Machine Learning Design Patterns

*Solutions to Common Challenges in Data
Preparation, Model Building, and MLOps*

*Valliappa Lakshmanan, Sara Robinson,
and Michael Munn*

Beijing · Boston · Farnham · Sebastopol · Tokyo

Machine Learning Design Patterns

by Valliappa Lakshmanan, Sara Robinson, and Michael Munn

Published by O'Reilly Media, Inc., 1005 Gravenstein Highway North, Sebastopol, CA 95472.

O'Reilly books may be purchased for educational, business, or sales promotional use. Online editions are also available for most titles (*http://oreilly.com*). For more information, contact our corporate/institutional sales department: 800-998-9938 or *corporate@oreilly.com*.

Acquisitions Editor: Rebecca Novack	**Indexer:** nSight, Inc.
Developmental Editor: Corbin Collins	**Interior Designer:** David Futato
Production Editor: Beth Kelly	**Cover Designer:** Karen Montgomery
Copyeditor: Charles Roumeliotis	**Illustrator:** Kate Dullea
Proofreader: Holly Bauer Forsyth	

October 2020:   First Edition

Revision History for the First Edition

2020-10-15:   First Release

See *http://oreilly.com/catalog/errata.csp?isbn=9781098115784* for release details.

978-1-098-11578-4

[LSI]

Table of Contents

Preface. ix

1. The Need for Machine Learning Design Patterns. 1
   What Are Design Patterns?   1
   How to Use This Book   3
   Machine Learning Terminology   3
      Models and Frameworks   4
      Data and Feature Engineering   6
      The Machine Learning Process   7
      Data and Model Tooling   8
      Roles   9
   Common Challenges in Machine Learning   11
      Data Quality   11
      Reproducibility   13
      Data Drift   14
      Scale   16
      Multiple Objectives   16
   Summary   17

2. Data Representation Design Patterns. 19
   Simple Data Representations   22
      Numerical Inputs   22
      Categorical Inputs   28
   Design Pattern 1: Hashed Feature   32
      Problem   32
      Solution   33
      Why It Works   34
      Trade-Offs and Alternatives   35

Design Pattern 2: Embeddings   39
    Problem   39
    Solution   41
    Why It Works   46
    Trade-Offs and Alternatives   48
Design Pattern 3: Feature Cross   52
    Problem   52
    Solution   53
    Why It Works   57
    Trade-Offs and Alternatives   58
Design Pattern 4: Multimodal Input   62
    Problem   62
    Solution   63
    Trade-Offs and Alternatives   65
Summary   77

3. Problem Representation Design Patterns. 79
Design Pattern 5: Reframing   80
    Problem   80
    Solution   80
    Why It Works   82
    Trade-Offs and Alternatives   85
Design Pattern 6: Multilabel   90
    Problem   90
    Solution   91
    Trade-Offs and Alternatives   93
Design Pattern 7: Ensembles   99
    Problem   99
    Solution   100
    Why It Works   104
    Trade-Offs and Alternatives   107
Design Pattern 8: Cascade   108
    Problem   109
    Solution   110
    Trade-Offs and Alternatives   114
Design Pattern 9: Neutral Class   117
    Problem   117
    Solution   118
    Why It Works   118
    Trade-Offs and Alternatives   120
Design Pattern 10: Rebalancing   122
    Problem   122

Solution	123
Trade-Offs and Alternatives	129
Summary	137

4. Model Training Patterns **139**

Typical Training Loop	139
Stochastic Gradient Descent	139
Keras Training Loop	140
Training Design Patterns	141
Design Pattern 11: Useful Overfitting	141
Problem	141
Solution	142
Why It Works	144
Trade-Offs and Alternatives	145
Design Pattern 12: Checkpoints	149
Problem	150
Solution	150
Why It Works	152
Trade-Offs and Alternatives	154
Design Pattern 13: Transfer Learning	161
Problem	161
Solution	162
Why It Works	169
Trade-Offs and Alternatives	172
Design Pattern 14: Distribution Strategy	175
Problem	175
Solution	175
Why It Works	181
Trade-Offs and Alternatives	183
Design Pattern 15: Hyperparameter Tuning	187
Problem	187
Solution	190
Why It Works	192
Trade-Offs and Alternatives	194
Summary	198

5. Design Patterns for Resilient Serving **201**

Design Pattern 16: Stateless Serving Function	201
Problem	203
Solution	205
Why It Works	207
Trade-Offs and Alternatives	209

Design Pattern 17: Batch Serving   213
  Problem   213
  Solution   214
  Why It Works   215
  Trade-Offs and Alternatives   217
Design Pattern 18: Continued Model Evaluation   220
  Problem   220
  Solution   221
  Why It Works   227
  Trade-Offs and Alternatives   227
Design Pattern 19: Two-Phase Predictions   232
  Problem   232
  Solution   234
  Trade-Offs and Alternatives   241
Design Pattern 20: Keyed Predictions   244
  Problem   244
  Solution   244
  Trade-Offs and Alternatives   247
Summary   248

6. Reproducibility Design Patterns. 249
Design Pattern 21: Transform   250
  Problem   250
  Solution   251
  Trade-Offs and Alternatives   252
Design Pattern 22: Repeatable Splitting   258
  Problem   258
  Solution   259
  Trade-Offs and Alternatives   260
Design Pattern 23: Bridged Schema   266
  Problem   266
  Solution   266
  Trade-Offs and Alternatives   271
Design Pattern 24: Windowed Inference   273
  Problem   273
  Solution   275
  Trade-Offs and Alternatives   277
Design Pattern 25: Workflow Pipeline   282
  Problem   283
  Solution   284
  Why It Works   289
  Trade-Offs and Alternatives   289

Design Pattern 26: Feature Store   295
  Problem   295
  Solution   296
  Why It Works   306
  Trade-Offs and Alternatives   308
Design Pattern 27: Model Versioning   310
  Problem   310
  Solution   311
  Trade-Offs and Alternatives   315
Summary   317

7. Responsible AI. 319
Design Pattern 28: Heuristic Benchmark   320
  Problem   320
  Solution   321
  Trade-Offs and Alternatives   324
Design Pattern 29: Explainable Predictions   326
  Problem   326
  Solution   327
  Trade-Offs and Alternatives   339
Design Pattern 30: Fairness Lens   343
  Problem   343
  Solution   345
  Trade-Offs and Alternatives   354
Summary   358

8. Connected Patterns. 359
Patterns Reference   359
Pattern Interactions   363
Patterns Within ML Projects   366
  ML Life Cycle   366
  AI Readiness   373
Common Patterns by Use Case and Data Type   377
  Natural Language Understanding   377
  Computer Vision   378
  Predictive Analytics   378
  Recommendation Systems   379
  Fraud and Anomaly Detection   380

Index. 383

Preface

Who Is This Book For?

Introductory machine learning books usually focus on the *what* and *how* of machine learning (ML). They then explain the mathematical aspects of new methods from AI research labs and teach how to use AI frameworks to implement these methods. This book, on the other hand, brings together hard-earned experience around the "why" that underlies the tips and tricks that experienced ML practitioners employ when applying machine learning to real-world problems.

We assume that you have prior knowledge of machine learning and data processing. This is not a fundamental textbook on machine learning. Instead, this book is for you if you are a data scientist, data engineer, or ML engineer who is looking for a second book on practical machine learning. If you already know the basics, this book will introduce you to a catalog of ideas, some of which you (an ML practitioner) may recognize, and give those ideas a name so that you can confidently reach for them.

If you are a computer science student headed for a job in industry, this book will round out your knowledge and prepare you for the professional world. It will help you learn how to build high-quality ML systems.

What's Not in the Book

This is a book that is primarily for ML engineers in the enterprise, not ML scientists in academia or industry research labs.

We purposefully do not discuss areas of active research—you will find very little here, for example, on machine learning model architecture (bidirectional encoders, or the attention mechanism, or short-circuit layers, for example) because we assume that you will be using a pre-built model architecture (such as ResNet-50 or GRUCell), not writing your own image classification or recurrent neural network.

Here are some concrete examples of areas that we intentionally stay away from because we believe that these topics are more appropriate for college courses and ML researchers:

ML algorithms

We do not cover the differences between random forests and neural networks, for example. This is covered in introductory machine learning textbooks.

Building blocks

We do not cover different types of gradient descent optimizers or activation functions. We recommend using Adam and ReLU—in our experience, the potential for improvements in performance by making different choices in these sorts of things tends to be minor.

ML model architectures

If you are doing image classification, we recommend that you use an off-the-shelf model like ResNet or whatever the latest hotness is at the time you are reading this. Leave the design of new image classification or text classification models to researchers who specialize in this problem.

Model layers

You won't find convolutional neural networks or recurrent neural networks in this book. They are doubly disqualified—first, for being a building block and second, for being something you can use off-the-shelf.

Custom training loops

Just calling `model.fit()` in Keras will fit the needs of practitioners.

In this book, we have tried to include only common patterns of the kind that machine learning engineers in enterprises will employ in their day-to-day work.

As an analogy, consider data structures. While a college course on data structures will delve into the implementations of different data structures, and a researcher on data structures will have to learn how to formally represent their mathematical properties, the practitioner can be more pragmatic. An enterprise software developer simply needs to know how to work effectively with arrays, linked lists, maps, sets, and trees. It is for a pragmatic practitioner in machine learning that this book is written.

Code Samples

We provide code for machine learning (sometimes in Keras/TensorFlow, and other times in scikit-learn or BigQuery ML) and data processing (in SQL) as a way to show how the techniques we are discussing are implemented in practice. All the code that is referenced in the book is part of our GitHub repository (*https://github.com/Google CloudPlatform/ml-design-patterns*), where you will find fully working ML models. We strongly encourage you to try out those code samples.

The code is secondary in importance to the concepts and techniques being covered. Our aim has been that the topic and principles should remain relevant regardless of changes to TensorFlow or Keras, and we can easily imagine updating the GitHub repository to include other ML frameworks, for example, while keeping the book text unchanged. Therefore, the book should be equally informative if your primary ML framework is PyTorch or even a non-Python framework like H20.ai or R. Indeed, we welcome your contributions to the GitHub repository of implementations of one or more of these patterns in your favorite ML framework.

If you have a technical question or a problem using the code examples, please send email to *bookquestions@oreilly.com*.

This book is here to help you get your job done. In general, if example code is offered with this book, you may use it in your programs and documentation. You do not need to contact us for permission unless you're reproducing a significant portion of the code. For example, writing a program that uses several chunks of code from this book does not require permission. Selling or distributing examples from O'Reilly books does require permission. Answering a question by citing this book and quoting example code does not require permission. Incorporating a significant amount of example code from this book into your product's documentation does require permission.

We appreciate, but generally do not require, attribution. An attribution usually includes the title, author, publisher, and ISBN. For example: "*Machine Learning Design Patterns* by Valliappa Lakshmanan, Sara Robinson, and Michael Munn (O'Reilly). Copyright 2021 Valliappa Lakshmanan, Sara Robinson, and Michael Munn, 978-1-098-11578-4." If you feel your use of code examples falls outside fair use or the permission given above, feel free to contact us at *permissions@oreilly.com*.

Conventions Used in This Book

The following typographical conventions are used in this book:

Italic
> Indicates new terms, URLs, email addresses, filenames, and file extensions.

`Constant width`
> Used for program listings, as well as within paragraphs to refer to program elements such as variable or function names, databases, data types, environment variables, statements, and keywords.

`Constant width bold`
> Shows commands or other text that should be typed literally by the user.

Constant width italic

Shows text that should be replaced with user-supplied values or by values determined by context.

This element signifies a tip or suggestion.

This element signifies a general note.

This element indicates a warning or caution.

O'Reilly Online Learning

 For more than 40 years, *O'Reilly Media* has provided technology and business training, knowledge, and insight to help companies succeed.

Our unique network of experts and innovators share their knowledge and expertise through books, articles, and our online learning platform. O'Reilly's online learning platform gives you on-demand access to live training courses, in-depth learning paths, interactive coding environments, and a vast collection of text and video from O'Reilly and 200+ other publishers. For more information, visit *http://oreilly.com*.

How to Contact Us

Please address comments and questions concerning this book to the publisher:

O'Reilly Media, Inc.
1005 Gravenstein Highway North
Sebastopol, CA 95472
800-998-9938 (in the United States or Canada)
707-829-0515 (international or local)
707-829-0104 (fax)

We have a web page for this book, where we list errata, examples, and any additional information. You can access this page at *https://oreil.ly/MLDP*.

Email *bookquestions@oreilly.com* to comment or ask technical questions about this book.

For news and information about our books and courses, visit *http://oreilly.com*.

Find us on Facebook: *http://facebook.com/oreilly*

Follow us on Twitter: *http://twitter.com/oreillymedia*

Watch us on YouTube: *http://youtube.com/oreillymedia*

Acknowledgments

A book like this would not be possible without the generosity of numerous Googlers, especially our colleagues in the Cloud AI, Solution Engineering, Professional Services, and Developer Relations teams. We are grateful to them for letting us observe, analyze, and question their solutions to the challenging problems they encountered in training, improving, and operationalizing ML models. Thanks to our managers, Karl Weinmeister, Steve Cellini, Hamidou Dia, Abdul Razack, Chris Hallenbeck, Patrick Cole, Louise Byrne, and Rochana Golani for fostering the spirit of openness within Google, giving us the freedom to catalog these patterns, and publish this book.

Salem Haykal, Benoit Dherin, and Khalid Salama reviewed every pattern and every chapter. Sal pointed out nuances we had missed, Benoit narrowed down our claims, and Khalid pointed us to relevant research. This book would be nowhere near as good without your inputs. Thank you! Amy Unruh, Rajesh Thallam, Robbie Haertel, Zhitao Li, Anusha Ramesh, Ming Fang, Parker Barnes, Andrew Zaldivar, James Wexler, Andrew Sellergren, and David Kanter reviewed parts of this book that align with their areas of expertise and made numerous suggestions on how the near-term roadmap would affect our recommendations. Nitin Aggarwal and Matthew Yeager brought a reader's eye to the manuscript and improved its clarity. Special thanks to

Rajesh Thallam for prototyping the design of the very last figure in Chapter 8. Any errors that remain are ours, of course.

O'Reilly is the publisher of choice for technical books, and the professionalism of our team illustrates why. Rebecca Novak shepherded us through putting together a compelling outline, Kristen Brown managed the entire content development with aplomb, Corbin Collins gave us helpful guidance at every stage, Elizabeth Kelly was a delight to work with during production, and Charles Roumeliotis brought a sharp eye to the copyediting. Thanks for all your help!

Michael: Thanks to my parents for always believing in me and encouraging my interests, both academic and otherwise. You will be able to appreciate as much as I do the surreptitious cover. To Phil, thank you for patiently bearing with my less-than-bearable schedule while working on this book. Now, I'mma be asleep.

Sara: Jon—you're a big reason this book exists. Thank you for encouraging me to write this, for always knowing how to make me laugh, appreciating my weirdness, and for believing in me especially when I didn't. To my parents, thank you for being my biggest fans since day one and encouraging my love of technology and writing for as long as I can remember. To Ally, Katie, Randi, and Sophie—thank you for being a constant source of light and laughter in these uncertain times.

Lak: I took on this book thinking I'd get to work on it while waiting in airports. COVID-19 made it so that much of the work was done at home. Thanks Abirami, Sidharth, and Sarada for all your forbearance as I hunkered down to write yet again. More hikes on weekends now!

The three of us are donating 100% of the royalties from this book to Girls Who Code (https://girlswhocode.com), an organization whose mission is to build a large pipeline of future female engineers. Diversity, equity, and inclusion are particularly important in machine learning to ensure that AI models don't perpetuate existing biases in human society.

## CHAPTER 1

The Need for Machine Learning Design Patterns

In engineering disciplines, design patterns capture best practices and solutions to commonly occurring problems. They codify the knowledge and experience of experts into advice that all practitioners can follow. This book is a catalog of machine learning design patterns that we have observed in the course of working with hundreds of machine learning teams.

What Are Design Patterns?

The idea of patterns, and a catalog of proven patterns, was introduced in the field of architecture by Christopher Alexander and five coauthors in a hugely influential book titled *A Pattern Language* (Oxford University Press, 1977). In their book, they catalog 253 patterns, introducing them this way:

> Each pattern describes a problem which occurs over and over again in our environment, and then describes the core of the solution to that problem, in such a way that you can use this solution a million times over, without ever doing it the same way twice.
>
> ...
>
> Each solution is stated in such a way that it gives the essential field of relationships needed to solve the problem, but in a very general and abstract way—so that you can solve the problem for yourself, in your own way, by adapting it to your preferences, and the local conditions at the place where you are making it.

For example, a couple of the patterns that incorporate human details when building a home are *Light on Two Sides of Every Room* and *Six-Foot Balcony*. Think of your favorite room in your home, and your least-favorite room. Does your favorite room have windows on two walls? What about your least-favorite room? According to Alexander:

> Rooms lit on two sides, with natural light, create less glare around people and objects; this lets us see things more intricately; and most important, it allows us to read in detail the minute expressions that flash across people's faces....

Having a name for this pattern saves architects from having to continually rediscover this principle. Yet where and how you get two light sources in any specific local condition is up to the architect's skill. Similarly, when designing a balcony, how big should it be? Alexander recommends 6 feet by 6 feet as being enough for 2 (mismatched!) chairs and a side table, and 12 feet by 12 feet if you want both a covered sitting space and a sitting space in the sun.

Erich Gamma, Richard Helm, Ralph Johnson, and John Vlissides brought the idea to software by cataloging 23 object-oriented design patterns in a 1994 book entitled *Design Patterns: Elements of Reusable Object-Oriented Software* (Addison-Wesley, 1995). Their catalog includes patterns such as Proxy, Singleton, and Decorator and led to lasting impact on the field of object-oriented programming. In 2005 the Association of Computing Machinery (ACM) awarded their annual Programming Languages Achievement Award to the authors, recognizing the impact of their work "on programming practice and programming language design."

Building production machine learning models is increasingly becoming an engineering discipline, taking advantage of ML methods that have been proven in research settings and applying them to business problems. As machine learning becomes more mainstream, it is important that practitioners take advantage of tried-and-proven methods to address recurring problems.

One benefit of our jobs in the customer-facing part of Google Cloud is that it brings us in contact with a wide variety of machine learning and data science teams and individual developers from around the world. At the same time, we each work closely with internal Google teams solving cutting-edge machine learning problems. Finally, we have been fortunate to work with the TensorFlow, Keras, BigQuery ML, TPU, and Cloud AI Platform teams that are driving the democratization of machine learning research and infrastructure. All this gives us a rather unique perspective from which to catalog the best practices we have observed these teams carrying out.

This book is a catalog of design patterns or repeatable solutions to commonly occurring problems in ML engineering. For example, the Transform pattern (Chapter 6) enforces the separation of inputs, features, and transforms and makes the transformations persistent in order to simplify moving an ML model to production. Similarly, Keyed Predictions, in Chapter 5, is a pattern that enables the large-scale distribution of batch predictions, such as for recommendation models.

For each pattern, we describe the commonly occurring problem that is being addressed and then walk through a variety of potential solutions to the problem, the trade-offs of these solutions, and recommendations for choosing between these solutions. Implementation code for these solutions is provided in SQL (useful if you are carrying out preprocessing and other ETL in Spark SQL, BigQuery, and so on), scikit-learn, and/or Keras with a TensorFlow backend.

How to Use This Book

This is a catalog of patterns that we have observed in practice, among multiple teams. In some cases, the underlying concepts have been known for many years. We don't claim to have invented or discovered these patterns. Instead, we hope to provide a common frame of reference and set of tools for ML practitioners. We will have succeeded if this book gives you and your team a vocabulary when talking about concepts that you already incorporate intuitively into your ML projects.

We don't expect you to read this book sequentially (although you can!). Instead, we anticipate that you will skim through the book, read a few sections more deeply than others, reference the ideas in conversations with colleagues, and refer back to the book when faced with problems you remember reading about. If you plan to skip around, we recommend that you start with Chapter 1 and Chapter 8 before dipping into individual patterns.

Each pattern has a brief problem statement, a canonical solution, an explanation of why the solution works, and a many-part discussion on tradeoffs and alternatives. We recommend that you read the discussion section with the canonical solution firmly in mind, so as to compare and contrast. The pattern description will include code snippets taken from the implementation of the canonical solution. The full code can be found in our GitHub repository (*https://github.com/GoogleCloudPlatform/ml-design-patterns*). We strongly encourage you to peruse the code as you read the pattern description.

Machine Learning Terminology

Because machine learning practitioners today may have different areas of primary expertise—software engineering, data analysis, DevOps, or statistics—there can be subtle differences in the way that different practitioners use certain terms. In this section, we define terminology that we use throughout the book.

Models and Frameworks

At its core, *machine learning* is a process of building models that learn from data. This is in contrast to traditional programming where we write explicit rules that tell programs how to behave. Machine learning *models* are algorithms that learn patterns from data. To illustrate this point, imagine we are a moving company and need to estimate moving costs for potential customers. In traditional programming, we might solve this with an if statement:

```
if num_bedrooms == 2 and num_bathrooms == 2:
    estimate = 1500
elif num_bedrooms == 3 and sq_ft > 2000:
    estimate = 2500
```

You can imagine how this will quickly get complicated as we add more variables (number of large furniture items, amount of clothing, fragile items, and so on) and try to handle edge cases. More to the point, asking for all this information ahead of time from customers can cause them to abandon the estimation process. Instead, we can train a machine learning model to estimate moving costs based on past data on previous households our company has moved.

Throughout the book, we primarily use feed-forward neural network models in our examples, but we'll also reference linear regression models, decision trees, clustering models, and others. *Feed-forward neural networks,* which we will commonly shorten as *neural networks*, are a type of machine learning algorithm whereby multiple layers, each with many neurons, analyze and process information and then send that information to the next layer, resulting in a final layer that produces a prediction as output. Though they are in no way identical, neural networks are often compared to the neurons in our brain because of the connectivity between nodes and the way they are able to generalize and form new predictions from the data they process. Neural networks with more than one *hidden layer* (layers other than the input and output layer) are classified as *deep learning* (see Figure 1-1).

Machine learning models, regardless of how they are depicted visually, are mathematical functions and can therefore be implemented from scratch using a numerical software package. However, ML engineers in industry tend to employ one of several open source frameworks designed to provide intuitive APIs for building models. The majority of our examples will use *TensorFlow*, an open source machine learning framework created by Google with a focus on deep learning models. Within the TensorFlow library, we'll be using the *Keras* API in our examples, which can be imported through `tensorflow.keras`. `Keras is a higher-level API for build` `ing neural networks`. While Keras supports many backends, we'll be using its TensorFlow backend. In other examples, we'll be using *scikit-learn*, *XGBoost*, and *PyTorch*, which are other popular open source frameworks that provide utilities for preparing your data, along with APIs for building linear and deep models. Machine learning continues to become more accessible, and one exciting development is the availability of machine learning models that can be expressed in SQL. We'll use *Big-Query ML* as an example of this, especially in situations where we want to combine data preprocessing and model creation.

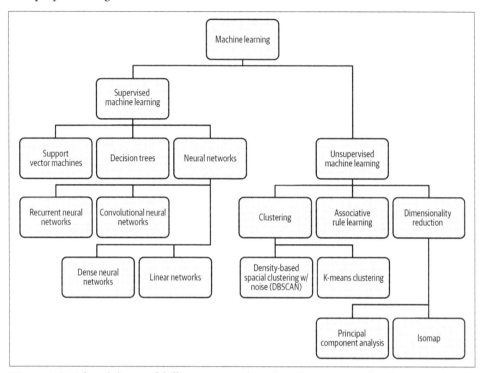

Figure 1-1. A breakdown of different types of machine learning, with a few examples of each. Note that although it is not included in this diagram, neural networks like autoencoders can also be used for unsupervised learning.

Conversely, neural networks with only an input and output layer are another subset of machine learning known as *linear models*. Linear models represent the patterns they've learned from data using a linear function. *Decision trees* are machine learning models that use your data to create a subset of paths with various branches. These branches approximate the results of different outcomes from your data. Finally, *clustering* models look for similarities between different subsets of your data and use these identified patterns to group data into clusters.

Machine learning problems (see Figure 1-1) can be broken into two types: supervised and unsupervised learning. *Supervised learning* defines problems where you know the ground truth label for your data in advance. For example, this could include labeling an image as "cat" or labeling a baby as being 2.3 kg at birth. You feed this labeled data to your model in hopes that it can learn enough to label new examples. With *unsupervised learning*, you do not know the labels for your data in advance, and the goal is to build a model that can find natural groupings of your data (called *clustering*), compress the information content (*dimensionality reduction*), or find association rules. The majority of this book will focus on supervised learning because the vast majority of machine learning models used in production are supervised.

With supervised learning, problems can typically be defined as either classification or regression. *Classification* models assign your input data a label (or labels) from a discrete, predefined set of categories. Examples of classification problems include determining the type of pet breed in an image, tagging a document, or predicting whether or not a transaction is fraudulent. *Regression* models assign continuous, numerical values to your inputs. Examples of regression models include predicting the duration of a bike trip, a company's future revenue, or the price of a product.

Data and Feature Engineering

Data is at the heart of any machine learning problem. When we talk about *datasets*, we're referring to the data used for training, validating, and testing a machine learning model. The bulk of your data will be *training data*: the data fed to your model during the training process. *Validation data* is data that is held out from your training set and used to evaluate how the model is performing after each training *epoch* (or pass through the training data). The performance of the model on the validation data is used to decide when to stop the training run, and to choose *hyperparameters*, such as the number of trees in a random forest model. *Test data* is data that is not used in the training process at all and is used to evaluate how the trained model performs. Performance reports of the machine learning model must be computed on the independent test data, rather than the training or validation tests. It's also important that the data be split in such a way that all three datasets (training, test, validation) have similar statistical properties.

The data you use to train your model can take many forms depending on the model type. We define *structured data* as numerical and categorical data. Numerical data includes integer and float values, and categorical data includes data that can be divided into a finite set of groups, like type of car or education level. You can also think of structured data as data you would commonly find in a spreadsheet. Throughout the book, we'll use the term *tabular data* interchangeably with structured data. *Unstructured data*, on the other hand, includes data that cannot be represented as neatly. This typically includes free-form text, images, video, and audio.

Numeric data can often be fed directly to a machine learning model, where other data requires various *data preprocessing* before it's ready to be sent to a model. This preprocessing step typically includes scaling numerical values, or converting nonnumerical data into a numerical format that can be understood by your model. Another term for preprocessing is *feature engineering*. We'll use these two terms interchangeably throughout the book.

There are various terms used to describe data as it goes through the feature engineering process. *Input* describes a single column in your dataset before it has been processed, and *feature* describes a single column *after* it has been processed. For example, a timestamp could be your input, and the feature would be day of the week. To convert the data from timestamp to day of the week, you'll need to do some data preprocessing. This preprocessing step can also be referred to as *data transformation*.

An *instance* is an item you'd like to send to your model for prediction. An instance could be a row in your test dataset (without the label column), an image you want to classify, or a text document to send to a sentiment analysis model. Given a set of features about the instance, the model will calculate a predicted value. In order to do that, the model is trained on *training examples*, which associate an instance with a *label*. A *training example* refers to a single instance (row) of data from your dataset that will be fed to your model. Building on the timestamp use case, a full training example might include: "day of week," "city," and "type of car." A *label* is the output column in your dataset—the item your model is predicting. *Label* can refer both to the target column in your dataset (also called a *ground truth label*) and the output given by your model (also called a *prediction*). A sample label for the training example outlined above could be "trip duration"—in this case, a float value denoting minutes.

Once you've assembled your dataset and determined the features for your model, *data validation* is the process of computing statistics on your data, understanding your schema, and evaluating the dataset to identify problems like drift and training-serving skew. Evaluating various statistics on your data can help you ensure the dataset contains a balanced representation of each feature. In cases where it's not possible to collect more data, understanding data balance will help you design your model to account for this. Understanding your schema involves defining the data type for each feature and identifying training examples where certain values may be incorrect or missing. Finally, data validation can identify inconsistencies that may affect the quality of your training and test sets. For example, maybe the majority of your training dataset contains *weekday* examples while your test set contains primarily *weekend* examples.

The Machine Learning Process

The first step in a typical machine learning workflow is *training*—the process of passing training data to a model so that it can learn to identify patterns. After training, the next step in the process is testing how your model performs on data outside of your training set. This is known as model *evaluation*. You might run training and evaluation multiple times, performing additional feature engineering and tweaking your model architecture. Once you are happy with your model's performance during evaluation, you'll likely want to serve your model so that others can access it to make predictions. We use the term *serving* to refer to accepting incoming requests and sending back predictions by deploying the model as a microservice. The serving infrastructure could be in the cloud, on-premises, or on-device.

The process of sending new data to your model and making use of its output is called *prediction*. This can refer both to generating predictions from local models that have not yet been deployed as well as getting predictions from deployed models. For deployed models, we'll refer both to online and batch prediction. *Online prediction* is used when you want to get predictions on a few examples in near real time. With online prediction, the emphasis is on low latency. *Batch prediction*, on the other hand, refers to generating predictions on a large set of data offline. Batch prediction jobs take longer than online prediction and are useful for precomputing predictions (such as in recommendation systems) and in analyzing your model's predictions across a large sample of new data.

The word *prediction* is apt when it comes to forecasting future values, such as in predicting the duration of a bicycle ride or predicting whether a shopping cart will be abandoned. It is less intuitive in the case of image and text classification models. If an ML model looks at a text review and outputs that the sentiment is positive, it's not really a "prediction" (there is no future outcome). Hence, you will also see word *inference* being used to refer to predictions. The statistical term inference is being repurposed here, but it's not really about reasoning.

Often, the processes of collecting training data, feature engineering, training, and evaluating your model are handled separately from the production pipeline. When this is the case, you'll reevaluate your solution whenever you decide you have enough additional data to train a new version of your model. In other situations, you may have new data being ingested continuously and need to process this data immediately before sending it to your model for training or prediction. This is known as *streaming*. To handle streaming data, you'll need a multistep solution for performing feature engineering, training, evaluation, and predictions. Such multistep solutions are called *ML pipelines*.

Data and Model Tooling

There are various Google Cloud products we'll be referencing that provide tooling for solving data and machine learning problems. These products are merely one option for implementing the design patterns referenced in this book and are not meant to be an exhaustive list. All of the products included here are serverless, allowing us to focus more on implementing machine learning design patterns instead of the infrastructure behind them.

BigQuery (*https://oreil.ly/7PnVj*) is an enterprise data warehouse designed for analyzing large datasets quickly with SQL. We'll use BigQuery in our examples for data collection and feature engineering. Data in BigQuery is organized by Datasets, and a Dataset can have multiple Tables. Many of our examples will use data from *Google Cloud Public Datasets* (*https://oreil.ly/AbTaJ*), a set of free, publicly available data hosted in BigQuery. Google Cloud Public Datasets consists of hundreds of different datasets, including NOAA weather data since 1929, Stack Overflow questions and answers, open source code from GitHub, natality data, and more. To build some of the models in our examples, we'll use *BigQuery Machine Learning* (*https://oreil.ly/_VjVz*) (or BigQuery ML). BigQuery ML is a tool for building models from data stored in BigQuery. With BigQuery ML, we can train, evaluate, and generate predictions on our models using SQL. It supports classification and regression models, along with unsupervised clustering models. It's also possible to import previously trained TensorFlow models to BigQuery ML for prediction.

Cloud AI Platform (*https://oreil.ly/90KLs*) includes a variety of products for training and serving custom machine learning models on Google Cloud. In our examples, we'll be using AI Platform Training and AI Platform Prediction. AI Platform Training provides infrastructure for training machine learning models on Google Cloud. With AI Platform Prediction, you can deploy your trained models and generate predictions on them using an API. Both services support TensorFlow, scikit-Learn, and XGBoost models, along with custom containers for models built with other frameworks. We'll also reference *Explainable AI* (*https://oreil.ly/lDocn*), a tool for interpreting the results of your model's predictions, available for models deployed to AI Platform.

Roles

Within an organization, there are many different job roles relating to data and machine learning. Below we'll define a few common ones referenced frequently throughout the book. This book is targeted primarily at data scientists, data engineers, and ML engineers, so let's start with those.

A *data scientist* is someone focused on collecting, interpreting, and processing datasets. They run statistical and exploratory analysis on data. As it relates to machine learning, a data scientist may work on data collection, feature engineering, model building, and more. Data scientists often work in Python or R in a notebook environment, and are usually the first to build out an organization's machine learning models.

A *data engineer* is focused on the infrastructure and workflows powering an organization's data. They might help manage how a company ingests data, data pipelines, and how data is stored and transferred. Data engineers implement infrastructure and pipelines around data.

Machine learning engineers do similar tasks to data engineers, but for ML models. They take models developed by data scientists, and manage the infrastructure and operations around training and deploying those models. ML engineers help build production systems to handle updating models, model versioning, and serving predictions to end users.

The smaller the data science team at a company and the more agile the team is, the more likely it is that the same person plays multiple roles. If you are in such a situation, it is very likely that you read the above three descriptions and saw yourself partially in all three categories. You might commonly start out a machine learning project as a data engineer and build data pipelines to operationalize the ingest of data. Then, you transition to the data scientist role and build the ML model(s). Finally, you put on the ML engineer hat and move the model to production. In larger organizations, machine learning projects may move through the same phases, but different teams might be involved in each phase.

Research scientists, data analysts, and developers may also build and use AI models, but these job roles are not a focus audience for this book.

Research scientists focus primarily on finding and developing new algorithms to advance the discipline of ML. This could include a variety of subfields within machine learning, like model architectures, natural language processing, computer vision, hyperparameter tuning, model interpretability, and more. Unlike the other roles discussed here, research scientists spend most of their time prototyping and evaluating new approaches to ML, rather than building out production ML systems.

Data analysts evaluate and gather insights from data, then summarize these insights for other teams within their organization. They tend to work in SQL and spreadsheets, and use business intelligence tools to create data visualizations to share their findings. Data analysts work closely with product teams to understand how their insights can help address business problems and create value. While data analysts focus on identifying trends in existing data and deriving insights from it, data scientists are concerned with using that data to generate future predictions and in automating or scaling out the generation of insights. With the increasing democratization of machine learning, data analysts can upskill themselves to become data scientists.

Developers are in charge of building production systems that enable end users to access ML models. They are often involved in designing the APIs that query models and return predictions in a user-friendly format via a web or mobile application. This could involve models hosted in the cloud, or models served on-device. Developers utilize the model serving infrastructure implemented by ML Engineers to build applications and user interfaces for surfacing predictions to model users.

Figure 1-2 illustrates how these different roles work together throughout an organization's machine learning model development process.

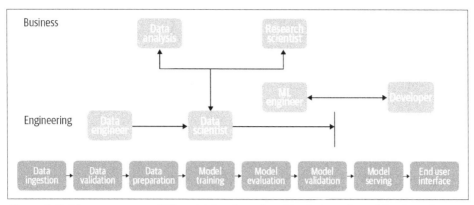

Figure 1-2. There are many different job roles related to data and machine learning, and these roles collaborate on the ML workflow, from data ingestion to model serving and the end user interface. For example, the data engineer works on data ingestion and data validation and collaborates closely with data scientists.

Common Challenges in Machine Learning

Why do we need a book about machine learning design patterns? The process of building out ML systems presents a variety of unique challenges that influence ML design. Understanding these challenges will help you, an ML practitioner, develop a frame of reference for the solutions introduced throughout the book.

Data Quality

Machine learning models are only as reliable as the data used to train them. If you train a machine learning model on an incomplete dataset, on data with poorly selected features, or on data that doesn't accurately represent the population using the model, your model's predictions will be a direct reflection of that data. As a result, machine learning models are often referred to as "garbage in, garbage out." Here we'll highlight four important components of data quality: accuracy, completeness, consistency, and timeliness.

Data *accuracy* refers to both your training data's features and the ground truth labels corresponding with those features. Understanding where your data came from and any potential errors in the data collection process can help ensure feature accuracy. After your data has been collected, it's important to do a thorough analysis to screen for typos, duplicate entries, measurement inconsistencies in tabular data, missing features, and any other errors that may affect data quality. Duplicates in your training dataset, for example, can cause your model to incorrectly assign more weight to these data points.

Accurate data labels are just as important as feature accuracy. Your model relies solely on the ground truth labels in your training data to update its weights and minimize loss. As a result, incorrectly labeled training examples can cause misleading model accuracy. For example, let's say you're building a sentiment analysis model and 25% of your "positive" training examples have been incorrectly labeled as "negative." Your model will have an inaccurate picture of what should be considered negative sentiment, and this will be directly reflected in its predictions.

To understand data *completeness*, let's say you're training a model to identify cat breeds. You train the model on an extensive dataset of cat images, and the resulting model is able to classify images into 1 of 10 possible categories ("Bengal," "Siamese," and so forth) with 99% accuracy. When you deploy your model to production, however, you find that in addition to uploading cat photos for classification, many of your users are uploading photos of dogs and are disappointed with the model's results. Because the model was trained only to identify 10 different cat breeds, this is all it knows how to do. These 10 breed categories are, essentially, the model's entire "world view." No matter what you send the model, you can expect it to slot it into one of these 10 categories. It may even do so with high confidence for an image that looks nothing like a cat. Additionally, there's no way your model will be able to return "not a cat" if this data and label weren't included in the training dataset.

Another aspect of data completeness is ensuring your training data contains a varied representation of each label. In the cat breed detection example, if all of your images are close-ups of a cat's face, your model won't be able to correctly identify an image of a cat from the side, or a full-body cat image. To look at a tabular data example, if you are building a model to predict the price of real estate in a specific city but only include training examples of houses larger than 2,000 square feet, your resulting model will perform poorly on smaller houses.

The third aspect of data quality is data *consistency*. For large datasets, it's common to divide the work of data collection and labeling among a group of people. Developing a set of standards for this process can help ensure consistency across your dataset, since each person involved in this will inevitably bring their own biases to the process. Like data completeness, data inconsistencies can be found in both data features and labels. For an example of inconsistent features, let's say you're collecting atmospheric data from temperature sensors. If each sensor has been calibrated to different standards, this will result in inaccurate and unreliable model predictions. Inconsistencies can also refer to data format. If you're capturing location data, some people may write out a full street address as "Main Street" and others may abbreviate it as "Main St." Measurement units, like miles and kilometers, can also differ around the world.

In regards to labeling inconsistencies, let's return to the text sentiment example. In this case, it's likely people will not always agree on what is considered positive and negative when labeling training data. To solve this, you can have multiple people labeling each example in your dataset, then take the most commonly applied label for each item. Being aware of potential labeler bias, and implementing systems to account for it, will ensure label consistency throughout your dataset. We'll explore the concept of bias in the "Design Pattern 30: Fairness Lens" on page 343 in Chapter 7.

Timeliness in data refers to the latency between when an event occurred and when it was added to your database. If you're collecting data on application logs, for example, an error log might take a few hours to show up in your log database. For a dataset recording credit card transactions, it might take one day from when the transaction occurred before it is reported in your system. To deal with timeliness, it's useful to record as much information as possible about a particular data point, and make sure that information is reflected when you transform your data into features for a machine learning model. More specifically, you can keep track of the timestamp of when an event occurred and when it was added to your dataset. Then, when performing feature engineering, you can account for these differences accordingly.

Reproducibility

In traditional programming, the output of a program is reproducible and guaranteed. For example, if you write a Python program that reverses a string, you know that an input of the word "banana" will always return an output of "ananab." Similarly, if there's a bug in your program causing it to incorrectly reverse strings containing numbers, you could send the program to a colleague and expect them to be able to reproduce the error with the same inputs you used (unless the bug has something to do with the program maintaining some incorrect internal state, differences in architecture such as floating point precision, or differences in execution such as threading).

Machine learning models, on the other hand, have an inherent element of randomness. When training, ML model weights are initialized with random values. These weights then converge during training as the model iterates and learns from the data. Because of this, the same model code given the same training data will produce slightly different results across training runs. This introduces a challenge of reproducibility. If you train a model to 98.1% accuracy, a repeated training run is not guaranteed to reach the same result. This can make it difficult to run comparisons across experiments.

In order to address this problem of repeatability, it's common to set the random seed value used by your model to ensure that the same randomness will be applied each time you run training. In TensorFlow, you can do this by running `tf.random.set_seed(value)` at the beginning of your program.

Additionally, in scikit-learn, many utility functions for shuffling your data also allow you to set a random seed value:

```
from sklearn.utils import shuffle
data = shuffle(data, random_state=value)
```

Keep in mind that you'll need to use the same data *and* the same random seed when training your model to ensure repeatable, reproducible results across different experiments.

Training an ML model involves several artifacts that need to be fixed in order to ensure reproducibility: the data used, the splitting mechanism used to generate datasets for training and validation, data preparation and model hyperparameters, and variables like the batch size and learning rate schedule.

Reproducibility also applies to machine learning framework dependencies. In addition to manually setting a random seed, frameworks also implement elements of randomness internally that are executed when you call a function to train your model. If this underlying implementation changes between different framework versions, repeatability is not guaranteed. As a concrete example, if one version of a framework's `train()` method makes 13 calls to `rand()`, and a newer version of the same framework makes 14 calls, using different versions between experiments will cause slightly different results, even with the same data and model code. Running ML workloads in containers and standardizing library versions can help ensure repeatability. Chapter 6 introduces a series of patterns for making ML processes reproducible.

Finally, reproducibility can refer to a model's training environment. Often, due to large datasets and complexity, many models take a significant amount of time to train. This can be accelerated by employing distribution strategies like data or model parallelism (see Chapter 5). With this acceleration, however, comes an added challenge of repeatability when you rerun code that makes use of distributed training.

Data Drift

While machine learning models typically represent a static relationship between inputs and outputs, data can change significantly over time. Data drift refers to the challenge of ensuring your machine learning models stay relevant, and that model predictions are an accurate reflection of the environment in which they're being used.

For example, let's say you're training a model to classify news article headlines into categories like "politics," "business," and "technology." If you train and evaluate your model on historical news articles from the 20th century, it likely won't perform as well on current data. Today, we know that an article with the word "smartphone" in the headline is probably about technology. However, a model trained on historical data would have no knowledge of this word. To solve for drift, it's important to continually update your training dataset, retrain your model, and modify the weight your model assigns to particular groups of input data.

To see a less-obvious example of drift, look at the NOAA dataset (*https://oreil.ly/ obzvn*) of severe storms in BigQuery. If we were training a model to predict the likelihood of a storm in a given area, we would need to take into account the way weather reporting has changed over time (*https://github.com/GoogleCloudPlatform/ml-design-patterns/blob/master/01_need_for_design_patterns/ml_challenges.ipynb*). We can see in Figure 1-3 that the total number of severe storms recorded has been steadily increasing since 1950.

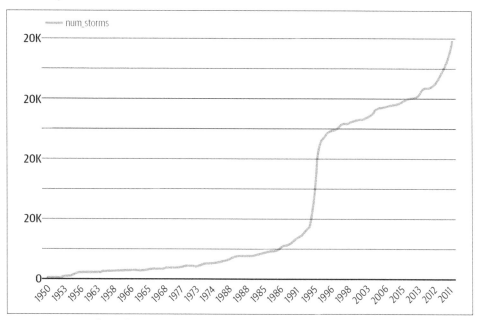

Figure 1-3. Number of severe storms reported in a year, as recorded by NOAA from 1950 to 2011.

From this trend, we can see that training a model on data before 2000 to generate predictions on storms today would lead to inaccurate predictions. In addition to the total number of reported storms increasing, it's also important to consider other factors that may have influenced the data in Figure 1-3. For example, the technology for observing storms has improved over time, most dramatically with the introduction of weather radars in the 1990s. In the context of features, this may mean that newer data contains more information about each storm, and that a feature available in today's data may not have been observed in 1950. Exploratory data analysis can help identify this type of drift and can inform the correct window of data to use for training. Section , "Design Pattern 23: Bridged Schema" on page 266 provides a way to handle datasets in which the availability of features improves over time.

Scale

The challenge of scaling is present throughout many stages of a typical machine learning workflow. You'll likely encounter scaling challenges in data collection and preprocessing, training, and serving. When ingesting and preparing data for a machine learning model, the size of the dataset will dictate the tooling required for your solution. It is often the job of data engineers to build out data pipelines that can scale to handle datasets with millions of rows.

For model training, ML engineers are responsible for determining the necessary infrastructure for a specific training job. Depending on the type and size of the dataset, model training can be time consuming and computationally expensive, requiring infrastructure (like GPUs) designed specifically for ML workloads. Image models, for instance, typically require much more training infrastructure than models trained entirely on tabular data.

In the context of model serving, the infrastructure required to support a team of data scientists getting predictions from a model prototype is entirely different from the infrastructure necessary to support a production model getting millions of prediction requests every hour. Developers and ML engineers are typically responsible for handling the scaling challenges associated with model deployment and serving prediction requests.

Most of the ML patterns in this book are useful without regard to organizational maturity. However, several of the patterns in Chapters 6 and 7 address resilience and reproducibility challenges in different ways, and the choice between them will often come down to the use case and the ability of your organization to absorb complexity.

Multiple Objectives

Though there is often a single team responsible for building a machine learning model, many teams across an organization will make use of the model in some way. Inevitably, these teams may have different ideas of what defines a successful model.

To understand how this may play out in practice, let's say you're building a model to identify defective products from images. As a data scientist, your goal may be to minimize your model's cross-entropy loss. The product manager, on the other hand, may want to reduce the number of defective products that are misclassified and sent to customers. Finally, the executive team's goal might be to increase revenue by 30%. Each of these goals vary in what they are optimizing for, and balancing these differing needs within an organization can present a challenge.

As a data scientist, you could translate the product team's needs into the context of your model by saying false negatives are five times more costly than false positives. Therefore, you should optimize for recall over precision to satisfy this when designing your model. You can then find a balance between the product team's goal of optimizing for precision and your goal of minimizing the model's loss.

When defining the goals for your model, it's important to consider the needs of different teams across an organization, and how each team's needs relate back to the model. By analyzing what each team is optimizing for before building out your solution, you can find areas of compromise in order to optimally balance these multiple objectives.

Summary

Design patterns are a way to codify the knowledge and experience of experts into advice that all practitioners can follow. The design patterns in this book capture best practices and solutions to commonly occurring problems in designing, building, and deploying machine learning systems. The common challenges in machine learning tend to revolve around data quality, reproducibility, data drift, scale, and having to satisfy multiple objectives.

We tend to use different ML design patterns at different stages of the ML life cycle. There are patterns that are useful in problem framing and assessing feasibility. The majority of patterns address either development or deployment, and quite a few patterns address the interplay between these stages.

## CHAPTER 2

Data Representation Design Patterns

At the heart of any machine learning model is a mathematical function that is defined to operate on specific types of data only. At the same time, real-world machine learning models need to operate on data that may not be directly pluggable into the mathematical function. The mathematical core of a decision tree, for example, operates on boolean variables. Note that we are talking here about the mathematical core of a decision tree—decision tree machine learning software will typically also include functions to learn an optimal tree from data and ways to read in and process different types of numeric and categorical data. The mathematical function (see Figure 2-1) that underpins a decision tree, however, operates on boolean variables and uses operations such as AND (&& in Figure 2-1) and OR (+ in Figure 2-1).

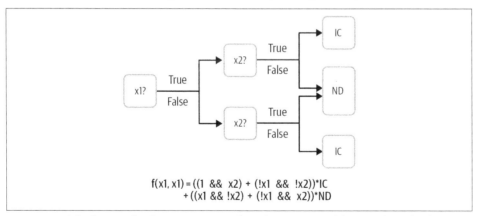

$$f(x1, x1) = ((1 \;\&\&\; x2) + (!x1 \;\&\&\; !x2))^*IC$$
$$+ ((x1 \;\&\&\; !x2) + (!x1 \;\&\&\; x2))^*ND$$

Figure 2-1. The heart of a decision tree machine learning model to predict whether or not a baby requires intensive care is a mathematical model that operates on boolean variables.

Suppose we have a decision tree to predict whether a baby will require intensive care (IC) or can be normally discharged (ND), and suppose that the decision tree takes as inputs two variables, *x1* and *x2*. The trained model might look something like Figure 2-1.

It is pretty clear that *x1* and *x2* need to be boolean variables in order for *f(x1, x2)* to work. Suppose that two of the pieces of information we'd like the model to consider when classifying a baby as requiring intensive care or not is the hospital that the baby is born in and the baby's weight. Can we use the hospital that a baby is born in as an input to the decision tree? No, because the hospital takes neither the value True nor the value False and cannot be fed into the && (AND) operator. It's mathematically not compatible. Of course, we can "make" the hospital value boolean by performing an operation such as:

```
x1 = (hospital IN France)
```

so that *x1* is True when the hospital is in France, and False if not. Similarly, a baby's weight cannot be fed directly into the model, but by performing an operation such as:

```
x1 = (babyweight < 3 kg)
```

we can use the hospital or the baby weight as an input to the model. This is an example of how input data (hospital, a complex object or baby weight, a floating point number) can be represented in the form (boolean) expected by the model. This is what we mean by *data representation*.

In this book, we will use the term *input* to represent the real-world data fed to the model (for example, the baby weight) and the term *feature* to represent the transformed data that the model actually operates on (for example, whether the baby weight is less than 3 kilograms). The process of creating features to represent the input data is called *feature engineering*, and so we can think of feature engineering as a way of selecting the data representation.

Of course, rather than hardcoding parameters such as the threshold value of 3 kilograms, we'd prefer the machine learning model to learn how to create each node by selecting the input variable and the threshold. Decision trees are an example of machine learning models that are capable of learning the data representation.[1] Many of the patterns that we look at in this chapter will involve similarly *learnable data representations*.

The *Embeddings* design pattern is the canonical example of a data representation that deep neural networks are capable of learning on their own. In an embedding, the learned representation is dense and lower-dimensional than the input, which could

---

1 Here, the learned data representation consists of baby weight as the input variable, the less than operator, and the threshold of 3 kg.

---

be sparse. The learning algorithm needs to extract the most salient information from the input and represent it in a more concise way in the feature. The process of learning features to represent the input data is called *feature extraction*, and we can think of learnable data representations (like embeddings) as automatically engineered features.

The data representation doesn't even need to be of a single input variable—an oblique decision tree, for example, creates a boolean feature by thresholding a linear combination of two or more input variables. A decision tree where each node can represent only one input variable reduces to a stepwise linear function, whereas an oblique decision tree where each node can represent a linear combination of input variables reduces to a piecewise linear function (see Figure 2-2). Considering how many steps will have to be learned to adequately represent the line, the piecewise linear model is simpler and faster to learn. An extension of this idea is the *Feature Cross* design pattern, which simplifies the learning of AND relationships between multivalued categorical variables.

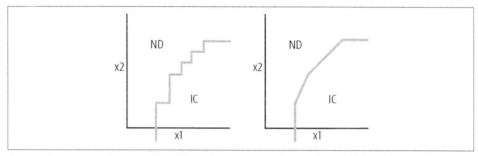

Figure 2-2. A decision tree classifier where each node can threshold only one input value (x1 or x2) will result in a stepwise linear boundary function, whereas an oblique tree classifier where a node can threshold a linear combination of input variables will result in a piecewise linear boundary function. The piecewise linear function requires fewer nodes and can achieve greater accuracy.

The data representation doesn't need to be learned or fixed—a hybrid is also possible. The *Hashed Feature* design pattern is deterministic, but doesn't require a model to know all the potential values that a particular input can take.

The data representations we have looked at so far are all one-to-one. Although we could represent input data of different types separately or represent each piece of data as just one feature, it can be more advantageous to use *Multimodal Input*. That is the fourth design pattern we will explore in this chapter.

Simple Data Representations

Before we delve into learnable data representations, feature crosses, and more, let's look at simpler data representations. We can think of these simple data representations as common *idioms* in machine learning—not quite patterns, but commonly employed solutions nevertheless.

Numerical Inputs

Most modern, large-scale machine learning models (random forests, support vector machines, neural networks) operate on numerical values, and so if our input is numeric, we can pass it through to the model unchanged.

Why scaling is desirable

Often, because the ML framework uses an optimizer that is tuned to work well with numbers in the [−1, 1] range, scaling the numeric values to lie in that range can be beneficial.

Why Scale Numeric Values to Lie in [−1, 1]?

Gradient descent optimizers require more steps to converge as the curvature of the loss function increases. This is because the derivatives of features with larger relative magnitudes will tend to be larger as well, and so lead to abnormal weight updates. The abnormally large weight updates will require more steps to converge and thereby increase the computation load.

"Centering" the data to lie in the [−1, 1] range makes the error function more spherical. Therefore, models trained with transformed data tend to converge faster and are therefore faster/cheaper to train. In addition, the [−1, 1] range offers the highest floating point precision.

A quick test with one of scikit-learn's built-in datasets can prove the point (this is an excerpt from this book's code repository (*https://github.com/GoogleCloudPlatform/ml-design-patterns/blob/master/02_data_representation/simple_data_representation.ipynb*)):

```
from sklearn import datasets, linear_model
diabetes_X, diabetes_y = datasets.load_diabetes(return_X_y=True)
raw = diabetes_X[:, None, 2]
max_raw = max(raw)
min_raw = min(raw)
scaled = (2*raw - max_raw - min_raw)/(max_raw - min_raw)

def train_raw():
    linear_model.LinearRegression().fit(raw, diabetes_y)
```

```
def train_scaled():
    linear_model.LinearRegression().fit(scaled, diabetes_y)

raw_time = timeit.timeit(train_raw, number=1000)
scaled_time = timeit.timeit(train_scaled, number=1000)
```

When we ran this, we got a nearly 9% improvement on this model which uses just one input feature. Considering the number of features in a typical machine learning model, the savings can add up.

Another important reason for scaling is that some machine learning algorithms and techniques are very sensitive to the relative magnitudes of the different features. For example, a k-means clustering algorithm that uses the Euclidean distance as its proximity measure will end up relying heavily on features with larger magnitudes. Lack of scaling also affects the efficacy of L1 or L2 regularization since the magnitude of weights for a feature depends on the magnitude of values of that feature, and so different features will be affected differently by regularization. By scaling all features to lie between [−1, 1], we ensure that there is not much of a difference in the relative magnitudes of different features.

Linear scaling

Four forms of scaling are commonly employed:

Min-max scaling
> The numeric value is linearly scaled so that the minimum value that the input can take is scaled to −1 and the maximum possible value to 1:
>
> ```
> x1_scaled = (2*x1 - max_x1 - min_x1)/(max_x1 - min_x1)
> ```
>
> The problem with min-max scaling is that the maximum and minimum value (max_x1 and min_x1) have to be estimated from the training dataset, and they are often outlier values. The real data often gets shrunk to a very narrow range in the [−1, 1] band.

Clipping (in conjunction with min-max scaling)
> Helps address the problem of outliers by using "reasonable" values instead of estimating the minimum and maximum from the training dataset. The numeric value is linearly scaled between these two reasonable bounds, then clipped to lie in the range [−1, 1]. This has the effect of treating outliers as −1 or 1.

Z-score normalization
> Addresses the problem of outliers without requiring prior knowledge of what the reasonable range is by linearly scaling the input using the mean and standard deviation estimated over the training dataset:
>
> ```
> x1_scaled = (x1 - mean_x1)/stddev_x1
> ```

The name of the method reflects the fact that the scaled value has zero mean and is normalized by the standard deviation so that it has unit variance over the training dataset. The scaled value is unbounded, but does lie between [−1, 1] the majority of the time (67%, if the underlying distribution is normal). Values outside this range get rarer the larger their absolute value gets, but are still present.

Winsorizing

Uses the empirical distribution in the training dataset to clip the dataset to bounds given by the 10th and 90th percentile of the data values (or 5th and 95th percentile, and so forth). The winsorized value is min-max scaled.

All the methods discussed so far scale the data linearly (in the case of clipping and winsorizing, linear within the typical range). Min-max and clipping tend to work best for uniformly distributed data, and Z-score tends to work best for normally distributed data. The impact of different scaling functions on the `mother_age` column in the baby weight prediction example is shown in Figure 2-3 (see the full code (*https://github.com/GoogleCloudPlatform/ml-design-patterns/blob/master/02_data_representation/simple_data_representation.ipynb*)).

Don't Throw Away "Outliers"

Note that we defined clipping as taking scaled values less than −1 and treating them as −1, and scaled values greater than 1 and treating them as 1. We don't simply discard such "outliers" because we expect that the machine learning model will encounter outliers like this in production. Take, for example, babies born to 50-year-old mothers. Because we don't have enough older mothers in our dataset, clipping ends up treating all mothers older than 45 (for example) as 45. This same treatment will be applied in production, and therefore, our model will be able to handle older mothers. The model would not learn to reflect outliers if we had simply thrown away all the training examples of babies born to mothers aged 50+!

Another way to think about this is that while it is acceptable to throw away *invalid input*, it is not acceptable to throw away *valid data*. Thus, we would be justified in throwing away rows where mother_age is negative because it's probably a data entry error. In production, validation of the input form will ensure that the admitting clerk has to reenter the mother's age. However, we are not justified in throwing away rows where mother_age is 50 because 50 is a perfectly valid input and we expect to encounter 50-year-old mothers once the model is deployed in production.

In Figure 2-3, note that minmax_scaled gets the x values into the desired range of [−1, 1] but continues to retain values at the extreme ends of the distribution where there are not enough examples. Clipping rolls up many of the problematic values, but requires getting the clipping thresholds exactly correct—here, the slow decline in the number of babies with mothers' ages above 40 poses problems in setting a hard threshold. Winsorizing, similar to clipping, requires getting the percentile thresholds exactly correct. Z-score normalization improves the range (but does not constrain values to be between [–1, 1]) and pushes the problematic values further out. Of these three methods, zero-norming works best for mother_age because the raw age values were somewhat of a bell curve. For other problems, min-max scaling, clipping, or winsorizing might be better.

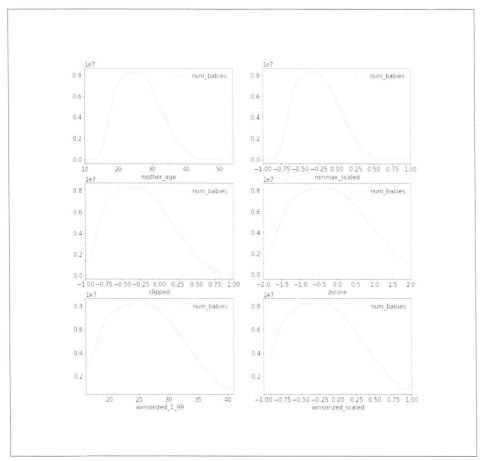

Figure 2-3. The histogram of mother_age in the baby weight prediction example is shown in the top-left panel, and different scaling functions (see the x-axis label) are shown in the remaining panels.

Nonlinear transformations

What if our data is skewed and neither uniformly distributed nor distributed like a bell curve? In that case, it is better to apply a *nonlinear transform* to the input before scaling it. One common trick is to take the logarithm of the input value before scaling it. Other common transformations include the sigmoid and polynomial expansions (square, square root, cube, cube root, and so on). We'll know that we have a good transformation function if the distribution of the transformed value becomes uniform or normally distributed.

Assume that we are building a model to predict the sales of a nonfiction book. One of the inputs to the model is the popularity of the Wikipedia page corresponding to the topic. The number of views of pages in Wikipedia is, however, highly skewed and occupies a large dynamic range (see the left panel of Figure 2-4: the distribution is highly skewed toward rarely viewed pages, but the most common pages are viewed tens of millions of times). By taking the logarithm of the views, then taking the fourth root of this log value and scaling the result linearly, we obtain something that is in the desired range and somewhat bell-shaped. For details of the code to query the Wikipedia data, apply these transformations, and generate this plot, refer to the GitHub repository (*https://github.com/GoogleCloudPlatform/ml-design-patterns/blob/master/02_data_representation/simple_data_representation.ipynb*) for this book.

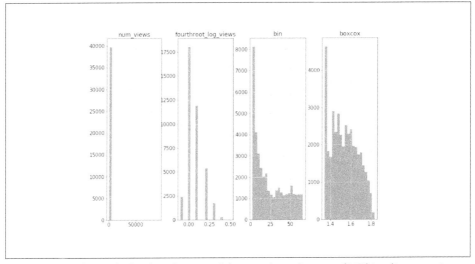

Figure 2-4. Left panel: the distribution of the number of views of Wikipedia pages is highly skewed and occupies a large dynamic range. The second panel demonstrates that problems can be addressed by transforming the number of views using the logarithm, a power function, and linear scaling in succession. The third panel shows the effect of histogram equalization and the fourth panel shows the effect of the Box-Cox transform.

It can be difficult to devise a linearizing function that makes the distribution look like a bell curve. An easier approach is to bucketize the number of views, choosing the bucket boundaries to fit the desired output distribution. A principled approach to choosing these buckets is to do *histogram equalization*, where the bins of the histogram are chosen based on quantiles of the raw distribution, (see the third panel of Figure 2-4). In the ideal situation, histogram equalization results in a uniform distribution (although not in this case, because of repeated values in the quantiles).

To carry out histogram equalization in BigQuery, we can do:

```
ML.BUCKETIZE(num_views, bins) AS bin
```

where the bins are obtained from:

```
APPROX_QUANTILES(num_views, 100) AS bins
```

See the notebook (*https://github.com/GoogleCloudPlatform/ml-design-patterns/blob/master/02_data_representation/simple_data_representation.ipynb*) in the code repository of this book for full details.

Another method to handle skewed distributions is to use a parametric transformation technique like the *Box-Cox transform*. Box-Cox chooses its single parameter, lambda, to control the "heteroscedasticity" so that the variance no longer depends on the magnitude. Here, the variance among rarely viewed Wikipedia pages will be much smaller than the variance among frequently viewed pages, and Box-Cox tries to equalize the variance across all ranges of the number of views. This can be done using Python's SciPy package:

```
traindf['boxcox'], est_lambda = (
    scipy.stats.boxcox(traindf['num_views']))
```

The parameter estimated over the training dataset (`est_lambda`) is then used to transform other values:

```
evaldf['boxcox'] = scipy.stats.boxcox(evaldf['num_views'], est_lambda)
```

Array of numbers

Sometimes, the input data is an array of numbers. If the array is of fixed length, data representation can be rather simple: flatten the array and treat each position as a separate feature. But often, the array will be of variable length. For example, one of the inputs to the model to predict the sales of a nonfiction book might be the sales of all previous books on the topic. An example input might be:

```
[2100, 15200, 230000, 1200, 300, 532100]
```

Obviously, the length of this array will vary in each row because there are different numbers of books published on different topics.

Common idioms to handle arrays of numbers include the following:

- Representing the input array in terms of its bulk statistics. For example, we might use the length (that is, count of previous books on topic), average, median, minimum, maximum, and so forth.
- Representing the input array in terms of its empirical distribution—i.e., by the 10th/20th/... percentile, and so on.
- If the array is ordered in a specific way (for example, in order of time or by size), representing the input array by the last three or some other fixed number of items. For arrays of length less than three, the feature is padded to a length of three with missing values.

All these end up representing the variable-length array of data as a fixed-length feature. We could have also formulated this problem as a time-series forecasting problem, as the problem of forecasting the sales of the next book on the topic based on the time history of sales of previous books. By treating the sales of previous books as an array input, we are assuming that the most important factors in predicting a book's sales are characteristics of the book itself (author, publisher, reviews, and so on) and not the temporal continuity of the sales amounts.

Categorical Inputs

Because most modern, large-scale machine learning models (random forests, support vector machines, neural networks) operate on numerical values, categorical inputs have to be represented as numbers.

Simply enumerating the possible values and mapping them to an ordinal scale will work poorly. Suppose that one of the inputs to the model that predicts the sales of a nonfiction book is the language that the book is written in. We can't simply create a mapping table like this:

Categorical input	Numeric feature
English	1.0
Chinese	2.0
German	3.0

This is because the machine learning model will then attempt to interpolate between the popularity of German and English books to get the popularity of the book in Chinese! Because there is no ordinal relationship between languages, we need to use a categorical to numeric mapping that allows the model to learn the market for books written in these languages independently.

One-hot encoding

The simplest method of mapping categorical variables while ensuring that the variables are independent is *one-hot encoding*. In our example, the categorical input variable would be converted into a three-element feature vector using the following mapping:

Categorical input	Numeric feature
English	[1.0, 0.0, 0.0]
Chinese	[0.0, 1.0, 0.0]
German	[0.0, 0.0, 1.0]

One-hot encoding requires us to know the *vocabulary* of the categorical input beforehand. Here, the vocabulary consists of three tokens (English, Chinese, and German), and the length of the resulting feature is the size of this vocabulary.

Dummy Coding or One-Hot Encoding?

Technically, a 2-element feature vector is enough to provide a unique mapping for a vocabulary of size 3:

Categorical input	Numeric feature
English	[0.0, 0.0]
Chinese	[1.0, 0.0]
German	[0.0, 1.0]

This is called *dummy coding*. Because dummy coding is a more compact representation, it is preferred in statistical models that perform better when the inputs are linearly independent.

Modern machine learning algorithms, though, don't require their inputs to be linearly independent and use methods such as L1 regularization to prune redundant inputs. The additional degree of freedom allows the framework to transparently handle a missing input in production as all zeros:

Categorical input	Numeric feature
English	[1.0, 0.0, 0.0]
Chinese	[0.0, 1.0, 0.0]
German	[0.0, 0.0, 1.0]
(missing)	[0.0, 0.0, 0.0]

Therefore, many machine learning frameworks often support only one-hot encoding.

In some circumstances, it can be helpful to treat a numeric input as categorical and map it to a one-hot encoded column:

When the numeric input is an index

For example, if we are trying to predict traffic levels and one of our inputs is the day of the week, we could treat the day of the week as numeric (1, 2, 3, …, 7), but it is helpful to recognize that the day of the week here is not a continuous scale but really just an index. It is better to treat it as categorical (Sunday, Monday, …, Saturday) because the indexing is arbitrary. Should the week start on Sunday (as in the USA), Monday (as in France), or Saturday (as in Egypt)?

When the relationship between input and label is not continuous

What should tip the scale toward treating day of the week as a categorical feature is that traffic levels on Friday are not affected by those on Thursday and Saturday.

When it is advantageous to bucket the numeric variable

In most cities, traffic levels depend on whether it is the weekend, and this can vary by location (Saturday and Sunday in most of the world, Thursday and Friday in some Islamic countries). It would be helpful to then treat day of the week as a boolean feature (weekend or weekday). Such a mapping where the number of distinct inputs (here, seven) is greater than the number of distinct feature values (here, two) is called bucketing. Commonly, bucketing is done in terms of ranges—for example, we might bucket `mother_age` into ranges that break at 20, 25, 30, etc. and treat each of these bins as categorical, but it should be realized that this loses the ordinal nature of `mother_age`.

When we want to treat different values of the numeric input as being independent when it comes to their effect on the label

For example, the weight of a baby depends on the plurality[2] of the delivery since twins and triplets tend to weigh less than single births. So, a lower-weight baby, if part of a triplet, might be healthier than a twin baby with the same weight. In this case, we might map the plurality to a categorical variable, since a categorical variable allows the model to learn independent tunable parameters for the different values of plurality. Of course, we can do this only if we have enough examples of twins and triplets in our dataset.

---

2 If twins, the plurality is 2. If triplets, the plurality is 3.

Array of categorical variables

Sometimes, the input data is an array of categories. If the array is of fixed length, we can treat each array position as a separate feature. But often, the array will be of variable length. For example, one of the inputs to the natality model might be the type of previous births to this mother:

```
[Induced, Induced, Natural, Cesarean]
```

Obviously, the length of this array will vary in each row because there are different numbers of older siblings for each baby.

Common idioms to handle arrays of categorical variables include the following:

- *Counting* the number of occurrences of each vocabulary item. So, the representation for our example would be [2, 1, 1] assuming that the vocabulary is Induced, Natural, and Cesarean (in that order). This is now a fixed-length array of numbers that can be flattened and used in positional order. If we have an array where an item can occur only once (for example, of languages a person speaks), or if the feature just indicates presence and not count (such as whether the mother has ever had a Cesarean operation), then the count at each position is 0 or 1, and this is called *multi-hot encoding*.

- To avoid large numbers, the *relative frequency* can be used instead of the count. The representation for our example would be [0.5, 0.25, 0.25] instead of [2, 1, 1]. Empty arrays (first-born babies with no previous siblings) are represented as [0, 0, 0]. In natural language processing, the relative frequency of a word overall is normalized by the relative frequency of documents that contain the word to yield TF-IDF (*https://oreil.ly/kNYHr*) (short for term frequency–inverse document frequency). TF-IDF reflects how unique a word is to a document.

- If the array is ordered in a specific way (e.g., in order of time), representing the input array by the last three items. Arrays shorter than three are padded with missing values.

- Representing the array by bulk statistics, e.g., the length of the array, the mode (most common entry), the median, the 10th/20th/… percentile, etc.

Of these, the counting/relative-frequency idiom is the most common. Note that both of these are a generalization of one-hot encoding—if the baby had no older siblings, the representation would be [0, 0, 0], and if the baby had one older sibling who was born in a natural birth, the representation would be [0, 1, 0].

Having seen simple data representations, let's discuss design patterns that help with data representation.

Design Pattern 1: Hashed Feature

The Hashed Feature design pattern addresses three possible problems associated with categorical features: incomplete vocabulary, model size due to cardinality, and cold start. It does so by grouping the categorical features and accepting the trade-off of collisions in the data representation.

Problem

One-hot encoding a categorical input variable requires knowing the vocabulary beforehand. This is not a problem if the input variable is something like the language a book is written in or the day of the week that traffic level is being predicted.

What if the categorical variable in question is something like the `hospital_id` of where the baby is born or the `physician_id` of the person delivering the baby? Categorical variables like these pose a few problems:

- Knowing the vocabulary requires extracting it from the training data. Due to random sampling, it is possible that the training data does not contain all the possible hospitals or physicians. The vocabulary might be *incomplete*.

- The categorical variables have *high cardinality*. Instead of having feature vectors with three languages or seven days, we have feature vectors whose length is in the thousands to millions. Such feature vectors pose several problems in practice. They involve so many weights that the training data may be insufficient. Even if we can train the model, the trained model will require a lot of space to store because the entire vocabulary is needed at serving time. Thus, we may not be able to deploy the model on smaller devices.

- After the model is placed into production, new hospitals might be built and new physicians hired. The model will be unable to make predictions for these, and so a separate serving infrastructure will be required to handle such *cold-start* problems.

 Even with simple representations like one-hot encoding, it is worth anticipating the cold-start problem and explicitly reserving all zeros for out-of-vocabulary inputs.

As a concrete example, let's take the problem of predicting the arrival delay of a flight. One of the inputs to the model is the departure airport. There were, at the time the dataset was collected, 347 airports in the United States:

```
SELECT
    DISTINCT(departure_airport)
FROM `bigquery-samples.airline_ontime_data.flights`
```

Some airports had as few as one to three flights over the entire time period, and so we expect that the training data vocabulary will be incomplete. 347 is large enough that the feature will be quite sparse, and it is certainly the case that new airports will get built. All three problems (incomplete vocabulary, high cardinality, cold start) will exist if we one-hot encode the departure airport.

The airline dataset, like the natality dataset and nearly all the other datasets that we use in this book for illustration, is a public dataset in BigQuery (*https://oreil.ly/lgcKA*), so you can try the query out. At the time we are writing this, 1 TB/month of querying is free, and there is a sandbox available so that you can use BigQuery up to this limit without putting down a credit card. We encourage you to bookmark our GitHub repository. For example, see the notebook (*https://github.com/GoogleCloud Platform/ml-design-patterns/blob/master/02_data_representation/hashed_feature.ipynb*) in GitHub for the full code.

Solution

The Hashed Feature design pattern represents a categorical input variable by doing the following:

1. Converting the categorical input into a unique string. For the departure airport, we can use the three-letter IATA code (*https://oreil.ly/B8nLw*) for the airport.

2. Invoking a deterministic (no random seeds or salt) and portable (so that the same algorithm can be used in both training and serving) hashing algorithm on the string.

3. Taking the remainder when the hash result is divided by the desired number of buckets. Typically, the hashing algorithm returns an integer that can be negative and the modulo of a negative integer is negative. So, the absolute value of the result is taken.

In BigQuery SQL, these steps are achieved like this:

```
ABS(MOD(FARM_FINGERPRINT(airport), numbuckets))
```

The `FARM_FINGERPRINT` function uses FarmHash, a family of hashing algorithms that is deterministic, well-distributed (*https://github.com/google/farmhash/blob/master/Understanding_Hash_Functions*), and for which implementations are available (*https://github.com/google/farmhash*) in a number of programming languages.

In TensorFlow, these steps are implemented by the `feature_column` function:

```
tf.feature_column.categorical_column_with_hash_bucket(
    airport, num_buckets, dtype=tf.dtypes.string)
```

For example, Table 2-1 shows the FarmHash of some IATA airport codes when hashed into 3, 10, and 1,000 buckets.

Table 2-1. The FarmHash of some IATA airport codes when hashed into different numbers of buckets

Row	departure_airport	hash3	hash10	hash1000
1	DTW	1	3	543
2	LBB	2	9	709
3	SNA	2	7	587
4	MSO	2	7	737
5	ANC	0	8	508
6	PIT	1	7	267
7	PWM	1	9	309
8	BNA	1	4	744
9	SAF	1	2	892
10	IPL	2	1	591

Why It Works

Assume that we have chosen to hash the airport code using 10 buckets (hash10 in Table 2-1). How does this address the problems we identified?

Out-of-vocabulary input

Even if an airport with a handful of flights is not part of the training dataset, its hashed feature value will be in the range [0–9]. Therefore, there is no resilience problem during serving—the unknown airport will get the predictions corresponding with other airports in the hash bucket. The model will not error out.

If we have 347 airports, an average of 35 airports will get the same hash bucket code if we hash it into 10 buckets. An airport that is missing from the training dataset will "borrow" its characteristics from the other similar ~35 airports in the hash bucket. Of course, the prediction for a missing airport won't be accurate (it is unreasonable to expect accurate predictions for unknown inputs), but it will be in the right range.

Choose the number of hash buckets by balancing the need to handle out-of-vocabulary inputs reasonably and the need to have the model accurately reflect the categorical input. With 10 hash buckets, ~35 airports get commingled. A good rule of thumb is to choose the number of hash buckets such that each bucket gets about five entries. In this case, that would mean that 70 hash buckets is a good compromise.

High cardinality

It's easy to see that the high cardinality problem is addressed as long as we choose a small enough number of hash buckets. Even if we have millions of airports or hospitals or physicians, we can hash them into a few hundred buckets, thus keeping the system's memory and model size requirements practical.

We don't need to store the vocabulary because the transformation code is independent of the actual data value and the core of the model only deals with num_buckets inputs, not the full vocabulary.

It is true that hashing is lossy—since we have 347 airports, an average of 35 airports will get the same hash bucket code if we hash it into 10 buckets. When the alternative is to discard the variable because it is too wide, though, a lossy encoding is an acceptable compromise.

Cold start

The cold-start situation is similar to the out-of-vocabulary situation. If a new airport gets added to the system, it will initially get the predictions corresponding to other airports in the hash bucket. As an airport gets popular, there will be more flights from that airport. As long as we periodically retrain the model, its predictions will start to reflect arrival delays from the new airport. This is discussed in more detail in the "Design Pattern 18: Continued Model Evaluation" on page 220 in Chapter 5.

By choosing the number of hash buckets such that each bucket gets about five entries, we can ensure that any bucket will have reasonable initial results.

Trade-Offs and Alternatives

Most design patterns involve some kind of a trade-off, and the Hashed Feature design pattern is no exception. The key trade-off here is that we lose model accuracy.

Bucket collision

The modulo part of the Hashed Feature implementation is a lossy operation. By choosing a hash bucket size of 100, we are choosing to have 3–4 airports share a bucket. We are explicitly compromising on the ability to accurately represent the data (with a fixed vocabulary and one-hot encoding) in order to handle out-of-vocabulary inputs, cardinality/model size constraints, and cold-start problems. It is not a free lunch. Do not choose Hashed Feature if you know the vocabulary beforehand, if the vocabulary size is relatively small (in the thousands is acceptable for a dataset with millions of examples), and if cold start is not a concern.

Note that we cannot simply increase the number of buckets to an extremely high number hoping to avoid collisions altogether. Even if we raise the number of buckets to 100,000 with only 347 airports, the probability that at least two airports share the same hash bucket is 45%—unacceptably high (see Table 2-2). Therefore, we should use Hashed Features only if we are willing to tolerate multiple categorical inputs sharing the same hash bucket value.

Table 2-2. The expected number of entries per bucket and the probability of at least one collision when IATA airport codes are hashed into different numbers of buckets

num_hash_buckets	entries_per_bucket	collision_prob
3	115.666667	1.000000
10	34.700000	1.000000
100	3.470000	1.000000
1000	0.347000	1.000000
10000	0.034700	0.997697
100000	0.003470	0.451739

Skew

The loss of accuracy is particularly acute when the distribution of the categorical input is highly skewed. Consider the case of the hash bucket that contains ORD (Chicago, one of the busiest airports in the world). We can find this using the following:

```
CREATE TEMPORARY FUNCTION hashed(airport STRING, numbuckets INT64) AS (
    ABS(MOD(FARM_FINGERPRINT(airport), numbuckets))
);

WITH airports AS (
SELECT
    departure_airport, COUNT(1) AS num_flights
FROM `bigquery-samples.airline_ontime_data.flights`
GROUP BY departure_airport
)

SELECT
    departure_airport, num_flights
FROM airports
WHERE hashed(departure_airport, 100) = hashed('ORD', 100)
```

The result shows that while there are ~3.6 million flights from ORD, there are only ~67,000 flights from BTV (Burlington, Vermont):

departure_airport	num_flights
ORD	3610491
BTV	66555
MCI	597761

This indicates that, for all practical purposes, the model will impute the long taxi times and weather delays that Chicago experiences to the municipal airport in

Burlington, Vermont! The model accuracy for BTV and MCI (Kansas City airport) will be quite poor because there are so many flights out of Chicago.

Aggregate feature

In cases where the distribution of a categorical variable is skewed or where the number of buckets is so small that bucket collisions are frequent, we might find it helpful to add an aggregate feature as an input to our model. For example, for every airport, we could find the probability of on-time flights in the training dataset and add it as a feature to our model. This allows us to avoid losing the information associated with individual airports when we hash the airport codes. In some cases, we might be able to avoid using the airport name as a feature entirely, since the relative frequency of on-time flights might be sufficient.

Hyperparameter tuning

Because of the trade-offs with bucket collision frequency, choosing the number of buckets can be difficult. It very often depends on the problem itself. Therefore, we recommend that you treat the number of buckets as a hyperparameter that is tuned:

```
- parameterName: nbuckets
    type: INTEGER
    minValue: 10
    maxValue: 20
    scaleType: UNIT_LINEAR_SCALE
```

Make sure that the number of buckets remains within a sensible range of the cardinality of the categorical variable being hashed.

Cryptographic hash

What makes the Hashed Feature lossy is the modulo part of the implementation. What if we were to avoid the modulo altogether? After all, the farm fingerprint has a fixed length (an INT64 is 64 bits), and so it can be represented using 64 feature values, each of which is 0 or 1. This is called *binary encoding*.

However, binary encoding does not solve the problem of out-of-vocabulary inputs or cold start (only the problem of high cardinality). In fact, the bitwise coding is a red herring. If we don't do a modulo, we can get a unique representation by simply encoding the three characters that form the IATA code (thus using a feature of length $3*26=78$). The problem with this representation is immediately obvious: airports whose names start with the letter O have nothing in common when it comes to their flight delay characteristics—the encoding has created a *spurious correlation* between airports that start with the same letter. The same insight holds in binary space as well. Because of this, we do not recommend binary encoding of farm fingerprint values.

Binary encoding of an MD5 hash will not suffer from this spurious correlation problem because the output of an MD5 hash is uniformly distributed, and so the resulting bits will be uniformly distributed. However, unlike the Farm Fingerprint algorithm, the MD5 hash is not deterministic and not unique—it is a one-way hash and will have many unexpected collisions.

In the Hashed Feature design pattern, we have to use a fingerprint hashing algorithm and not a cryptographic hashing algorithm. This is because the goal of a fingerprint function is to produce a deterministic and unique value. If you think about it, this is a key requirement of preprocessing functions in machine learning, since we need to apply the same function during model serving and get the same hashed value. A fingerprint function does not produce a uniformly distributed output. Cryptographic algorithms such as MD5 or SHA1 do produce uniformly distributed output, but they are not deterministic and are purposefully made to be computationally expensive. Therefore, a cryptographic hash is not usable in a feature engineering context where the hashed value computed for a given input during prediction has to be the same as the hash computed during training, and where the hash function should not slow down the machine learning model.

The reason that MD5 is not deterministic is that a "salt" is typically added to the string to be hashed. The salt is a random string added to each password (*https://oreil.ly/cv7PS*) to ensure that even if two users happen to use the same password, the hashed value in the database will be different. This is needed to thwart attacks based on "rainbow tables," which are attacks that rely on dictionaries of commonly chosen passwords and that compare the hash of the known password against hashes in the database. As computational power has increased, it is possible to carry out a brute-force attack on every possible salt as well, and so modern cryptographic implementations do their hashing in a loop to increase the computational expense. Even if we were to turn off the salt and reduce the number of iterations to one, the MD5 hash is only one way. It won't be unique.

The bottom line is that we need to use a fingerprint hashing algorithm, and we need to modulo the resulting hash.

Order of operations

Note that we do the modulo first, and then the absolute value:

```
CREATE TEMPORARY FUNCTION hashed(airport STRING, numbuckets INT64) AS (
    ABS(MOD(FARM_FINGERPRINT(airport), numbuckets))
);
```

The order of `ABS`, `MOD`, and `FARM_FINGERPRINT` in the preceding snippet is important because the range of `INT64` is not symmetric. Specifically, its range is between -9,223,372,036,854,775,808 and 9,223,372,036,854,775,807 (both inclusive). So, if we were to do:

```
ABS(FARM_FINGERPRINT(airport))
```

we would run into a rare and likely unreproducible overflow error if the `FARM_FINGERPRINT` operation happened to return -9,223,372,036,854,775,808 since its absolute value can not be represented using an `INT64`!

Empty hash buckets

Although unlikely, there is a remote possibility that even if we choose 10 hash buckets to represent 347 airports, one of the hash buckets will be empty. Therefore, when using hashed feature columns, it may be beneficial (*https://oreil.ly/xlwAH*) to also use L2 regularization so that the weights associated with an empty bucket will be driven to near-zero. This way, if an out-of-vocabulary airport does fall into an empty bucket, it will not cause the model to become numerically unstable.

Design Pattern 2: Embeddings

Embeddings are a learnable data representation that map high-cardinality data into a lower-dimensional space in such a way that the information relevant to the learning problem is preserved. Embeddings are at the heart of modern-day machine learning and have various incarnations throughout the field.

Problem

Machine learning models systematically look for patterns in data that capture how the properties of the model's input features relate to the output label. As a result, the data representation of the input features directly affects the quality of the final model. While handling structured, numeric input is fairly straightforward, the data needed to train a machine learning model can come in myriad varieties, such as categorical features, text, images, audio, time series, and many more. For these data representations, we need a meaningful numeric value to supply our machine learning model so these features can fit within the typical training paradigm. Embeddings provide a way to handle some of these disparate data types in a way that preserves similarity between items and thus improves our model's ability to learn those essential patterns.

One-hot encoding is a common way to represent categorical input variables. For example, consider the plurality input in the natality dataset.[3] This is a categorical

---

3 This dataset is available in BigQuery: *bigquery-public-data.samples.natality*.

input that has six possible values: ['Single(1)', 'Multiple(2+)', 'Twins(2)', 'Triplets(3)', 'Quadruplets(4)', 'Quintuplets(5)']. We can handle this categorical input using a one-hot encoding that maps each potential input string value to a unit vector in R^6, as shown in Table 2-3.

Table 2-3. An example of one-hot encoding categorical inputs for the natality dataset

Plurality	One-hot encoding
Single(1)	[1,0,0,0,0,0]
Multiple(2+)	[0,1,0,0,0,0]
Twins(2)	[0,0,1,0,0,0]
Triplets(3)	[0,0,0,1,0,0]
Quadruplets(4)	[0,0,0,0,1,0]
Quintuplets(5)	[0,0,0,0,0,1]

When encoded in this way, we need six dimensions to represent each of the different categories. Six dimensions may not be so bad, but what if we had many, many more categories to consider?

For example, what if our dataset consisted of customers' view history of our video database and our task is to suggest a list of new videos given customers' previous video interactions? In this scenario, the customer_id field could have millions of unique entries. Similarly, the video_id of previously watched videos could contain thousands of entries as well. One-hot encoding *high-cardinality* categorical features like video_ids or customer_ids as inputs to a machine learning model leads to a sparse matrix that isn't well suited for a number of machine learning algorithms.

The second problem with one-hot encoding is that it treats the categorical variables as being *independent*. However, the data representation for twins should be close to the data representation for triplets and quite far away from the data representation for quintuplets. A multiple is most likely a twin, but could be a triplet. As an example, Table 2-4 shows an alternate representation of the plurality column in a lower dimension that captures this *closeness* relationship.

Table 2-4. Using lower dimensionality embeddings to represent the plurality column in the natality dataset.

Plurality	Candidate encoding
Single(1)	[1.0,0.0]
Multiple(2+)	[0.0,0.6]
Twins(2)	[0.0,0.5]
Triplets(3)	[0.0,0.7]
Quadruplets(4)	[0.0,0.8]
Quintuplets(5)	[0.0,0.9]

These numbers are arbitrary of course. But is it possible to learn the best possible representation of the plurality column using just two dimensions for the natality problem? That is the problem that the Embeddings design pattern solves.

The same problem of high cardinality and dependent data also occurs in images and text. Images consist of thousands of pixels, which are not independent of one another. Natural language text is drawn from a vocabulary in the tens of thousands of words, and a word like walk is closer to the word run than to the word book.

Solution

The Embeddings design pattern addresses the problem of representing high-cardinality data densely in a lower dimension by passing the input data through an embedding layer that has trainable weights. This will map the high-dimensional, categorical input variable to a real-valued vector in some low-dimensional space. The weights to create the dense representation are learned as part of the optimization of the model (see Figure 2-5). In practice, these embeddings end up capturing closeness relationships in the input data.

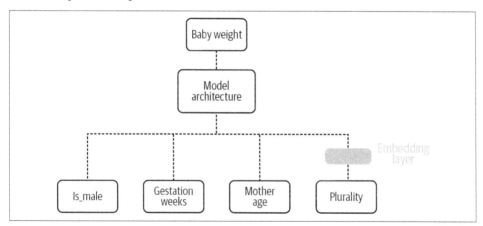

Figure 2-5. The weights of an embedding layer are learned as parameters during training.

Because embeddings capture closeness relationships in the input data in a lower-dimensional representation, we can use an embedding layer as a replacement for clustering techniques (e.g., customer segmentation) and dimensionality reduction methods like principal components analysis (PCA). Embedding weights are determined in the main model training loop, thus saving the need to cluster or do PCA beforehand.

The weights in the embedding layer would be learned as part of the gradient descent procedure when training the natality model.

At the end of training, the weights of the embedding layer might be such that the encoding for the categorical variables is as shown in Table 2-5.

Table 2-5. One-hot and learned encodings for the plurality column in the natality dataset

Plurality	One-hot encoding	Learned encoding
Single(1)	[1,0,0,0,0,0]	[0.4, 0.6]
Multiple(2+)	[0,1,0,0,0,0]	[0.1, 0.5]
Twins(2)	[0,0,1,0,0,0]	[-0.1, 0.3]
Triplets(3)	[0,0,0,1,0,0]	[-0.2, 0.5]
Quadruplets(4)	[0,0,0,0,1,0]	[-0.4, 0.3]
Quintuplets(5)	[0,0,0,0,0,1]	[-0.6, 0.5]

The embedding maps a sparse, one-hot encoded vector to a dense vector in R^2.

In TensorFlow, we first construct a categorical feature column for the feature, then wrap that in an embedding feature column. For example, for our plurality feature, we would have:

```
plurality = tf.feature_column.categorical_column_with_vocabulary_list(
        'plurality', ['Single(1)', 'Multiple(2+)', 'Twins(2)',
'Triplets(3)', 'Quadruplets(4)', 'Quintuplets(5)'])
plurality_embed = tf.feature_column.embedding_column(plurality, dimension=2)
```

The resulting feature column (`plurality_embed`) is used as input to the downstream nodes of the neural network instead of the one-hot encoded feature column (`plurality`).

Text embeddings

Text provides a natural setting where it is advantageous to use an embedding layer. Given the cardinality of a vocabulary (often on the order of tens of thousands of words), one-hot encoding each word isn't practical. This would create an incredibly large (high-dimensional) and sparse matrix for training. Also, we'd like similar words to have embeddings close by and unrelated words to be far away in embedding space. Therefore, we use a dense word embedding to vectorize the discrete text input before passing to our model.

To implement a text embedding in Keras, we first create a tokenization for each word in our vocabulary, as shown in Figure 2-6. Then we use this tokenization to map to an embedding layer, similar to how it was done for the plurality column.

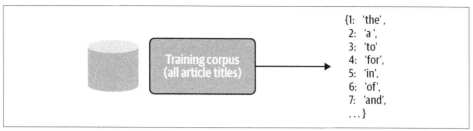

Figure 2-6. The tokenizer creates a lookup table that maps each word to an index.

The tokenization is a lookup table that maps each word in our vocabulary to an index. We can think of this as a one-hot encoding of each word where the tokenized index is the location of the nonzero element in the one-hot encoding. This requires a full pass over the entire dataset (let's assume these consist of titles of articles[4]) to create the lookup table and can be done in Keras. The complete code can be found in the repository (*https://github.com/GoogleCloudPlatform/ml-design-patterns/blob/master/02_data_representation/embeddings.ipynb*) for this book:

```
from tensorflow.keras.preprocessing.text import Tokenizer

tokenizer = Tokenizer()
tokenizer.fit_on_texts(titles_df.title)
```

Here we can use the `Tokenizer` class in the *keras.preprocessing.text* library. The call to `fit_on_texts` creates a lookup table that maps each of the words found in our titles to an index. By calling `tokenizer.index_word`, we can examine this lookup table directly:

```
tokenizer.index_word
{1: 'the',
 2: 'a',
 3: 'to',
 4: 'for',
 5: 'in',
 6: 'of',
 7: 'and',
 8: 's',
 9: 'on',
 10: 'with',
 11: 'show',
 ...
```

---

4 This dataset is available in BigQuery: *bigquery-public-data.hacker_news.stories.*

We can then invoke this mapping with the `texts_to_sequences` method of our tokenizer. This maps each sequence of words in the text input being represented (here, we assume that they are titles of articles) to a sequence of tokens corresponding to each word as in Figure 2-7:

```
integerized_titles = tokenizer.texts_to_sequences(titles_df.title)
```

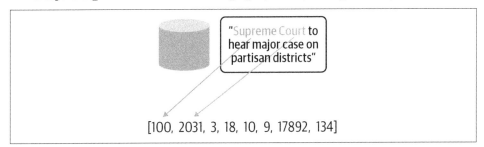

Figure 2-7. Using the tokenizer, each title is mapped to a sequence of integer index values.

The tokenizer contains other relevant information that we will use later for creating an embedding layer. In particular, `VOCAB_SIZE` captures the number of elements of the index lookup table and `MAX_LEN` contains the maximum length of the text strings in the dataset:

```
VOCAB_SIZE = len(tokenizer.index_word)
MAX_LEN = max(len(sequence) for sequence in integerized_titles)
```

Before creating the model, it is necessary to preprocess the titles in the dataset. We'll need to pad the elements of our title to feed into the model. Keras has the helper functions `pad_sequence` for that on the top of the tokenizer methods. The function `create_sequences` takes both titles as well as the maximum sentence length as input and returns a list of the integers corresponding to our tokens padded to the sentence maximum length:

```
from tensorflow.keras.preprocessing.sequence import pad_sequences

def create_sequences(texts, max_len=MAX_LEN):
    sequences = tokenizer.texts_to_sequences(texts)
    padded_sequences = pad_sequences(sequences,
                                     max_len,
                                     padding='post')
    return padded_sequences
```

Next, we'll build a deep neural network (DNN) model in Keras that implements a simple embedding layer to transform the word integers into dense vectors. The Keras `Embedding` layer can be thought of as a map from the integer indices of specific words to dense vectors (their embeddings). The dimensionality of the embedding is determined by `output_dim`. The argument `input_dim` indicates the size of the vocabulary, and `input_shape` indicates the length of input sequences. Since here we have padded the titles before passing to the model, we set `input_shape=[MAX_LEN]`:

```
model = models.Sequential([layers.Embedding(input_dim=VOCAB_SIZE + 1,
                                            output_dim=embed_dim,
                                            input_shape=[MAX_LEN]),
                          layers.Lambda(lambda x: tf.reduce_mean(x,axis=1)),
                          layers.Dense(N_CLASSES, activation='softmax')])
```

Note that we need to put a custom Keras Lambda layer in between the embedding layer and the dense softmax layer to average the word vectors returned by the embedding layer. This is the average that's fed to the dense softmax layer. By doing so, we create a model that is simple but that loses information about the word order, creating a model that sees sentences as a "bag of words."

Image embeddings

While text deals with very sparse input, other data types, such as images or audio, consist of dense, high-dimensional vectors, usually with multiple channels containing raw pixel or frequency information. In this setting, an Embedding captures a relevant, low-dimensional representation of the input.

For image embeddings, a complex convolutional neural network—like Inception or ResNet—is first trained on a large image dataset, like ImageNet, containing millions of images and thousands of possible classification labels. Then, the last softmax layer is removed from the model. Without the final softmax classifier layer, the model can be used to extract a feature vector for a given input. This feature vector contains all the relevant information of the image so it is essentially a low-dimensional embedding of the input image.

Similarly, consider the task of image captioning, that is, generating a textual caption of a given image, shown in Figure 2-8.

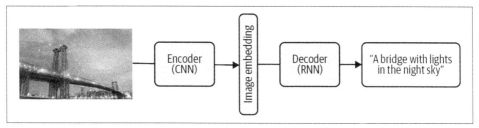

Figure 2-8. For the image translation task, the encoder produces a low-dimensional embedding representation of the image.

By training this model architecture on a massive dataset of image/caption pairs, the encoder learns an efficient vector representation for images. The decoder learns how to translate this vector to a text caption. In this sense, the encoder becomes an Image2Vec embedding machine.

Why It Works

The embedding layer is just another hidden layer of the neural network. The weights are then associated to each of the high-cardinality dimensions, and the output is passed through the rest of the network. Therefore, the weights to create the embedding are learned through the process of gradient descent just like any other weights in the neural network. This means that the resulting vector embeddings represent the most efficient low-dimensional representation of those feature values with respect to the learning task.

While this improved embedding ultimately aids the model, the embeddings themselves have inherent value and allow us to gain additional insight into our dataset.

Consider again the customer video dataset. By only using one-hot encoding, any two separate users, user_i and user_j, will have the same similarity measure. Similarly, the dot product or cosine similarity for any two distinct six-dimensional one-hot encodings of birth plurality would have zero similarity. This makes sense since the one-hot encoding is essentially telling our model to treat any two different birth pluralities as separate and unrelated. For our dataset of customers and video watches, we lose any notion of similarity between customers or videos. But this doesn't feel quite right. Two different customers or videos likely do have similarities between them. The same goes for birth plurality. The occurrence of quadruplets and quintuplets likely affects the birthweight in a statistically similar way as opposed to single child birthweights (see Figure 2-9).

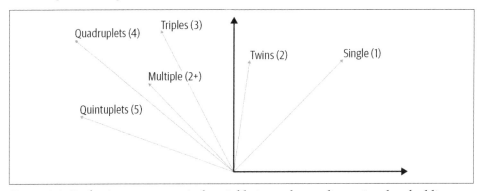

Figure 2-9. By forcing our categorical variable into a lower-dimensional embedding space, we can also learn relationships between the different categories.

When computing the similarity of plurality categories as one-hot encoded vectors, we obtain the identity matrix since each category is treated as a distinct feature (see Table 2-6).

Table 2-6. When features are one-hot encoded, the similarity matrix is just the identity matrix

	Single(1)	Multiple(2+)	Twins(2)	Triplets(3)	Quadruplets(4)	Quintuplets(5)
Single(1)	1	0	0	0	0	0
Multiple(2+)	-	1	0	0	0	0
Twins(2)	-	-	1	0	0	0
Triplets(3)	-	-	-	1	0	0
Quadruplets(4)	-	-	-	-	1	0
Quintuplets(5)	-	-	-	-	-	1

However, once the plurality is embedded into two dimensions, the similarity measure becomes nontrivial, and important relationships between the different categories emerge (see Table 2-7).

Table 2-7. When the features are embedded in two dimensions, the similarity matrix gives us more information

	Single(1)	Multiple(2+)	Twins(2)	Triplets(3)	Quadruplets(4)	Quintuplets(5)
Single(1)	1	0.92	0.61	0.57	0.06	0.1
Multiple(2+)	-	1	0.86	0.83	0.43	0.48
Twins(2)	-		1	0.99	0.82	0.85
Triplets(3)	-			1	0.85	0.88
Quadruplets(4)	-				1	0.99
Quintuplets(5)	-	-	-	-	-	1

Thus, a learned embedding allows us to extract inherent similarities between two separate categories and, given there is a numeric vector representation, we can precisely quantify the similarity between two categorical features.

This is easy to visualize with the natality dataset, but the same principle applies when dealing with `customer_ids` embedded into 20-dimensional space. When applied to our customer dataset, embeddings allow us to retrieve similar customers to a given `customer_id` and make suggestions based on similarity, such as which videos they are likely to watch, as shown in Figure 2-10. Furthermore, these user and item embeddings can be combined with other features when training a separate machine learning model. Using pre-trained embeddings in machine learning models is referred to as *transfer learning*.

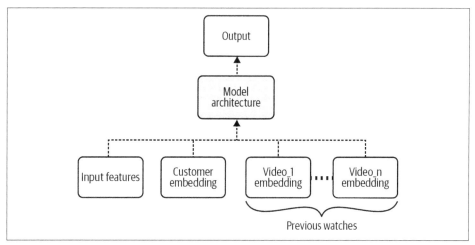

Figure 2-10. By learning a low-dimensional, dense embedding vector for each customer and video, an embedding-based model is able to generalize well with less of a manual feature engineering burden.

Trade-Offs and Alternatives

The main trade-off with using an embedding is the compromised representation of the data. There is a loss of information involved in going from a high-cardinality representation to a lower-dimensional representation. In return, we gain information about closeness and context of the items.

Choosing the embedding dimension

The exact dimensionality of the embedding space is something that we choose as a practitioner. So, should we choose a large or small embedding dimension? Of course, as with most things in machine learning, there is a trade-off. The lossiness of the representation is controlled by the size of the embedding layer. By choosing a very small output dimension of an embedding layer, too much information is forced into a small vector space and context can be lost. On the other hand, when the embedding dimension is too large, the embedding loses the learned contextual importance of the features. At the extreme, we're back to the problem encountered with one-hot encoding. The optimal embedding dimension is often found through experimentation, similar to choosing the number of neurons in a deep neural network layer.

If we're in a hurry, one rule of thumb is to use the fourth root (*https://oreil.ly/ywFco*) of the total number of unique categorical elements while another is that the embedding dimension should be approximately 1.6 times the square root (*https://oreil.ly/github-fastai-2-blob-fastai-2-tabular-model-py*) of the number of unique elements in the category, and no less than 600. For example, suppose we wanted to use an embedding layer to encode a feature that has 625 unique values. Using the first rule of thumb, we would choose an embedding dimension for plurality of 5, and using the second rule of thumb, we'd choose 40. If we are doing hyperparameter tuning, it might be worth searching within this range.

Autoencoders

Training embeddings in a supervised way can be hard because it requires a lot of labeled data. For an image classification model like Inception to be able to produce useful image embeddings, it is trained on ImageNet, which has 14 million labeled images. Autoencoders provide one way to get around this need for a massive labeled dataset.

The typical autoencoder architecture, shown in Figure 2-11, consists of a bottleneck layer, which is essentially an embedding layer. The portion of the network before the bottleneck (the "encoder") maps the high-dimensional input into a lower-dimensional embedding layer, while the latter network (the "decoder") maps that representation back to a higher dimension, typically the same dimension as the original. The model is typically trained on some variant of a reconstruction error, which forces the model's output to be as similar as possible to the input.

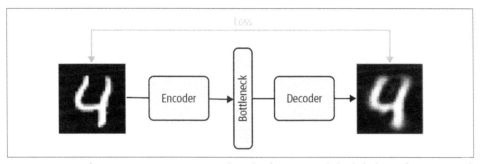

Figure 2-11. When training an autoencoder, the feature and the label are the same and the loss is the reconstruction error. This allows the autoencoder to achieve nonlinear dimension reduction.

Because the input is the same as the output, no additional labels are needed. The encoder learns an optimal nonlinear dimension reduction of the input. Similar to how PCA achieves linear dimension reduction, the bottleneck layer of an autoencoder is able to obtain nonlinear dimension reduction through the embedding.

This allows us to break a hard machine learning problem into two parts. First, we use all the unlabeled data we have to go from high cardinality to lower cardinality by using autoencoders as an *auxiliary learning task*. Then, we solve the actual image classification problem for which we typically have much less labeled data using the embedding produced by the auxiliary autoencoder task. This is likely to boost model performance, because now the model only has to learn the weights for the lower-dimension setting (i.e., it has to learn fewer weights).

In addition to image autoencoders, recent work (*https://oreil.ly/ywFco*) has focused on applying deep learning techniques for structured data. TabNet is a deep neural network specifically designed to learn from tabular data and can be trained in an unsupervised manner. By modifying the model to have an encoder-decoder structure, TabNet works as an autoencoder on tabular data, which allows the model to learn embeddings from structured data via a feature transformer.

Context language models

Is there an auxiliary learning task that works for text? Context language models like Word2Vec and masked language models like Bidirectional Encoding Representations from Transformers (BERT) change the learning task to a problem so that there is no scarcity of labels.

Word2Vec is a well-known method for constructing an embedding using shallow neural networks and combining two techniques—Continuous Bag of Words (CBOW) and a skip-gram model—applied to a large corpus of text, such as Wikipedia. While the goal of both models is to learn the context of a word by mapping input word(s) to the target word(s) with an intermediate embedding layer, an auxiliary goal is achieved that learns low-dimensional embeddings that best capture the context of words. The resulting word embeddings learned through Word2Vec capture the semantic relationships between words so that, in the embedding space, the vector representations maintain meaningful distance and directionality (Figure 2-12).

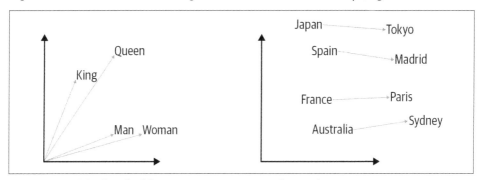

Figure 2-12. Word embeddings capture semantic relationships.

BERT is trained using a masked language model and next sentence prediction. For a masked language model, words are randomly masked from text and the model guesses what the missing word(s) are. Next sentence prediction is a classification task where the model predicts whether or not two sentences followed each other in the original text. So any corpus of text is suitable as a labeled dataset. BERT was initially trained on all of the English Wikipedia and BooksCorpus. Despite learning on these auxiliary tasks, the learned embeddings from BERT or Word2Vec have proven very powerful when used on other downstream training tasks. The word embeddings learned by Word2Vec are the same regardless of the sentence where the word appears. However, the BERT word embeddings are contextual, meaning the embedding vector is different depending on the context of how the word is used.

A pre-trained text embedding, like Word2Vec, NNLM, GLoVE, or BERT, can be added to a machine learning model to process text features in conjunction with structured inputs and other learned embeddings from our customer and video dataset (Figure 2-13).

Ultimately, embeddings learn to preserve information relevant to the prescribed training task. In the case of image captioning, the task is to learn how the context of the elements of an image relates to text. In the autoencoder architecture, the label is the same as the feature, so the dimension reduction of the bottleneck attempts to learn everything with no specific context of what is important.

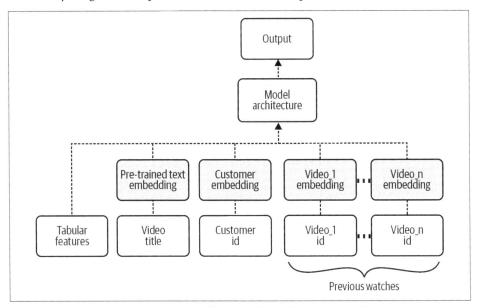

Figure 2-13. A pre-trained text embedding can be added to a model to process text features.

Embeddings in a data warehouse

Machine learning on structured data is best carried out directly in SQL on a data warehouse. This avoids the need to export data out of the warehouse and mitigates problems with data privacy and security.

Many problems, however, require a mix of structured data and natural language text or image data. In data warehouses, natural language text (such as reviews) is stored directly as columns, and images are typically stored as URLs to files in a cloud storage bucket. In these cases, it simplifies later machine learning to additionally store the embeddings of the text columns or of the images as array-type columns. Doing so will enable the easy incorporation of such unstructured data into machine learning models.

To create text embeddings, we can load a pre-trained model such as Swivel from TensorFlow Hub into BigQuery. The full code is on GitHub (*https://github.com/ GoogleCloudPlatform/ml-design-patterns/blob/master/02_data_representation/text_ embeddings.ipynb*):

```
CREATE OR REPLACE MODEL advdata.swivel_text_embed
OPTIONS(model_type='tensorflow', model_path='gs://BUCKET/swivel/*')
```

Then, use the model to transform the natural language text column into an embedding array and store the embedding lookup into a new table:

```
CREATE OR REPLACE TABLE advdata.comments_embedding AS
SELECT
  output_0 as comments_embedding,
  comments
FROM ML.PREDICT(MODEL advdata.swivel_text_embed,(
  SELECT comments, LOWER(comments) AS sentences
  FROM `bigquery-public-data.noaa_preliminary_severe_storms.wind_reports`
))
```

It is now possible to join against this table to get the text embedding for any comment. For image embeddings, we can similarly transform image URLs into embeddings and load them into the data warehouse.

Precomputing features in this manner is an example of the "Design Pattern 26: Feature Store" on page 295 (see Chapter 6).

Design Pattern 3: Feature Cross

The Feature Cross design pattern helps models learn relationships between inputs faster by explicitly making each combination of input values a separate feature.

Problem

Consider the dataset in Figure 2-14 and the task of creating a binary classifier that separates the + and – labels.

Using only the x_1 and x_2 coordinates, it is not possible to find a linear boundary that separates the + and – classes.

This means that to solve this problem, we have to make the model more complex, perhaps by adding more layers to the model. However, a simpler solution exists.

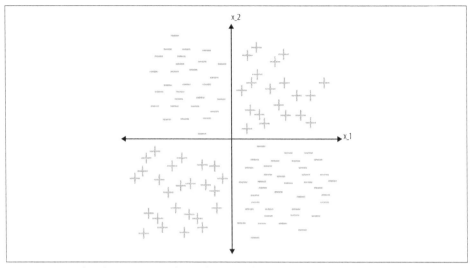

Figure 2-14. This dataset is not linearly separable using only x_1 and x_2 as inputs.

Solution

In machine learning, feature engineering is the process of using domain knowledge to create new features that aid the machine learning process and increase the predictive power of our model. One commonly used feature engineering technique is creating a feature cross.

A feature cross is a synthetic feature formed by concatenating two or more categorical features in order to capture the interaction between them. By joining two features in this way, it is possible to encode nonlinearity into the model, which can allow for predictive abilities beyond what each of the features would have been able to provide individually. Feature crosses provide a way to have the ML model learn relationships between the features faster. While more complex models like neural networks and trees can learn feature crosses on their own, using feature crosses explicitly can allow us to get away with training just a linear model. Consequently, feature crosses can speed up model training (less expensive) and reduce model complexity (less training data is needed).

To create a feature column for the dataset above, we can bucketize x_1 and x_2 each into two buckets, depending on their sign. This converts x_1 and x_2 into categorical features. Let A denote the bucket where x_1 >= 0 and B the bucket where x_1 < 0. Let C denote the bucket where x_2 >= 0 and D the bucket where x_2 < 0 (Figure 2-15).

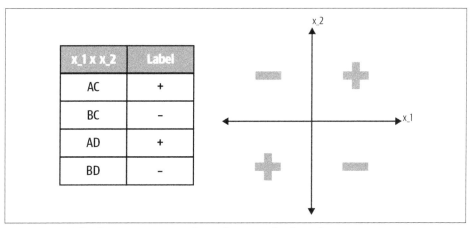

Figure 2-15. The feature cross introduces four new boolean features.

A feature cross of these bucketized features introduces four new boolean features for our model:

> AC where x_1 >= 0 and x_2 >= 0
>
> BC where x_1 < 0 and x_2 >= 0
>
> AD where x_1 >= 0 and x_2 < 0
>
> BD where x_1 < 0 and x_2 < 0

Each of these four boolean features (AC, BC, AD, and BD) would get its own weight when training the model. This means we can treat each quadrant as its own feature. Since the original dataset was split perfectly by the buckets we created, a feature cross of A and B is able to linearly separate the dataset.

But this is just an illustration. What about real-world data? Consider a public dataset of yellow cab rides in New York City (see Table 2-8).[5]

Table 2-8. A preview of the public New York City taxi dataset in BigQuery

pickup_datetime	pickuplon	pickuplat	dropofflon	dropofflat	passengers	fare_amount
2014-05-17 15:15:00 UTC	-73.99955	40.7606	-73.99965	40.72522	1	31
2013-12-09 15:03:00 UTC	-73.99095	40.749772	-73.870807	40.77407	1	34.33
2013-04-18 08:48:00 UTC	-73.973102	40.785075	-74.011462	40.708307	1	29
2009-11-05 06:47:00 UTC	-73.980313	40.744282	-74.015285	40.711458	1	14.9
2009-05-21 09:47:06 UTC	-73.901887	40.764021	-73.901795	40.763612	1	12.8

---

5 The feature_cross.ipynb notebook in the book's repository (*https://github.com/GoogleCloudPlatform/ml-design-patterns/blob/master/02_data_representation/feature_cross.ipynb*) of this book will help you follow the discussion better.

This dataset contains information on taxi rides in New York City with features such as the timestamp of pickup, the pickup and drop-off latitude and longitude, and number of passengers. The label here is `fare_amount`, the cost of the taxi ride. Which feature crosses might be relevant for this dataset?

There could be many. Let's consider the `pickup_datetime`. From this feature, we can use information about the ride's hour and day of the week. Each of these is a categorical variable, and certainly both contain predictive power in determining the price of a taxi ride. For this dataset, it makes sense to consider a feature cross of `day_of_week` and `hour_of_day` since it's reasonable to assume that taxi rides at 5pm on Monday should be treated differently than taxi rides at 5 p.m. on Friday (see Table 2-9).

Table 2-9. A preview of the data we're using to create a feature cross: the day of week and hour of day columns

day_of_week	hour_of_day
Sunday	00
Sunday	01
...	...
Saturday	23

A feature cross of these two features would be a 168-dimensional one-hot encoded vector (24 hours × 7 days = 168) with the example "Monday at 5 p.m." occupying a single index denoting (`day_of_week` is Monday concatenated with `hour_of_day` is 17).

While the two features are important on their own, allowing for a feature cross of hour_of_day and day_of_week makes it easier for a taxi fare prediction model to recognize that end-of-the-week rush hour influences the taxi ride duration and thus the taxi fare in its own way.

Feature cross in BigQuery ML

To create the feature cross in BigQuery, we can use the function `ML.FEATURE_CROSS` and pass in a STRUCT of the features `day_of_week` and `hour_of_day`:

```
ML.FEATURE_CROSS(STRUCT(day_of_week,hour_of_week)) AS day_X_hour
```

The STRUCT clause creates an ordered pair of the two features. If our software framework doesn't support a feature cross function, we can get the same effect using string concatenation:

```
CONCAT(CAST(day_of_week AS STRING),
       CAST(hour_of_week AS STRING)) AS day_X_hour
```

A complete training example for the natality problem is shown below, with a feature cross of the is_male and plurality columns used as a feature; see the full code in this book's repository (*https://github.com/GoogleCloudPlatform/ml-design-patterns/blob/ master/02_data_representation/feature_cross.ipynb*):

```
CREATE OR REPLACE MODEL babyweight.natality_model_feat_eng
TRANSFORM(weight_pounds,
    is_male,
    plurality,
    gestation_weeks,
    mother_age,
    CAST(mother_race AS string) AS mother_race,
    ML.FEATURE_CROSS(
            STRUCT(
                is_male,
                plurality)
        ) AS gender_X_plurality)
OPTIONS
  (MODEL_TYPE='linear_reg',
   INPUT_LABEL_COLS=['weight_pounds'],
   DATA_SPLIT_METHOD="NO_SPLIT") AS
SELECT
  *
FROM
    babyweight.babyweight_data_train
```

 TrThe Transform pattern (see Chapter 6) is being used here when engineering features of the natality model. This also allows the model to "remember" to carry out the feature cross of the input data fields during prediction.

When we have enough data, the Feature Cross pattern allows models to become simpler. On the natality dataset, the RMSE for the evaluation set for a linear model with the Feature Cross pattern is 1.056. Alternatively, training a deep neural network in BigQuery ML on the same dataset with no feature crosses yields an RMSE of 1.074. There is a slight improvement in our performance despite using a much simpler linear model, and the training time is also drastically reduced.

Feature crosses in TensorFlow

To implement a feature cross using the features is_male and plurality in Tensor-Flow, we use tf.feature_column.crossed_column. The method crossed_column takes two arguments: a list of the feature keys to be crossed and the hash bucket size. Crossed features will be hashed according to hash_bucket_size so it should be large enough to comfortably decrease the likelihood of collisions. Since the is_male input can take 3 values (True, False, Unknown) and the plurality input can take 6 values (Single(1), Twins(2), Triplets(3), Quadruplets(4), Quintuplets(5), Multiple(2+)), there are 18 possible (is_male, plurality) pairs. If we set hash_bucket_size to 1,000, we can be 85% sure there are no collisions.

Finally, to use a crossed column in a DNN model, we need to wrap it either in an indicator_column or an embedding_column depending on whether we want to one-hot encode it or represent it in a lower dimension (see the "Design Pattern 2: Embeddings" on page 39 in this chapter):

```
gender_x_plurality = fc.crossed_column(["is_male", "plurality"],
                                       hash_bucket_size=1000)
crossed_feature = fc.embedding_column(gender_x_plurality, dimension=2)
```

or

```
gender_x_plurality = fc.crossed_column(["is_male", "plurality"],
                                       hash_bucket_size=1000)
crossed_feature = fc.indicator_column(gender_x_plurality)
```

Why It Works

Feature crosses provide a valuable means of feature engineering. They provide more complexity, more expressivity, and more capacity to simple models. Think again about the crossed feature of is_male and plurality in the natality dataset. This Feature Cross pattern allows the model to treat twin males separately from female twins and separately from triplet males and separately from single females and so on. When we use an indicator_column, the model is able to treat each of the resulting crosses as an independent variable, essentially adding 18 additional binary categorical features to the model (see Figure 2-16).

Feature crosses scale well to massive data. While adding extra layers to a deep neural network could potentially provide enough nonlinearity to learn how pairs (is_male, plurality) behave, this drastically increases the training time. On the natality dataset, we observed that a linear model with a feature cross trained in BigQuery ML performs comparably with a DNN trained without a feature cross. However, the linear model trains substantially faster.

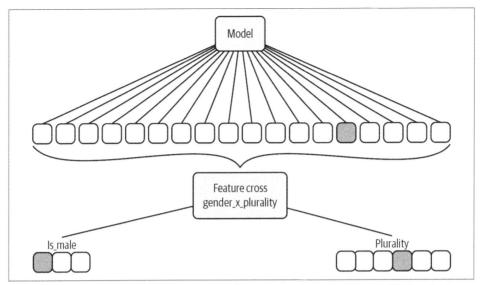

Figure 2-16. A feature cross between is_male and plurality creates an additional 18 binary features in our ML model.

Table 2-10 compares the training time in BigQuery ML and evaluation loss for both a linear model with a feature cross of (is_male, plurality) and a deep neural network without any feature cross.

Table 2-10. A comparison of BigQuery ML training metrics for models with and without feature crosses

Model type	Incl. feature cross	Training time (minutes)	Eval. loss (RMSE)
Linear	Yes	0.42	1.05
DNN	No	48	1.07

A simple linear regression achieves comparable error on the evaluation set but trains one hundred times faster. Combining feature crosses with massive data is an alternative strategy for learning complex relationships in training data.

Trade-Offs and Alternatives

While we discussed feature crosses as a way of handling categorical variables, they can be applied, with a bit of preprocessing, to numerical features also. Feature crosses cause sparsity in models and are often used along with techniques that counteract that sparsity.

Handling numerical features

We would never want to create a feature cross with a continuous input. Remember, if one input takes m possible values and another input takes n possible values, then the feature cross of the two would result in m*n elements. A numeric input is dense, taking a continuum of values. It would be impossible to enumerate all possible values in a feature cross of continuous input data.

Instead, if our data is continuous, then we can bucketize the data to make it categorical before applying a feature cross. For example, latitude and longitude are continuous inputs, and it makes intuitive sense to create a feature cross using these inputs since location is determined by an ordered pair of latitude and longitude. However, instead of creating a feature cross using the raw latitude and longitude, we would bin these continuous values and cross the binned_latitude and the binned_longitude:

```
import tensorflow.feature_column as fc

# Create a bucket feature column for latitude.
latitude_as_numeric = fc.numeric_column("latitude")
lat_bucketized = fc.bucketized_column(latitude_as_numeric,
                                      lat_boundaries)
# Create a bucket feature column for longitude.
longitude_as_numeric = fc.numeric_column("longitude")
lon_bucketized = fc.bucketized_column(longitude_as_numeric,
                                      lon_boundaries)

# Create a feature cross of latitude and longitude
lat_x_lon = fc.crossed_column([lat_bucketized, lon_bucketized],
                              hash_bucket_size=nbuckets**4)

crossed_feature = fc.indicator_column(lat_x_lon)
```

Handling high cardinality

Because the cardinality of resulting categories from a feature cross increases multiplicatively with respect to the cardinality of the input features, feature crosses lead to sparsity in our model inputs. Even with the day_of_week and hour_of_day feature cross, a feature cross would be a sparse vector of dimension 168 (see Figure 2-17).

It can be useful to pass a feature cross through an Embedding layer (see the "Design Pattern 2: Embeddings" on page 39 in this chapter) to create a lower-dimensional representation, as shown in Figure 2-18.

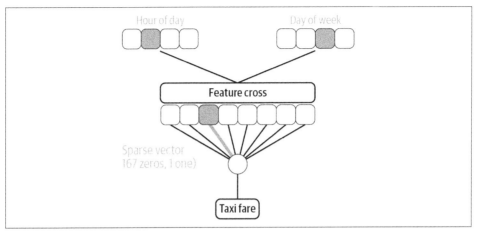

Figure 2-17. A feature cross of day_of_week and hour_of_day produces a sparse vector of dimension 168.

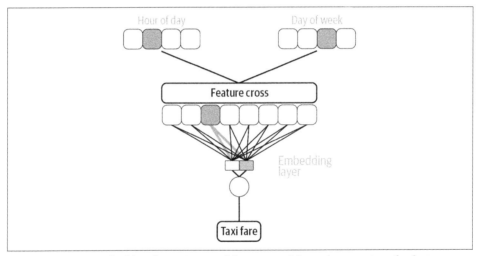

Figure 2-18. An embedding layer is a useful way to address the sparsity of a feature cross.

Because the Embeddings design pattern allows us to capture closeness relationships, passing the feature cross through an embedding layer allows the model to generalize how certain feature crosses coming from pairs of hour and day combinations affect the output of the model. In the example of latitude and longitude above, we could have used an embedding feature column in place of the indicator column:

```
crossed_feature = fc.embedding_column(lat_x_lon, dimension=2)
```

Need for regularization

When crossing two categorical features both with large cardinality, we produce a cross feature with multiplicative cardinality. Naturally, given more categories for an individual feature, the number of categories in a feature cross can increase dramatically. If this gets to the point where individual buckets have too few items, it will hinder the model's ability to generalize. Think of the latitude and longitude example. If we were to take very fine buckets for latitude and longitude, then a feature cross would be so precise it would allow the model to memorize every point on the map. However, if that memorization was based on just a handful of examples, the memorization would actually be an overfit.

To illustrate, take the example of predicting the taxi fare in New York given the pickup and dropoff locations and the time of pickup:[6]

```
CREATE OR REPLACE MODEL mlpatterns.taxi_l2reg
TRANSFORM(
  fare_amount
, ML.FEATURE_CROSS(STRUCT(CAST(EXTRACT(DAYOFWEEK FROM pickup_datetime)
                   AS STRING) AS dayofweek,
                           CAST(EXTRACT(HOUR FROM pickup_datetime)
                   AS STRING) AS hourofday), 2) AS day_hr
, CONCAT(
    ML.BUCKETIZE(pickuplon, GENERATE_ARRAY(-78, -70, 0.01)),
    ML.BUCKETIZE(pickuplat, GENERATE_ARRAY(37, 45, 0.01)),
    ML.BUCKETIZE(dropofflon, GENERATE_ARRAY(-78, -70, 0.01)),
    ML.BUCKETIZE(dropofflat, GENERATE_ARRAY(37, 45, 0.01))
  ) AS pickup_and_dropoff
)
OPTIONS(input_label_cols=['fare_amount'],
        model_type='linear_reg', l2_reg=0.1)
AS
SELECT * FROM mlpatterns.taxi_data
```

There are two feature crosses here: one in time (of day of week and hour of day) and the other in space (of the pickup and dropoff locations). The location, in particular, is very high cardinality, and it is likely that some of the buckets will have very few examples.

For this reason, it is advisable to pair feature crosses with L1 regularization, which encourages sparsity of features, or L2 regularization, which limits overfitting. This allows our model to ignore the extraneous noise generated by the many synthetic features and combat overfitting. Indeed, on this dataset, the regularization improves the RMSE slightly, by 0.3%.

---

6 Full code is in *02_data_representation/feature_cross.ipynb* in the code repository of this book.

As a related point, when choosing which features to combine for a feature cross, we would not want to cross two features that are highly correlated. We can think of a feature cross as combining two features to create an ordered pair. In fact, the term "cross" of "feature cross" refers to the Cartesian product. If two features are highly correlated, then the "span" of their feature cross doesn't bring any new information to the model. As an extreme example, suppose we had two features, x_1 and x_2, where $x\_2 = 5^*x\_1$. Bucketing values for x_1 and x_2 by their sign and creating a feature cross will still produce four new boolean features. However, due to the dependence of x_1 and x_2, two of those four features are actually empty, and the other two are precisely the two buckets created for x_1.

Design Pattern 4: Multimodal Input

The Multimodal Input design pattern addresses the problem of representing different types of data or data that can be expressed in complex ways by concatenating all the available data representations.

Problem

Typically, an input to a model can be represented as a number or as a category, an image, or free-form text. Many off-the-shelf models are defined for specific types of input only—a standard image classification model such as Resnet-50, for example, does not have the ability to handle inputs other than images.

To understand the need for multimodal inputs, let's say we've got a camera capturing footage at an intersection to identify traffic violations. We want our model to handle both image data (camera footage) and some metadata about when the image was captured (time of day, day of week, weather, etc.), as depicted in Figure 2-19.

This problem also occurs when training a structured data model where one of the inputs is free-form text. Unlike numerical data, images and text cannot be fed directly into a model. As a result, we'll need to represent image and text inputs in a way our model can understand (usually using the Embeddings design pattern), then combine these inputs with other tabular[7] features. For example, we might want to predict a restaurant patron's rating based on their review text and other attributes such as what they paid and whether it was lunch or dinner (see Figure 2-20).

---

7 We use the term "tabular data" to refer to numerical and categorical inputs, but not free-form text. You can think of tabular data as anything you might commonly find in a spreadsheet. For example, values like age, type of car, price, or number of hours worked. Tabular data does not include free-form text like descriptions or reviews.

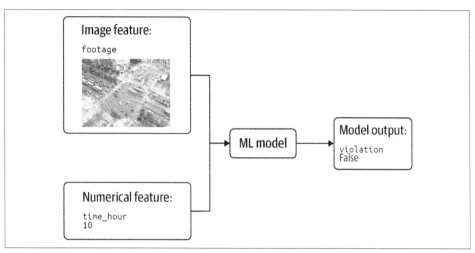

Figure 2-19. Model combining image and numerical features to predict whether footage of an intersection depicts a traffic violation.

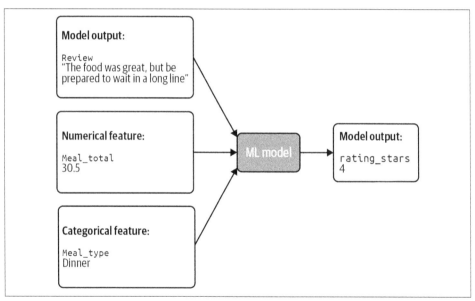

Figure 2-20. Model combining free-form text input with tabular data to predict the rating of a restaurant review.

Solution

To start, let's take the example above with text from a restaurant review combined with tabular metadata about the meal referenced by the review. We'll first combine the numerical and categorical features. There are three possible options for meal_type, so we can turn this into a one-hot encoding and will represent dinner as [0, 0, 1]. With this categorical feature represented as an array, we can now combine it with meal_total by adding the price of the meal as the fourth element of the array: [0, 0, 1, 30.5].

The Embeddings design pattern is a common approach to encoding text for machine learning models. If our model had only text, we could represent it as an embedding layer using the following tf.keras code:

```
from tensorflow.keras import Sequential
from tensorflow.keras.layers import Embedding

model = Sequential()
model.add(Embedding(batch_size, 64, input_length=30))
```

Here, we need to flatten the embedding[8] in order to concatenate with the meal_type and meal_total:

```
model.add(Flatten())
```

We could then use a series of Dense layers to transform that very large array[9] into smaller ones, ending with our output that is an array of, say, three numbers:

```
model.add(Dense(3, activation="relu"))
```

We now need to concatenate these three numbers, which form the sentence embedding of the review with the earlier inputs: [0, 0, 1, 30.5, 0.75, -0.82, 0.45].

To do this, we'll use the Keras functional API and apply the same steps. Layers built with the functional API are callable, enabling us to chain them together starting with an Input layer.[10] To make use of this, we'll first define both our embedding and tabular layers:

```
embedding_input = Input(shape=(30,))
embedding_layer = Embedding(batch_size, 64)(embedding_input)
embedding_layer = Flatten()(embedding_layer)
embedding_layer = Dense(3, activation='relu')(embedding_layer)

tabular_input = Input(shape=(4,))
tabular_layer = Dense(32, activation='relu')(tabular_input)
```

Note that we've defined the Input pieces of both of these layers as their own variables. This is because we need to pass Input layers when we build a Model with the

---

8 When we pass an encoded 30-word array to our model, the Keras layer will transform it into a 64-dimensional embedding representation, so we'll have a [64×30] matrix representing the review.

9 The starting point is an array that is 1,920 numbers.

10 See *02_data_representation/mixed_representation.ipynb* in the code repository of this book for the full model code.

functional API. Next, we'll create a concatenated layer, feed that into our output layer, and finally create the model by passing in the original Input layers we defined above:

```
merged_input = keras.layers.concatenate([embedding_layer, tabular_layer])
merged_dense = Dense(16)(merged_input)
output = Dense(1)(merged_dense)

model = Model(inputs=[embedding_input, tabular_input], outputs=output)
merged_dense = Dense(16, activation='relu')(merged_input)
output = Dense(1)(merged_dense)

model = Model(inputs=[embedding_input, tabular_input], outputs=output)
```

Now we have a single model that accepts the multimodal input.

Trade-Offs and Alternatives

As we just saw, the Multimodal Input design pattern explores how to represent *different input formats* in the same model. In addition to mixing different *types* of data, we may also want to represent the *same data in different ways* to make it easier for our model to identify patterns. For example, we may have a ratings field that is on an ordinal scale of 1 star to 5 stars, and treat that ratings field as both numeric and categorical. Here, we are referring to *multimodal inputs* as both:

- Combining different types of data, like images + metadata
- Representing complex data in multiple ways

We'll start by exploring how tabular data can be represented in different ways, and then we'll look at text and image data.

Tabular data multiple ways

To see how we can represent tabular data in different ways for the same model, let's return to the restaurant review example. We'll imagine instead that rating is an *input* to our model and we're trying to predict the review's usefulness (how many people liked the review). As an input, the rating can be represented both as an integer value ranging from 1 to 5 and as a categorical feature. To represent rating categorically, we can bucket it. The way we bucket the data is up to us and dependent on our dataset and use case. To keep things simple, let's say we want to create two buckets: "good" and "bad." The "good" bucket includes ratings of 4 and 5, and "bad" includes 3 and below. We can then create a boolean value to encode the rating buckets and concatenate both the integer and boolean into a single array (full code is on GitHub (*https://github.com/GoogleCloudPlatform/ml-design-patterns/blob/master/02_data_representation/mixed_representation.ipynb*)).

Here's what this might look like for a small dataset with three data points:

```
rating_data = [2, 3, 5]

def good_or_bad(rating):
  if rating > 3:
    return 1
  else:
    return 0

rating_processed = []

for i in rating_data:
  rating_processed.append([i, good_or_bad(i)])
```

The resulting feature is a two-element array consisting of the integer rating and its boolean representation:

```
[[2, 0], [3, 0], [5, 1]]
```

If we had instead decided to create more than two buckets, we would one-hot encode each input and append this one-hot array to the integer representation.

The reason it's useful to represent rating in two ways is because the value of rating as measured by 1 to 5 stars does not necessarily increase linearly. Ratings of 4 and 5 are very similar, and ratings of 1 to 3 most likely indicate that the reviewer was dissatisfied. Whether you give something you dislike 1, 2, or 3 stars is often related to your review tendencies rather than the review itself. Despite this, it's still useful to keep the more granular information present in the star rating, which is why we encode it in two ways.

Additionally, consider features with a larger range than 1 to 5, like the distance between a reviewer's home and a restaurant. If someone drives two hours to go to a restaurant, their review may be more critical than someone coming from across the street. In this case, we might have outlier values, and so it would make sense to both threshold the numeric distance representation at something like 50 km and to include a separate categorical representation of distance. The categorical feature could be bucketed into "in state," "in country," and "foreign."

Multimodal representation of text

Both text and images are unstructured and require more transformations than tabular data. Representing them in various formats can help our models extract more patterns. We'll build on our discussion of text models in the preceding section by looking at different approaches for representing text data. Then we'll introduce images and dive into a few options for representing image data in ML models.

Text data multiple ways.   Given the complex nature of text data, there are many ways to extract meaning from it. The Embeddings design pattern enables a model to group

---

similar words together, identify relationships between words, and understand syntactic elements of text. While representing text through word embeddings most closely mirrors how humans innately understand language, there are additional text representations that can maximize our model's ability to perform a given prediction task. In this section, we'll look at the bag of words approach to representing text, along with extracting tabular features from text.

To demonstrate text data representation, we'll be referencing a dataset that contains the text of millions of questions and answers from Stack Overflow,[11] along with metadata about each post. For example, the following query will give us a subset of questions tagged as either "keras," "matplotlib," or "pandas," along with the number of answers each question received:

```
SELECT
  title,
  answer_count,
  REPLACE(tags, "|", ",") as tags
FROM
  `bigquery-public-data.stackoverflow.posts_questions`
WHERE
  REGEXP_CONTAINS( tags, r"(?:keras|matplotlib|pandas)")
```

The query results in the following output:

Row	title	answer_count	tags
1	Building a new column in a pandas dataframe by matching string values in a list	6	python,python-2.7,pandas,replace,nested-loops
2	Extracting specific selected columns to new DataFrame as a copy	6	python,pandas,chained-assignment
3	Where do I call the BatchNormalization function in Keras?	7	python,keras,neural-network,data-science,batch-normalization
4	Using Excel like solver in Python or SQL	8	python,sql,numpy,pandas,solver

When representing text using the bag of words (BOW) approach, we imagine each text input to our model as a bag of Scrabble tiles, with each tile containing a single word instead of a letter. BOW does not preserve the order of our text, but it does detect the presence or absence of certain words in each piece of text we send to our model. This approach is a type of multi-hot encoding where each text input is converted into an array of 1s and 0s. Each index in this BOW array corresponds to a word from our vocabulary.

---

11 This dataset is available in BigQuery: *bigquery-public-data.stackoverflow.posts_questions*.

How Bag of Words Works

The first step in BOW encoding is choosing our vocabulary size, which will include the top N most frequently occurring words in our text corpus. In theory, our vocabulary size could be equal to the number of unique words in our entire dataset. However, this would lead to very large input arrays of mostly zeros, since many words could be unique to a single question. Instead, we'll want to choose a vocabulary size small enough to include key, recurring words that convey meaning for our prediction task, but big enough that our vocabulary isn't limited to words found in nearly every question (like "the," "is," "and," etc.).

Each input to our model will then be an array the size of our vocabulary. This BOW representation therefore entirely disregards words that aren't included in our vocabulary. There isn't a magic number or percentage for choosing vocabulary size—it's helpful to try a few and see which performs best on our model.

To understand BOW encoding, let's first look at a simplified example. For this example, let's say we're predicting the tag of a Stack Overflow question from a list of three possible tags: "pandas," "keras," and "matplotlib." To keep things simple, assume our vocabulary consists of only the 10 words listed below:

```
dataframe
layer
series
graph
column
plot
color
axes
read_csv
activation
```

This list is our *word index*, and every input we feed into our model will be a 10-element array where each index corresponds with one of the words listed above. For example, a 1 in the first index of an input array means a particular question contains the word *dataframe*. To understand BOW encoding from the perspective of our model, imagine we're learning a new language and the 10 words above are the only words we know. Every "prediction" we make will be based solely on the presence or absence of these 10 words and will disregard any words outside this list.

Therefore, given question title, "How to plot dataframe bar graph," how will we transform it into a BOW representation? First, let's take note of the words in this sentence that appear in our vocabulary: *plot*, *dataframe*, and *graph*. The other words in this sentence will be ignored by the bag of words approach. Using our word index above, this sentence becomes:

```
[ 1 0 0 1 0 1 0 0 0 0 ]
```

Note that the 1s in this array correspond with the indices of *dataframe*, *graph*, and *plot*, respectively. To summarize, Figure 2-21 shows how we transformed our input from raw text to a BOW-encoded array based on our vocabulary.

Keras has some utility methods for encoding text as a bag of words, so we don't need to write the code for identifying the top words from our text corpus and encoding raw text into multi-hot arrays from scratch.

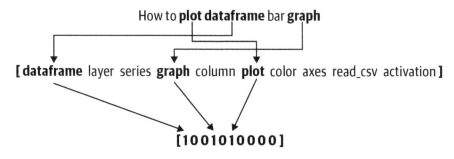

Figure 2-21. Raw input text → identifying words present in this text from our vocabulary → transforming to a multi-hot BOW encoding.

Given that there are two different approaches for representing text (Embedding and BOW), which approach should we choose for a given task? As with many aspects of machine learning, this depends on our dataset, the nature of our prediction task, and the type of model we're planning to use.

Embeddings add an extra layer to our model and provide extra information about word meaning that is not available from the BOW encoding. However, embeddings require training (unless we can use a pre-trained embedding for our problem). While a deep learning model may achieve higher accuracy, we can also try using BOW encoding in a linear regression or decision-tree model using frameworks like scikit-learn or XGBoost. Using BOW encoding with a simpler model type can be useful for fast prototyping or to verify that the prediction task we've chosen will work on our dataset. Unlike embeddings, BOW doesn't take into account the order or meaning of words in a text document. If either of these are important to our prediction task, embeddings may be the best approach.

There may also be benefits to building a deep model that combines *both* bag of words *and* text embedding representations to extract more patterns from our data. To do this, we can use the Multimodal Input approach, except that instead of concatenating text and tabular features, we can concatenate the Embedding and BOW representations (see code on GitHub (*https://github.com/GoogleCloudPlatform/ml-design-patterns/blob/master/02_data_representation/mixed_representation.ipynb*)). Here, the shape of our Input layer would be the vocabulary size of the BOW representation. Some benefits of representing text in multiple ways include:

- BOW encoding provides strong signals for the most significant words present in our vocabulary, while embeddings can identify relationships between words in a much larger vocabulary.

- If we have text that switches between languages, we can build embeddings (or BOW encodings) for each one and concatenate them.

- Embeddings can encode the frequency of words in text, where the BOW treats the presence of each word as a boolean value. Both representations are valuable.

- BOW encoding can identify patterns between reviews that all contain the word "amazing," while an embedding can learn to correlate the phrase "not amazing" with a below-average review. Again, both of these representations are valuable.

Extracting tabular features from text. In addition to encoding raw text data, there are often other characteristics of text that can be represented as tabular features. Let's say we are building a model to predict whether or not a Stack Overflow question will get a response. Various factors about the text but unrelated to the exact words themselves may be relevant to training a model on this task. For example, maybe the length of a question or the presence of a question mark influences the likelihood of an answer. However, when we create an embedding, we usually truncate the words to a certain length. The actual length of a question is lost in that data representation. Similarly, punctuation is often removed. We can use the Multimodal Input design pattern to bring back this lost information to the model.

In the following query, we'll extract some tabular features from the title field of the Stack Overflow dataset to predict whether or not a question will get an answer:

```
SELECT
  LENGTH(title) AS title_len,
  ARRAY_LENGTH(SPLIT(title, " ")) AS word_count,
  ENDS_WITH(title, "?") AS ends_with_q_mark,
IF
  (answer_count > 0,
    1,
    0) AS is_answered,
FROM
  `bigquery-public-data.stackoverflow.posts_questions`
```

This results in the following:

Row	title_len	word_count	ends_with_q_mark	is_answered
1	84	14	true	0
2	104	16	false	0
3	85	19	true	1

Row	title_len	word_count	ends_with_q_mark	is_answered
4	88	14	false	1
5	17	3	false	1

In addition to these features extracted directly from a question's title, we could also represent *metadata* about the question as features. For example, we could add features representing the number of tags the question had and the day of the week it was posted. We could then combine these tabular features with our encoded text and feed both representations into our model using Keras's Concatenate layer to combine the BOW-encoded text array with the tabular metadata describing our text.

Multimodal representation of images

Similar to our analysis of embeddings and BOW encoding for text, there are many ways to represent image data when preparing it for an ML model. Like raw text, images cannot be fed directly into a model and need to be transformed into a numerical format that the model can understand. We'll start by discussing some common approaches to representing image data: as pixel values, as sets of tiles, and as sets of windowed sequences. The Multimodal Input design pattern provides a way to use more than one representation of an image in our model.

Images as pixel values.   At their core, images are arrays of pixel values. A black and white image, for example, contains pixel values ranging from 0 to 255. We could therefore represent a 28×28-pixel black-and-white image in a model as a 28×28 array with integer values ranging from 0 to 255. In this section, we'll be referencing the MNIST dataset, a popular ML dataset that includes images of handwritten digits.

With the `Sequential` API, we can represent our MNIST images of pixel values using a Flatten layer, which flattens the image into a one-dimensional 784 (28 * 28) element array:

```
layers.Flatten(input_shape=(28, 28))
```

For color images, this gets more complex. Each pixel in an RGB color image has three values—one for red, green, and blue. If our images in the example above were instead color, we'd add a third dimension to the model's `input_shape` such that it would be:

```
layers.Flatten(input_shape=(28, 28, 3))
```

While representing images as arrays of pixel values works well for simple images like the grayscale ones in the MNIST dataset, it starts to break down when we introduce images with more edges and shapes throughout. When a network is fed with all of the pixels in an image at once, it's hard for it to focus on smaller areas of adjacent pixels that contain important information.

Images as tiled structures. We need a way to represent more complex, real-world images that will enable our model to extract meaningful details and understand patterns. If we feed the network only small pieces of an image at a time, it'll be more likely to identify things like spatial gradients and edges present in neighboring pixels. A common model architecture for accomplishing this is a *convolutional neural network* (CNN).

Convolutional Neural Network Layers

Take a look at Figure 2-22. In this example, we've got a 4×4 grid where each square represents pixel values on our image. We then use max pooling to take the largest value of each grid and generate a resulting, smaller matrix. By dividing our image into a grid of tiles, our model is able to extract key insights from each region of an image at different levels of granularity.

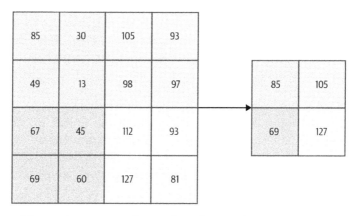

Figure 2-22. Max pooling on a single 4×4 slice of image data.

Figure 2-22 uses a *kernel size* of (2, 2). Kernel size refers to the size of each chunk of our image. The number of spaces our filter moves before creating its next chunk, also known as *stride*, is 2. Because our stride is equal to the size of our kernel, the chunks created *do not overlap*.

While this tiling method preserves more detail than representing images as arrays of pixel values, quite a bit of information is lost after each pooling step. In the diagram above, the next pooling step would produce a scalar value of 8, taking our matrix from 4 ×4 to a single value in just two steps. In a real-world image, you can imagine how this might bias a model to focus on areas with dominant pixel values while losing important details that may surround these areas.

How can we build on this idea of splitting images into smaller chunks, while still preserving important details in images? We'll do this by making these chunks *overlap*. If the example in Figure 2-22 had instead used a stride of 1, the output would instead be a 3×3 matrix (Figure 2-23).

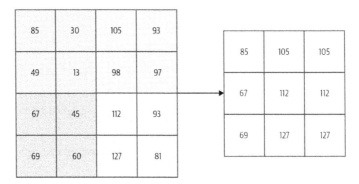

Figure 2-23. Using overlapping windows for max pooling on a 4×4 pixel grid.

We could then transform this into a 2×2 grid (Figure 2-24).

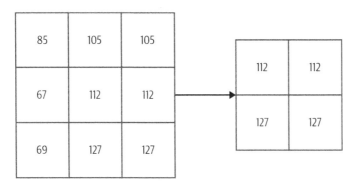

Figure 2-24. Transforming the 3×3 grid into 2×2 with sliding windows and max pooling.

We end with a final scalar value of 127. While the end value is the same, you can see how the intermediate steps preserved more detail from the original matrix.

Keras provides convolution layers to build models that split images into smaller, windowed chunks. Let's say we're building a model to classify 28×28 color images as either "dog" or "cat." Since these images are color, each image will be represented as a 28×28×3-dimensional array, since each pixel has three color channels. Here's how we'd define the inputs to this model using a convolution layer and the Sequential API:

```
Conv2D(filters=16, kernel_size=3, activation='relu', input_shape=(28,28,3))
```

In this example, we're dividing our input images into 3×3 chunks before passing them through a max pooling layer. Building a model architecture that splits images into chunks of sliding windows allows our model to recognize more granular details in an image like edges and shapes.

Combining different image representations. In addition, as with the bag of words and text embedding, it may be useful to represent the same image data in multiple ways. Again, we can accomplish this with the Keras functional API.

Here's how we'd combine our pixel values with the sliding window representation using the Keras Concatenate layer:

```
# Define image input layer (same shape for both pixel and tiled
# representation)
image_input = Input(shape=(28,28,3))

# Define pixel representation
pixel_layer = Flatten()(image_input)

# Define tiled representation
tiled_layer = Conv2D(filters=16, kernel_size=3,
                     activation='relu')(image_input)
tiled_layer = MaxPooling2D()(tiled_layer)
tiled_layer = tf.keras.layers.Flatten()(tiled_layer)

# Concatenate into a single layer
merged_image_layers = keras.layers.concatenate([pixel_layer, tiled_layer])
```

To define a model that accepts that multimodal input representation, we can then feed our concatenated layer into our output layer:

```
merged_dense = Dense(16, activation='relu')(merged_image_layers)
merged_output = Dense(1)(merged_dense)

model = Model(inputs=image_input, outputs=merged_output)
```

Choosing which image representation to use or whether to use multimodal representations depends largely on the type of image data we're working with. In general, the more detailed our images, the more likely it is that we'll want to represent them as tiles or sliding windows of tiles. For the MNIST dataset, representing images as pixel values alone may suffice. With complex medical images, on the other hand, we may see increased accuracy by combining multiple representations. Why combine multiple image representations? Representing images as pixel values allows the model to identify higher-level focus points in an image like dominant, high-contrast objects. Tiled representations, on the other hand, help models identify more granular, lower-contrast edges and shapes.

Using images with metadata.   Earlier we discussed different types of metadata that might be associated with text, and how to extract and represent this metadata as tabular features for our model. We can also apply this concept to images. To do this, let's return to the example referenced in Figure 2-19 of a model using footage of an intersection to predict whether or not it contains a traffic violation. Our model can extract many patterns from the traffic images on their own, but there may be other data available that could improve our model's accuracy. For example, maybe certain behavior (e.g., a right turn on red) is not permitted during rush hour but is OK at other times of day. Or maybe drivers are more likely to violate traffic laws in bad weather. If we're collecting image data from multiple intersections, knowing the location of our image might also be useful to our model.

We've now identified three additional tabular features that could enhance our image model:

- Time of day
- Weather
- Location

Next, let's think about possible representations for each of these features. We could represent time as an integer indicating the *hour* of the day. This might help us identify patterns associated with high-traffic times like rush hour. In the context of this model, it may be more useful to know whether or not it was dark when the image was taken. In this case, we could represent time as a boolean feature.

Weather can also be represented in various ways, as both numeric and categorical values. We could include temperature as a feature, but in this case, visibility might be more useful. Another option for representing weather is through a categorical variable indicating the presence of rain or snow.

If we're collecting data from many locations, we'd likely want to encode this as a feature as well. This would make most sense as a categorical feature, and could even be multiple features (city, country, state, etc.) depending on how many locations we're collecting footage from.

For this example, let's say we'd like to use the following tabular features:

- Time as hour of the day (integer)
- Visibility (float)
- Inclement weather (categorical: rain, snow, none)
- Location ID (categorical with five possible locations)

Here's what a subset of this dataset might look like for the three examples:

```
data = {
    'time': [9,10,2],
    'visibility': [0.2, 0.5, 0.1],
    'inclement_weather': [[0,0,1], [0,0,1], [1,0,0]],
    'location': [[0,1,0,0,0], [0,0,0,1,0], [1,0,0,0,0]]
}
```

We could then combine these tabular features into a single array for each example, so that our model's input shape would be 10. The input array for the first example would look like the following:

```
[9, 0.2, 0, 0, 1, 0, 1, 0, 0, 0]
```

We could feed this input into a Dense fully connected layer, and the output of our model would be a single value between 0 and 1 indicating whether or not the instance contains a traffic violation. To combine this with our image data, we'll use a similar approach to what we discussed for text models. First, we'd define a convolution layer to handle our image data, then a Dense layer to handle our tabular data, and finally we'd concatenate both into a single output.

This approach is outlined in Figure 2-25.

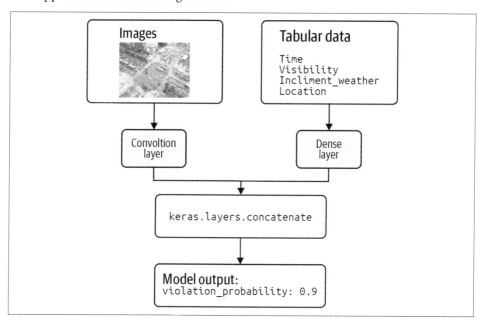

Figure 2-25. Concatenating layers to handle image and tabular metadata features.

Multimodal feature representations and model interpretability

Deep learning models are inherently difficult to explain. If we build a model that achieves 99% accuracy, we still don't know exactly *how* our model is making predictions and consequently, if the way it's making those predictions is correct. For example, let's say we train a model on images of petri dishes taken in a lab that achieves a high accuracy score. These images also contain annotations from the scientist who took the pictures. What we don't know is that the model is incorrectly using the annotations to make its predictions, rather than the contents of the petri dishes.

There are several techniques for explaining image models that can highlight the pixels that signaled a model's prediction. When we combine multiple data representations in a single model, however, these features become dependent on one another. As a result, it can be difficult to explain how the model is making predictions. Explainability is covered in Chapter 7.

Summary

In this chapter, we learned different approaches to representing data for our model. We started by looking at how to handle numerical inputs, and how scaling these inputs can speed up model training time and improve accuracy. Then we explored how to do feature engineering on categorical inputs, specifically with one-hot encoding and using arrays of categorical values.

Throughout the rest of the chapter, we discussed four design patterns for representing data. The first was the *Hashed Feature* design pattern, which involves encoding categorical inputs as unique strings. We explored a few different approaches to hashing using the airport dataset in BigQuery. The second pattern we looked at in this chapter was *Embeddings*, a technique for representing high-cardinality data such as inputs with many possible categories or text data. Embeddings represent data in multidimensional space, where the dimension is dependent on our data and prediction task. Next we looked at *Feature Crosses*, an approach that joins two features to extract relationships that may not have been easily captured by encoding the features on their own. Finally, we looked at *Multimodal Input* representations by addressing the problem of how to combine inputs of different types into the same model, and how a single feature can be represented multiple ways.

This chapter focused on preparing *input* data for our models. In the next chapter, we'll focus on model *output* by diving into different approaches for representing our prediction task.

## CHAPTER 3

Problem Representation Design Patterns

Chapter 2 looked at design patterns that catalog the myriad ways in which inputs to machine learning models can be represented. This chapter looks at different types of machine learning problems and analyzes how the model architectures vary depending on the problem.

The input and the output types are two key factors impacting the model architecture. For instance, the output in supervised machine learning problems can vary depending on whether the problem being solved is a classification or regression problem. Special neural network layers exist for specific types of input data: convolutional layers for images, speech, text, and other data with spatiotemporal correlation, recurrent networks for sequential data, and so on. A huge literature has arisen around special techniques such as max pooling, attention, and so forth on these types of layers. In addition, special classes of solutions have been crafted for commonly occurring problems like recommendations (such as matrix factorization) or time-series forecasting (for example, ARIMA). Finally, a group of simpler models together with common idioms can be used to solve more complex problems—for example, text generation often involves a classification model whose outputs are postprocessed using a beam search algorithm.

To limit our discussion and stay away from areas of active research, we will ignore patterns and idioms associated with specialized machine learning domains. Instead, we will focus on regression and classification and examine patterns with problem representation in just these two types of ML models.

The *Reframing* design pattern takes a solution that is intuitively a regression problem and poses it as a classification problem (and vice versa). The *Multilabel* design pattern handles the case that training examples can belong to more than one class. The *Cascade* design pattern addresses situations where a machine learning problem can be profitably broken into a series (or cascade) of ML problems. The *Ensemble* design pattern solves a problem by training multiple models and aggregating their responses. The *Neutral Class* design pattern looks at how to handle situations where experts disagree. The *Rebalancing* design pattern recommends approaches to handle highly skewed or imbalanced data.

Design Pattern 5: Reframing

The Reframing design pattern refers to changing the representation of the output of a machine learning problem. For example, we could take something that is intuitively a regression problem and instead pose it as a classification problem (and vice versa).

Problem

The first step of building any machine learning solution is framing the problem. Is this a supervised learning problem? Or unsupervised? What are the features? If it is a supervised problem, what are the labels? What amount of error is acceptable? Of course, the answers to these questions must be considered in context with the training data, the task at hand, and the metrics for success.

For example, suppose we wanted to build a machine learning model to predict future rainfall amounts in a given location. Starting broadly, would this be a regression or classification task? Well, since we're trying to predict rainfall amount (for example, 0.3 cm), it makes sense to consider this as a time-series forecasting problem: given the current and historical climate and weather patterns, what amount of rainfall should we expect in a given area in the next 15 minutes? Alternately, because the label (the amount of rainfall) is a real number, we could build a regression model. As we start to develop and train our model, we find (perhaps not surprisingly) that weather prediction is harder than it sounds. Our predicted rainfall amounts are all off because, for the same set of features, it sometimes rains 0.3 cm and other times it rains 0.5 cm. What should we do to improve our predictions? Should we add more layers to our network? Or engineer more features? Perhaps more data will help? Maybe we need a different loss function?

Any of these adjustments could improve our model. But wait. Is regression the only way we can pose this task? Perhaps we can reframe our machine learning objective in a way that improves our task performance.

Solution

The core issue here is that rainfall is probabilistic. For the same set of features, it sometimes rains 0.3 cm and other times it rains 0.5 cm. Yet, even if a regression model were able to learn the two possible amounts, it is limited to predicting only a single number.

Instead of trying to predict the amount of rainfall as a regression task, we can reframe our objective as a classification problem. There are different ways this can be accomplished. One approach is to model a discrete probability distribution, as shown in Figure 3-1. Instead of predicting rainfall as a real-valued output, we model the output as a multiclass classification giving the probability that the rainfall in the next 15 minutes is within a certain range of rainfall amounts.

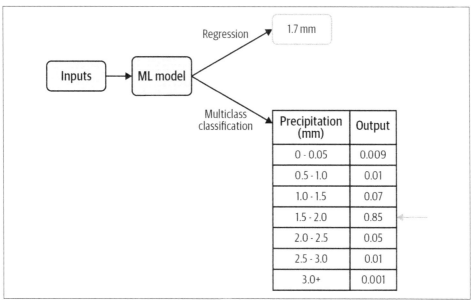

Figure 3-1. *Instead of predicting precipitation as a regression output, we can instead model discrete probability distribution using a multiclass classification.*

Both the regression approach and this reframed-as-classification approach give a prediction of the rainfall for the next 15 minutes. However, the classification approach allows the model to capture the probability distribution of rainfall of different quantities instead of having to choose the mean of the distribution. Modeling a distribution in this way is advantageous since precipitation does not exhibit the typical bell-shaped curve of a normal distribution and instead follows a Tweedie distribution (*https://oreil.ly/C8JfK*), which allows for a preponderance of points at zero. Indeed, that's the approach taken in a Google Research paper (*https://oreil.ly/PGAEw*) that predicts precipitation rates in a given location using a 512-way categorical distribution. Other reasons that modeling a distribution can be advantageous is when the distribution is bimodal, or even when it is normal but with a large variance. A recent paper that beats all benchmarks at predicting protein folding structure (*https://oreil.ly/-Hi3k*) also predicts the distance between amino acids as a 64-way classification problem where the distances are bucketized into 64 bins.

Another reason to reframe a problem is when the objective is better in the other type of model. For example, suppose we are trying to build a recommendation system for videos. A natural way to frame this problem is as a classification problem of predicting whether a user is likely to watch a certain video. This framing, however, can lead to a recommendation system that prioritizes click bait. It might be better to reframe this into a regression problem of predicting the fraction of the video that will be watched.

Why It Works

Changing the context and reframing the task of a problem can help when building a machine learning solution. Instead of learning a single real number, we relax our prediction target to be instead a discrete probability distribution. We lose a little precision due to bucketing, but gain the expressiveness of a full probability density function (PDF). The discretized predictions provided by the classification model are more adept at learning a complex target than the more rigid regression model.

An added advantage of this classification framing is that we obtain posterior probability distribution of our predicted values, which provides more nuanced information. For example, suppose the learned distribution is bimodal. By modeling a classification as a discrete probability distribution, the model is able to capture the bimodal structure of the predictions, as Figure 3-2 illustrates. Whereas, if only predicting a single numeric value, this information would be lost. Depending on the use case, this could make the task easier to learn and substantially more advantageous.

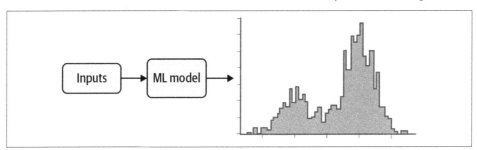

Figure 3-2. Reframing a classification task to model a probability distribution allows the predictions to capture bimodal output. The prediction is not limited to a single value as in a regression.

Capturing uncertainty

Let's look again at the natality dataset and the task of predicting baby weight. Since baby weight is a positive real value, this is intuitively a regression problem. However, notice that for a given set of inputs, `weight_pounds` (the label) can take many different values. We see that the distribution of babies' weights for a specific set of input values (male babies born to 25-year-old mothers at 38 weeks) approximately follows a normal distribution centered at about 7.5 pounds. The code to produce the graph in Figure 3-3 can be found in the repository (*https://github.com/GoogleCloudPlatform/ ml-design-patterns/03_problem_representation/reframing.ipynb*) for this book.

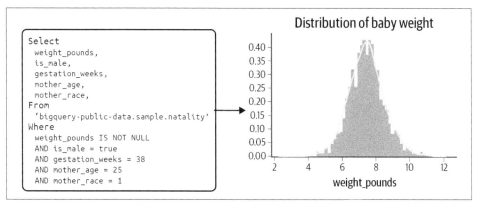

Figure 3-3. Given a specific set of inputs (for example, male babies born to 25-year-old mothers at 38 weeks) the weight_pounds variable takes a range of values, approximately following a normal distribution centered at 7.5 lbs.

However, notice the width of the distribution—even though the distribution peaks at 7.5 pounds, there is a nontrivial likelihood (actually 33%) that a given baby is less than 6.5 pounds or more than 8.5 pounds! The width of this distribution indicates the irreducible error inherent to the problem of predicting baby weight. Indeed, the best root mean square error we can obtain on this problem, if we frame it as a regression problem, is the standard deviation of the distribution seen in Figure 3-3.

If we frame this as a regression problem, we'd have to state the prediction result as 7.5 +/- 1.0 (or whatever the standard deviation is). Yet, the width of the distribution will vary for different combinations of inputs, and so learning the width is another machine learning problem in and of itself. For example, at the 36th week, for mothers of the same age, the standard deviation is 1.16 pounds. *Quantiles regression*, covered later in the pattern discussion, tries to do exactly this but in a nonparametric way.

Had the distribution been multimodal (with multiple peaks), the case for reframing the problem as a classification would be even stronger. However, it is helpful to realize that because of the law of large numbers, as long as we capture all of the relevant inputs, many of the distributions we will encounter on large datasets will be bell-shaped, although other distributions are possible. The wider the bell curve, and the more this width varies at different values of inputs, the more important it is to capture uncertainty and the stronger the case for reframing the regression problem as a classification one.

By reframing the problem, we train the model as a multiclass classification that learns a discrete probability distribution for the given training examples. These discretized predictions are more flexible in terms of capturing uncertainty and better able to approximate the complex target than a regression model. At inference time, the model then predicts a collection of probabilities corresponding to these potential outputs. That is, we obtain a discrete PDF giving the relative likelihood of any specific weight. Of course, care has to be taken here—classification models can be hugely uncalibrated (such as the model being overly confident and wrong).

Changing the objective

In some scenarios, reframing a classification task as a regression could be beneficial. For example, suppose we had a large movie database with customer ratings on a scale from 1 to 5, for all movies that the user had watched and rated. Our task is to build a machine learning model that will be used to serve recommendations to our users.

Viewed as a classification task, we could consider building a model that takes as input a user_id, along with that user's previous video watches and ratings, and predicts which movie from our database to recommend next. However, it is possible to reframe this problem as a regression. Instead of the model having a categorical output corresponding to a movie in our database, our model could instead carry out multitask learning, with the model learning a number of key characteristics (such as income, customer segment, and so on) of users who are likely to watch a given movie.

Reframed as a regression task, the model now predicts the user-space representation for a given movie. To serve recommendations, we choose the set of movies that are closest to the known characteristics of a user. In this way, instead of the model providing the probability that a user will like a movie as in a classification, we would get a cluster of movies that have been watched by users like this user.

By reframing the classification problem of recommending movies to be a regression of user characteristics, we gain the ability to easily adapt our recommendation model to recommend trending videos, or classic movies, or documentaries without having to train a separate classification model each time.

This type of model approach is also useful when the numerical representation has an intuitive interpretation; for example, a latitude and longitude pair can be used instead of urban area predictions. Suppose we wanted to predict which city will experience the next viral outbreak or which New York neighborhood will have a real estate pricing surge. It could be easier to predict the latitude and longitude and choose the city or neighborhood closest to that location, rather than predicting the city or neighborhood itself.

Trade-Offs and Alternatives

There is rarely just one way to frame a problem, and it is helpful to be aware of any trade-offs or alternatives of a given implementation. For example, bucketizing the output values of a regression is an approach to reframing the problem as a classification task. Another approach is multitask learning that combines both tasks (classification and regression) into a single model using multiple prediction heads. With any reframing technique, being aware of data limitations or the risk of introducing label bias is important.

Bucketized outputs

The typical approach to reframing a regression task as a classification is to bucketize the output values. For example, if our model is to be used to indicate when a baby might need critical care upon birth, the categories in Table 3-1 could be sufficient.

Table 3-1. Bucketized outputs for baby weight

Category	Description
High birth weight	More than 8.8 lbs
Average birth weight	Between 5.5 lbs and 8.8 lbs
Low birth weight	Between 3.31 lbs and 5.5 lbs
Very low birth weight	Less than 3.31 lbs

Our regression model now becomes a multiclass classification. Intuitively, it is easier to predict one out of four possible categorical cases than to predict a single value from the continuum of real numbers—just as it would be easier to predict a binary 0 versus 1 target for is_underweight instead of four separate categories high_weight versus avg_weight versus low_weight versus very_low_weight. By using categorical outputs, our model is incentivized less for getting arbitrarily close to the actual output value since we've essentially changed the output label to a range of values instead of a single real number.

In the notebook (*https://github.com/GoogleCloudPlatform/ml-design-patterns/blob/ master/03_problem_representation/reframing.ipynb*) accompanying this section, we train both a regression and a multiclass classification model. The regression model achieves an RMSE of 1.3 on the validation set while the classification model has an accuracy of 67%. Comparing these two models is difficult since one evaluation metric is RMSE and the other is accuracy. In the end, the design decision is governed by the use case. If medical decisions are based on bucketed values, then our model should be a classification using those buckets. However, if a more precise prediction of baby weight is needed, then it makes sense to use the regression model.

Other ways of capturing uncertainty

There are other ways to capture uncertainty in regression. A simple approach is to carry out quantile regression. For example, instead of predicting just the mean, we can estimate the conditional 10th, 20th, 30th, ..., 90th percentile of what needs to be predicted. Quantile regression is an extension of linear regression. Reframing, on the other hand, can work with more complex machine learning models.

Another, more sophisticated approach is to use a framework like TensorFlow Probability (*https://oreil.ly/AEtLG*) to carry out regression. However, we have to explicitly model the distribution of the output. For example, if the output is expected to be normally distributed around a mean that's dependent on the inputs, the model's output layer would be:

```
tfp.layers.DistributionLambda(lambda t: tfd.Normal(loc=t, scale=1))
```

On the other hand, if we know the variance increases with the mean, we might be able to model it using the lambda function. Reframing, on the other hand, doesn't require us to model the posterior distribution.

 When training any machine learning model, the data is key. More complex relationships typically require more training data examples to find those elusive patterns. With that in mind, it is important to consider how data requirements compare for regression or classification models. A common rule of thumb for classification tasks is that we should have 10 times the number of model features for each label category. For a regression model, the rule of thumb is 50 times the number of model features. Of course, these numbers are just rough heuristics and not precise. However, the intuition is that regression tasks typically require more training examples. Furthermore, this need for massive data only increases with the complexity of the task. Thus, there could be data limitations that should be considered when considering the type of model used or, in the case of classification, the number of label categories.

Precision of predictions

When thinking of reframing a regression model as a multiclass classification, the width of the bins for the output label governs the precision of the classification model. In the case of our baby weight example, if we needed more precise information from the discrete probability density function, we would need to increase the number of bins of our categorical model. Figure 3-4 shows how the discrete probability distributions would look as either a 4-way or 10-way classification.

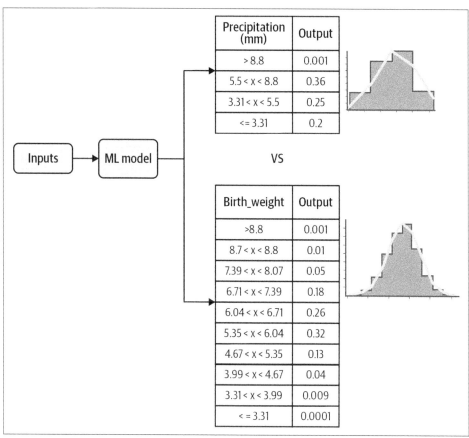

Precipitation (mm)	Output
> 8.8	0.001
5.5 < x < 8.8	0.36
3.31 < x < 5.5	0.25
<= 3.31	0.2

VS

Birth_weight	Output
>8.8	0.001
8.7 < x < 8.8	0.01
7.39 < x < 8.07	0.05
6.71 < x < 7.39	0.18
6.04 < x < 6.71	0.26
5.35 < x < 6.04	0.32
4.67 < x < 5.35	0.13
3.99 < x < 4.67	0.04
3.31 < x < 3.99	0.009
< = 3.31	0.0001

Figure 3-4. The precision of the multiclass classification is controlled by the width of the bins for the label.

The sharpness of the PDF indicates the precision of the task as a regression. A sharper PDF indicates a smaller standard deviation of the output distribution while a wider PDF indicates a larger standard deviation and thus more variance. For a very sharp density function, it's better to stick with a regression model (see Figure 3-5).

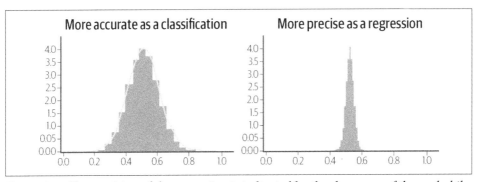

Figure 3-5. The precision of the regression is indicated by the sharpness of the probability density function for a fixed set of input values.

Restricting the prediction range

Another reason to reframe the problem is when it is essential to restrict the range of the prediction output. Let's say, for example, that realistic output values for a regression problem are in the range [3, 20]. If we train a regression model where the output layer is a linear activation function, there is always the possibility that the model predictions will fall outside this range. One way to limit the range of the output is to reframe the problem.

Make the activation function of the last-but-one layer a sigmoid function (which is typically associated with classification) so that it is in the range [0,1] and have the last layer scale these values to the desired range:

```
MIN_Y =  3
MAX_Y = 20
input_size = 10
inputs = keras.layers.Input(shape=(input_size,))
h1 = keras.layers.Dense(20, 'relu')(inputs)
h2 = keras.layers.Dense(1, 'sigmoid')(h1)  # 0-1 range
output = keras.layers.Lambda(
            lambda y : (y*(MAX_Y-MIN_Y) + MIN_Y))(h2) # scaled
model = keras.Model(inputs, output)
```

We can verify (see the notebook (*https://github.com/GoogleCloudPlatform/ml-design-patterns/blob/master/03_problem_representation/reframing.ipynb*) on GitHub for full code) that this model now emits numbers in the range [3, 20]. Note that because the output is a sigmoid, the model will never actually hit the minimum and maximum of the range, and only get quite close to it. When we trained the model above on some random data, we got values in the range [3.03, 19.99].

Label bias

Recommendation systems like matrix factorization can be reframed in the context of neural networks, both as a regression or classification. One advantage to this change of context is that a neural network framed as a regression or classification model can incorporate many more additional features outside of just the user and item embeddings learned in matrix factorization. So it can be an appealing alternative.

However, it is important to consider the nature of the target label when reframing the problem. For example, suppose we reframed our recommendation model to a classification task that predicts the likelihood a user will click on a certain video thumbnail. This seems like a reasonable reframing since our goal is to provide content a user will select and watch. But be careful. This change of label is not actually in line with our prediction task. By optimizing for user clicks, our model will inadvertently promote click bait and not actually recommend content of use to the user.

Instead, a more advantageous label would be video watch time, reframing our recommendation as a regression instead. Or perhaps we can modify the classification objective to predict the likelihood that a user will watch at least half the video clip. There is often more than one suitable approach, and it is important to consider the problem holistically when framing a solution.

> Be careful when changing the label and training task of your machine learning model, as it can inadvertently introduce label bias into your solution. Consider again the example of video recommendation we discussed in "Why It Works" on page 82.

Multitask learning

One alternative to reframing is multitask learning. Instead of trying to choose between regression or classification, do both! Generally speaking, multitask learning refers to any machine learning model in which more than one loss function is optimized. This can be accomplished in many different ways, but the two most common forms of multi task learning in neural networks is through hard parameter sharing and soft parameter sharing.

Parameter sharing refers to the parameters of the neural network being shared between the different output tasks, such as regression and classification. Hard parameter sharing occurs when the hidden layers of the model are shared between all the output tasks. In soft parameter sharing, each label has its own neural network with its own parameters, and the parameters of the different models are encouraged to be similar through some form of regularization. Figure 3-6 shows the typical architecture for hard parameter sharing and soft parameter sharing.

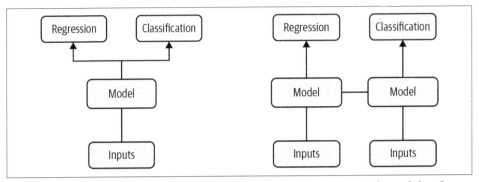

Figure 3-6. Two common implementations of multitask learning are through hard parameter sharing and soft parameter sharing.

In this context, we could have two heads to our model: one to predict a regression output and another to predict classification output. For example, this paper (*https://oreil.ly/sIjsF*) trains a computer vision model using a classification output of softmax probabilities together with a regression output to predict bounding boxes. They show that this approach achieves better performance than related work that trains networks separately for the classification and localization tasks. The idea is that through parameter sharing, the tasks are learned simultaneously and the gradient updates from the two loss functions inform both outputs and result in a more generalizable model.

Design Pattern 6: Multilabel

The Multilabel design pattern refers to problems where we can assign *more than one* label to a given training example. For neural networks, this design requires changing the activation function used in the final output layer of the model and choosing how our application will parse model output. Note that this is different from *multiclass* classification problems, where a single example is assigned exactly one label from a group of many (> 1) possible classes. You may also hear the Multilabel design pattern referred to as *multilabel, multiclass classification* since it involves choosing more than one label from a group of more than one possible class. When discussing this pattern, we'll focus primarily on neural networks.

Problem

Often, model prediction tasks involve applying a single classification to a given training example. This prediction is determined from N possible classes where N is greater than 1. In this case, it's common to use softmax as the activation function for the output layer. Using softmax, the output of our model is an N-element array, where the sum of all the values adds up to 1. Each value indicates the probability that a particular training example is associated with the class at that index.

For example, if our model is classifying images as cats, dogs, or rabbits, the softmax output might look like this for a given image: [.89, .02, .09]. This means our model is predicting an 89% chance the image is a cat, 2% chance it's a dog, and 9% chance it's a rabbit. Because each image can have *only one possible label* in this scenario, we can take the argmax (index of the highest probability) to determine our model's predicted class. The less-common scenario is when each training example can be assigned *more than one* label, which is what this pattern addresses.

The Multilabel design pattern exists for models trained on all data modalities. For image classification, in the earlier cat, dog, rabbit example, we could instead use training images that each depicted *multiple* animals, and could therefore have multiple labels. For text models, we can imagine a few scenarios where text can be labeled with multiple tags. Using the dataset of Stack Overflow questions on BigQuery as an example, we could build a model to predict the tags associated with a particular question. As an example, the question "How do I plot a pandas DataFrame?" could be tagged as "Python," "pandas," and "visualization." Another multilabel text classification example is a model that identifies toxic comments. For this model, we might want to flag comments with multiple toxicity labels. A comment could therefore be labeled both "hateful" and "obscene."

This design pattern can also apply to tabular datasets. Imagine a healthcare dataset with various physical characteristics for each patient, like height, weight, age, blood pressure, and more. This data could be used to predict the presence of multiple conditions. For example, a patient could show risk of both heart disease and diabetes.

Solution

The solution for building models that can assign *more than one label* to a given training example is to use the *sigmoid* activation function in our final output layer. Rather than generating an array where all values sum to 1 (as in softmax), each *individual* value in a sigmoid array is a float between 0 and 1. That is to say, when implementing the Multilabel design pattern, our label needs to be multi-hot encoded. The length of the multi-hot array corresponds with the number of classes in our model, and each output in this label array will be a sigmoid value.

Building on the image example above, let's say our training dataset included images with more than one animal. The sigmoid output for an image that contained a cat and a dog but not a rabbit might look like the following: [.92, .85, .11]. This output means the model is 92% confident the image contains a cat, 85% confident it contains a dog, and 11% confident it contains a rabbit.

A version of this model for 28×28-pixel images with sigmoid output might look like this, using the Keras `Sequential` API:

```
model = keras.Sequential([
    keras.layers.Flatten(input_shape=(28, 28)),
    keras.layers.Dense(128, activation='relu'),
    keras.layers.Dense(3, activation='sigmoid')
])
```

The main difference in output between the sigmoid model here and the softmax example in the Problem section is that the softmax array is guaranteed to contain three values that sum to 1, whereas the sigmoid output will contain three values, each between 0 and 1.

Sigmoid Versus Softmax Activation

Sigmoid is a nonlinear, continuous, and differentiable activation function that takes the outputs of each neuron in the previous layer in the ML model and squashes the value of those outputs between 0 and 1. Figure 3-7 shows what the sigmoid function looks like.

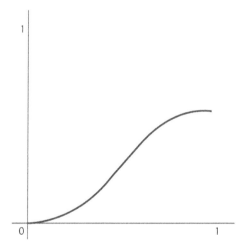

Figure 3-7. A sigmoid function.

While sigmoid takes a single value as input and provides a single value as output, softmax takes an array of values as input and transforms it into an array of probabilities that sum to 1. The input to the softmax function could be the output of N sigmoids.

In a multiclass classification problem where each example can only have one label, use softmax as the last layer to get a probability distribution. In the Multilabel pattern, it's acceptable for the output array to not sum to 1 since we're evaluating the probability of each individual label.

Following are sample sigmoid and softmax output arrays:

```
sigmoid = [.8, .9, .2, .5]
softmax = [.7, .1, .15, .05]
```

Trade-Offs and Alternatives

There are several special cases to consider when following the Multilabel design pattern and using sigmoid output. Next, we'll explore how to structure models that have two possible label classes, how to make sense of sigmoid results, and other important considerations for Multilabel models.

Sigmoid output for models with two classes

There are two types of models where the output can belong to two possible classes:

- Each training example can be assigned *only one* class. This is also called *binary classification* and is a special type of multiclass classification problem.

- Some training examples could belong to *both* classes. This is a type of *multilabel classification* problem.

Figure 3-8 shows the distinction between these classifications.

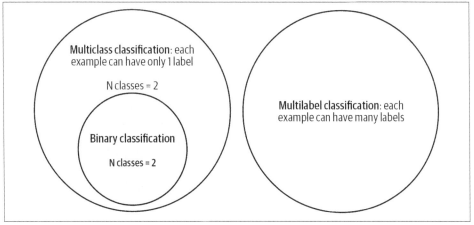

Figure 3-8. Understanding the distinction between multiclass, multilabel, and binary classification problems.

The first case (binary classification) is unique in that it is the only type of single-label classification problem where we would consider using sigmoid as our activation function. For nearly any other multiclass classification problem (for example, classifying text into one of five possible categories), we would use softmax. However, when we only have two classes, softmax is redundant. Take for example a model that predicts whether or not a specific transaction is fraudulent. Had we used a softmax output in this example, here's what a fraudulent model prediction might look like:

```
[.02, .98]
```

In this example, the first index corresponds with "not fraudulent" and the second index corresponds with "fraudulent." This is redundant because we could also represent this with a single scalar value, and thus use a sigmoid output. The same prediction could be represented as simply .98. Because each input can only be assigned a single class, we can infer from this output of .98 that the model has predicted a 98% chance of fraud and a 2% chance of nonfraud.

Therefore, for binary classification models, it is optimal to use an output shape of 1 with a sigmoid activation function. Models with a single output node are also more efficient, since they will have fewer trainable parameters and will likely train faster. Here is what the output layer of a binary classification model would look like:

```
keras.layers.Dense(1, activation='sigmoid')
```

For the second case where a training example could belong to *both possible classes* and fits into the Multilabel design pattern, we'll also want to use sigmoid, this time with a two-element output:

```
keras.layers.Dense(2, activation='sigmoid')
```

Which loss function should we use?

Now that we know when to use sigmoid as an activation function in our model, how should we choose which loss function to use with it? For the binary classification case where our model has a one-element output, use binary cross-entropy loss. In Keras, we provide a loss function when we compile our model:

```
model.compile(loss='binary_crossentropy', optimizer='adam',
    metrics=['accuracy'])
```

Interestingly, we also use binary cross-entropy loss for multilabel models with sigmoid output. This is because, as shown in Figure 3-9, a multilabel problem with three classes is essentially three smaller binary classification problems.

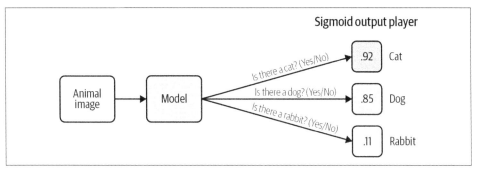

Figure 3-9. *Understanding the Multilabel pattern by breaking down the problem into smaller binary classification tasks.*

Parsing sigmoid results

To extract the predicted label for a model with softmax output, we can simply take the argmax (highest value index) of the output array to get the predicted class. Parsing sigmoid outputs is less straightforward. Instead of taking the class with the highest predicted probability, we need to evaluate the probability of each class in our output layer and consider the probability *threshold* for our use case. Both of these choices are largely dependent on the end user application of our model.

> By threshold, we're referring to the probability we're comfortable with for confirming an input belongs to a particular class. For example, if we're building a model to classify different types of animals in images, we might be comfortable saying an image has a cat even if the model is only 80% confident the image contains a cat. Alternatively, if we're building a model that's making healthcare predictions, we'll likely want the model to be closer to 99% confident before confirming a specific medical condition is present or not. While thresholding is something we'll need to consider for any type of classification model, it's especially relevant to the Multilabel design pattern since we'll need to determine thresholds for each class and they may be different.

To look at a specific example, let's take the Stack Overflow dataset in BigQuery and use it to build a model that predicts the tags associated with a Stack Overflow question given its title. We'll limit our dataset to questions that contain only five tags to keep things simple:

```
SELECT
  title,
  REPLACE(tags, "|", ",") as tags
FROM
  `bigquery-public-data.stackoverflow.posts_questions`
WHERE
```

```
    REGEXP_CONTAINS(tags,
    r"(?:keras|tensorflow|matplotlib|pandas|scikit-learn)")
```

The output layer of our model would look like the following (full code for this section is available in the GitHub repository (*https://github.com/GoogleCloudPlatform/ml-design-patterns/blob/master/03_problem_representation/multilabel.ipynb*)):

```
    keras.layers.Dense(5, activation='sigmoid')
```

Let's take the Stack Overflow question "What is the definition of a non-trainable parameter?" as an input example. Assuming our output indices correspond with the order of tags in our query, an output for that question might look like this:

```
    [.95, .83, .02, .08, .65]
```

Our model is 95% confident this question should be tagged Keras, and 83% confident it should be tagged TensorFlow. When evaluating model predictions, we'll need to iterate over every element in the output array and determine how we want to display those results to our end users. If 80% is our threshold for all tags, we'd show `Keras` *and* `TensorFlow` associated with this question. Alternatively, maybe we want to encourage users to add as many tags as possible and we want to show options for any tag with prediction confidence above 50%.

For examples like this one, where the goal is primarily to suggest possible tags with less emphasis on getting the tag *exactly* right, a typical rule of thumb is to use `n_specific_tag / n_total_examples` as a threshold for each class. Here, `n_specific_tag` is the number of examples with one tag in the dataset (for example, "pandas"), and `n_total_examples` is the total number of examples in the training set across all tags. This ensures that the model is doing better than guessing a certain label based on its occurrence in the training dataset.

 For a more precise approach to thresholding, consider using S-Cut or optimizing for your model's F-measure. Details on both can be found in this paper (*https://oreil.ly/oyR57*). Calibrating the per-label probabilities is often helpful as well, especially when there are thousands of labels and you want to consider the top K of them (this is common in search and ranking problems).

As you've seen, multilabel models provide more flexibility in how we parse predictions and require us to think carefully about the output for each class.

Dataset considerations

When dealing with single-label classification tasks, we can ensure our dataset is balanced by aiming for a relatively equal number of training examples for each class. Building a balanced dataset is more nuanced for the Multilabel design pattern.

Taking the Stack Overflow dataset example, there will likely be many questions tagged as both `TensorFlow` and `Keras`. But there will also be questions about Keras that have nothing to do with TensorFlow. Similarly, we might see questions about plotting data that is tagged with both `matplotlib` and `pandas`, and questions about data preprocessing that are tagged both `pandas` and `scikit-learn`. In order for our model to learn what is unique to each tag, we'll want to ensure the training dataset consists of varied combinations of each tag. If the majority of `matplotlib` questions in our dataset are also tagged `pandas`, the model won't learn to classify `matplotlib` on its own. To account for this, think about the different relationships between labels that might be present in our model and count the number of training examples that belong to each overlapping combination of labels.

When exploriing relationships between labels in our dataset, we may also encounter hierarchical labels. ImageNet (*https://oreil.ly/0VXtc*), the popular image classification dataset, contains thousands of labeled images and is often used as a starting point for transfer learning on image models. All of the labels used in ImageNet are hierarchical, meaning all images have at least one label, and many images have more specific labels that are part of a hierarchy. Here's an example of one label hierarchy in ImageNet:

animal → invertebrate → arthropod → arachnid → spider

Depending on the size and nature of the dataset, there are two common approaches for handling hierarchical labels:

- Use a flat approach and put every label in the same output array regardless of hierarchy, making sure you have enough examples of each "leaf node" label.
- Use the Cascade design pattern. Build one model to identify higher-level labels. Based on the high-level classification, send the example to a separate model for a more specific classification task. For example, we might have an initial model that labels images as "Plant," "Animal," or "Person." Depending on which labels the first model applies, we'll send the image to different model(s) to apply more granular labels like "succulent" or "barbary lion."

The flat approach is more straightforward than following the Cascade design pattern since it only requires one model. However, this might cause the model to lose information about more detailed label classes since there will naturally be more training examples with the higher-level labels in our dataset.

Inputs with overlapping labels

The Multilabel design pattern is also useful in cases where input data occasionally has overlapping labels. Let's take an image model that's classifying clothing items for a catalog as an example. If we have multiple people labeling each image in the training dataset, one labeler may label an image of a skirt as "maxi skirt," while another identifies it as "pleated skirt." Both are correct. However, if we build a multiclass classification model on this data, passing it multiple examples of the same image with different labels, we'll likely encounter situations where the model labels similar images differently when making predictions. Ideally, we want a model that labels this image as both "maxi skirt" and "pleated skirt" as seen in Figure 3-10, rather than sometimes predicting only one of these labels.

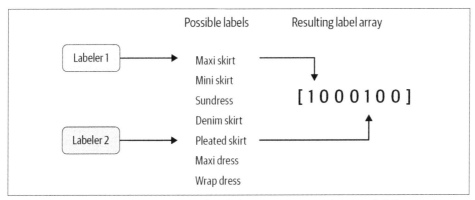

Figure 3-10. Using input from multiple labelers to create overlapping labels in cases where multiple descriptions of an item are correct.

The Multilabel design pattern solves this by allowing us to associate both overlapping labels with an image. In cases with overlapping labels where we have multiple labelers evaluating each image in our training dataset, we can choose the maximum number of labels we'd like labelers to assign to a given image, then take the most commonly chosen tags to associate with an image during training. The threshold for "most commonly chosen tags" will depend on our prediction task and the number of human labelers we have. For example, if we have 5 labelers evaluating every image and 20 possible tags for each image, we might encourage labelers to give each image 3 tags. From this list of 15 label "votes" per image, we could then choose the 2 to 3 with the most votes from the labelers. When evaluating this model, we need to take note of the average prediction confidence the model returns for each label and use this to iteratively improve our dataset and label quality.

One versus rest

Another technique for handling Multilabel classification is to train multiple binary classifiers instead of one multilabel model. This approach is called *one versus rest*. In the case of the Stack Overflow example where we want to tag questions as Tensor-Flow, Python, and pandas, we'd train an individual classifier for each of these three tags: Python or not, TensorFlow or not, and so forth. Then we'd choose a confidence threshold and tag the original input question with tags from each binary classifier above some threshold.

The benefit of one versus rest is that we can use it with model architectures that can only do binary classification, like SVMs. It may also help with rare categories since the model will be performing only one classification task at a time on each input, and it is possible to apply the Rebalancing design pattern. The disadvantage of this approach is the added complexity of training many different classifiers, requiring us to build our application in a way that generates predictions from each of these models rather than having just one.

To summarize, use the Multilabel design pattern when your data falls into any of the following classification scenarios:

- A single training example can be associated with mutually exclusive labels.
- A single training example can have many hierarchical labels.
- Labelers describe the same item in different ways, and each interpretation is accurate.

When implementing a multilabel model, ensure combinations of overlapping labels are well represented in your dataset, and consider the threshold values you're willing to accept for each possible label in your model. Using a sigmoid output layer is the most common approach for building models that can handle multilabel classification. Additionally, sigmoid output can also be applied to binary classification tasks where a training example can have only one out of two possible labels.

Design Pattern 7: Ensembles

The Ensembles design pattern refers to techniques in machine learning that combine multiple machine learning models and aggregate their results to make predictions. Ensembles can be an effective means to improve performance and produce predictions that are better than any single model.

Problem

Suppose we've trained our baby weight prediction model, engineering special features and adding additional layers to our neural network so that the error on our training set is nearly zero. Excellent, you say! However, when we look to use our model in production at the hospital or evaluate performance on the hold out test set, our predictions are all wrong. What happened? And, more importantly, how can we fix it?

No machine learning model is perfect. To better understand where and how our model is wrong, the error of an ML model can be broken down into three parts: the irreducible error, the error due to bias, and the error due to variance. The irreducible error is the inherent error in the model resulting from noise in the dataset, the framing of the problem, or bad training examples, like measurement errors or confounding factors. Just as the name implies, we can't do much about *irreducible error*.

The other two, the bias and the variance, are referred to as the *reducible error*, and here is where we can influence our model's performance. In short, the bias is the model's inability to learn enough about the relationship between the model's features and labels, while the variance captures the model's inability to generalize on new, unseen examples. A model with high bias oversimplifies the relationship and is said to be *underfit*. A model with high variance has learned too much about the training data and is said to be *overfit*. Of course, the goal of any ML model is to have low bias and low variance, but in practice, it is hard to achieve both. This is known as the bias–variance trade-off. We can't have our cake and eat it too. For example, increasing model complexity decreases bias but increases variance, while decreasing model complexity decreases variance but introduces more bias.

Recent work (*https://oreil.ly/PxUvs*) suggests that when using modern machine learning techniques such as large neural networks with high capacity, this behavior is valid only up to a point. In observed experiments, there is an "interpolation threshold" beyond which very high capacity models are able to achieve zero training error as well as low error on unseen data. Of course, we need much larger datasets in order to avoid overfitting on high-capacity models.

Is there a way to mitigate this bias–variance trade-off on small- and medium-scale problems?

Solution

Ensemble methods are meta-algorithms that combine several machine learning models as a technique to decrease the bias and/or variance and improve model performance. Generally speaking, the idea is that combining multiple models helps to improve the machine learning results. By building several models with different inductive biases and aggregating their outputs, we hope to get a model with better performance. In this section, we'll discuss some commonly used ensemble methods, including bagging, boosting, and stacking.

Bagging

Bagging (short for bootstrap aggregating) is a type of parallel ensembling method and is used to address high variance in machine learning models. The bootstrap part of bagging refers to the datasets used for training the ensemble members. Specifically, if there are k submodels, then there are k separate datasets used for training each submodel of the ensemble. Each dataset is constructed by randomly sampling (with replacement) from the original training dataset. This means there is a high probability that any of the k datasets will be missing some training examples, but also any dataset will likely have repeated training examples. The aggregation takes place on the output of the multiple ensemble model members—either an average in the case of a regression task or a majority vote in the case of classification.

A good example of a bagging ensemble method is the random forest: multiple decision trees are trained on randomly sampled subsets of the entire training data, then the tree predictions are aggregated to produce a prediction, as shown in Figure 3-11.

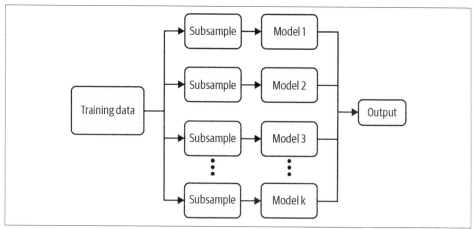

Figure 3-11. Bagging is good for decreasing variance in machine learning model output.

Popular machine learning libraries have implementations of bagging methods. For example, to implement a random Forest regression in scikit-learn to predict baby weight from our natality dataset:

```
from sklearn.ensemble import RandomForestRegressor

# Create the model with 50 trees
RF_model = RandomForestRegressor(n_estimators=50,
                                 max_features='sqrt',
                                 n_jobs=-1, verbose = 1)

# Fit on training data
RF_model.fit(X_train, Y_train)
```

Model averaging as seen in bagging is a powerful and reliable method for reducing model variance. As we'll see, different ensemble methods combine multiple submodels in different ways, sometimes using different models, different algorithms, or even different objective functions. With bagging, the model and algorithms are the same. For example, with random forest, the submodels are all short decision trees.

Boosting

Boosting is another Ensemble technique. However, unlike bagging, boosting ultimately constructs an ensemble model with *more* capacity than the individual member models. For this reason, boosting provides a more effective means of reducing bias than variance. The idea behind boosting is to iteratively build an ensemble of models where each successive model focuses on learning the examples the previous model got wrong. In short, boosting iteratively improves upon a sequence of weak learners taking a weighted average to ultimately yield a strong learner.

At the start of the boosting procedure, a simple base model f_0 is selected. For a regression task, the base model could just be the average target value: f_0 = np.mean(Y_train). For the first iteration step, the residuals delta_1 are measured and approximated via a separate model. This residual model can be anything, but typically it isn't very sophisticated; we'd often use a weak learner like a decision tree. The approximation provided by the residual model is then added to the current prediction, and the process continues.

After many iterations, the residuals tend toward zero and the prediction gets better and better at modeling the original training dataset. Notice that in Figure 3-12 the residuals for each element of the dataset decrease with each successive iteration.

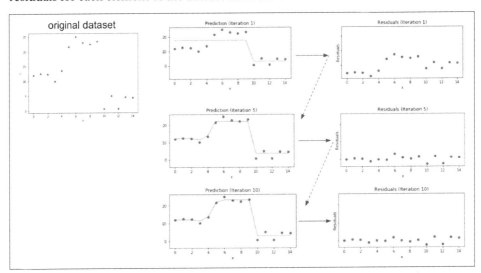

Figure 3-12. Boosting converts weak learners into strong learners by iteratively improving the model prediction.

Some of the more well-known boosting algorithms are AdaBoost, Gradient Boosting Machines, and XGBoost, and they have easy-to-use implementations in popular machine learning frameworks like scikit-learn or TensorFlow.

The implementation in scikit-learn is also straightforward:

```
from sklearn.ensemble import GradientBoostingRegressor

# Create the Gradient Boosting regressor
GB_model = GradientBoostingRegressor(n_estimators=1,
                                     max_depth=1,
                                     learning_rate=1,
                                     criterion='mse')

# Fit on training data
GB_model.fit(X_train, Y_train)
```

Stacking

Stacking is an ensemble method that combines the outputs of a collection of models to make a prediction. The initial models, which are typically of different model types, are trained to completion on the full training dataset. Then, a secondary meta-model is trained using the initial model outputs as features. This second meta-model learns how to best combine the outcomes of the initial models to decrease the training error and can be any type of machine learning model.

To implement a stacking ensemble, we first train all the members of the ensemble on the training dataset. The following code calls a function, fit_model, that takes as arguments a model and the training dataset inputs X_train and label Y_train. This way *members* is a list containing all the trained models in our ensemble. The full code for this example can be found in the code repository (*https://github.com/GoogleCloud Platform/ml-design-patterns/blob/master/03_problem_representation/ensemble_meth ods.ipynb*) for this book:

```
members = [model_1, model_2, model_3]

# fit and save models
n_members = len(members)

for i in range(n_members):
    # fit model
    model = fit_model(members[i])
    # save model
    filename = 'models/model_' + str(i + 1) + '.h5'
    model.save(filename, save_format='tf')
    print('Saved {}\n'.format(filename))
```

These submodels are incorporated into a larger stacking ensemble model as individual inputs. Since these input models are trained alongside the secondary ensemble model, we fix the weights of these input models. This can be done by setting layer.trainable to False for the ensemble member models:

```
for i in range(n_members):
    model = members[i]
```

```
for layer in model.layers:
    # make not trainable
    layer.trainable = False
    # rename to avoid 'unique layer name' issue
    layer._name = 'ensemble_' + str(i+1) + '_' + layer.name
```

We create the ensemble model stitching together the components using the Keras functional API:

```
member_inputs = [model.input for model in members]

# concatenate merge output from each model
member_outputs = [model.output for model in members]
merge = layers.concatenate(member_outputs)
hidden = layers.Dense(10, activation='relu')(merge)
ensemble_output = layers.Dense(1, activation='relu')(hidden)
ensemble_model = Model(inputs=member_inputs, outputs=ensemble_output)

# plot graph of ensemble
tf.keras.utils.plot_model(ensemble_model, show_shapes=True,
                          to_file='ensemble_graph.png')

# compile
ensemble_model.compile(loss='mse', optimizer='adam', metrics=['mse'])
```

In this example, the secondary model is a dense neural network with two hidden layers. Through training, this network learns how to best combine the results of the ensemble members when making predictions.

Why It Works

Model averaging methods like bagging work because typically the individual models that make up the ensemble model will not all make the same errors on the test set. In an ideal situation, each individual model is off by a random amount, so when their results are averaged, the random errors cancel out, and the prediction is closer to the correct answer. In short, there is wisdom in the crowd.

Boosting works well because the model is punished more and more according to the residuals at each iteration step. With each iteration, the ensemble model is encouraged to get better and better at predicting those hard-to-predict examples. Stacking works because it combines the best of both bagging and boosting. The secondary model can be thought of as a more sophisticated version of model averaging.

Bagging

More precisely, suppose we've trained k neural network regression models and average their results to create an ensemble model. If each model has error error_i on each example, where error_i is drawn from a zero-mean multivariate normal distribution with variance var and covariance cov, then the ensemble predictor will have an error:

```
ensemble_error = 1./k * np.sum([error_1, error_2,...,error_k])
```

If the errors error_i are perfectly correlated so that cov = var, then the mean square error of the ensemble model reduces to var. In this case, model averaging doesn't help at all. On the other extreme, if the errors error_i are perfectly uncorrelated, then cov = 0 and the mean square error of the ensemble model is var/k. So, the expected square error decreases linearly with the number k of models in the ensemble.[1] To summarize, on average, the ensemble will perform at least as well as any of the individual models in the ensemble. Furthermore, if the models in the ensemble make independent errors (for example, cov = 0), then the ensemble will perform significantly better. Ultimately, the key to success with bagging is model diversity.

This also explains why bagging is typically less effective for more stable learners like k-nearest neighbors (kNN), naive Bayes, linear models, or support vector machines (SVMs) since the size of the training set is reduced through bootstrapping. Even when using the same training data, neural networks can reach a variety of solutions due to random weight initializations or random mini-batch selection or different hyperparameters, creating models whose errors are partially independent. Thus, model averaging can even benefit neural networks trained on the same dataset. In fact, one recommended solution to fix the high variance of neural networks is to train multiple models and aggregate their predictions.

Boosting

The boosting algorithm works by iteratively improving the model to reduce the prediction error. Each new weak learner corrects for the mistakes of the previous prediction by modeling the residuals delta_i of each step. The final prediction is the sum of the outputs from the base learner and each of the successive weak learners, as shown in Figure 3-13.

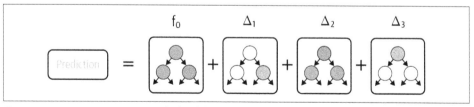

Figure 3-13. Boosting iteratively builds a strong learner from a sequence of weak learners that model the residual error of the previous iteration.

---

1 For the explicit computation of these values, see Ian Goodfellow, Yoshua Bengio, and Aaron Courville, *Deep Learning* (Cambridge, MA: MIT Press, 2016), Ch. 7.

Thus, the resulting ensemble model becomes successively more and more complex, having more capacity than any one of its members. This also explains why boosting is particularly good for combating high bias. Recall, the bias is related to the model's tendency to be underfit. By iteratively focusing on the hard-to-predict examples, boosting effectively decreases the bias of the resulting model.

Stacking

Stacking can be thought of as an extension of simple model averaging where we train k models to completion on the training dataset, then average the results to determine a prediction. Simple model averaging is similar to bagging, but the models in the ensemble could be of different types, while for bagging, the models are of the same type. More generally, we could modify the averaging step to take a weighted average, for example, to give more weight to one model in our ensemble over the others, as shown in Figure 3-14.

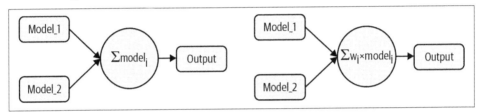

Figure 3-14. The simplest form of model averaging averages the outputs of two or more different machine learning models. Alternatively, the average could be replaced with a weighted average where the weight might be based on the relative accuracy of the models.

You can think of stacking as a more advanced version of model averaging, where instead of taking an average or weighted average, we train a second machine learning model on the outputs to learn how best to combine the results to the models in our ensemble to produce a prediction as shown in Figure 3-15. This provides all the benefits of decreasing variance as with bagging techniques but also controls for high bias.

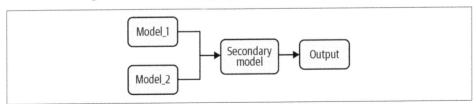

Figure 3-15. Stacking is an ensemble learning technique that combines the outputs of several different ML models as the input to a secondary ML model that makes predictions.

Trade-Offs and Alternatives

Ensemble methods have become quite popular in modern machine learning and have played a large part in winning well-known challenges, perhaps most notably the Netflix Prize (*https://oreil.ly/ybZ28*). There is also a lot of theoretical evidence to back up the success demonstrated on these real-world challenges.

Increased training and design time

One downside to ensemble learning is increased training and design time. For example, for a stacked ensemble model, choosing the ensemble member models can require its own level of expertise and poses its own questions: Is it best to reuse the same architectures or encourage diversity? If we do use different architectures, which ones should we use? And how many? Instead of developing a single ML model (which can be a lot of work on its own!), we are now developing k models. We've introduced an additional amount of overhead in our model development, not to mention maintenance, inference complexity, and resource usage if the ensemble model is to go into production. This can quickly become impractical as the number of models in the ensemble increases.

Popular machine learning libraries, like scikit-learn and TensorFlow, provide easy-to-use implementations for many common bagging and boosting methods, like random forest, AdaBoost, gradient boosting, and XGBoost. However, we should carefully consider whether the increased overhead associated with an ensemble method is worth it. Always compare accuracy and resource usage against a linear or DNN model. Note that distilling (see "Design Pattern 11: Useful Overfitting" on page 141) an ensemble of neural networks can often reduce complexity and improve performance.

Dropout as bagging

Techniques like dropout provide a powerful and effective alternative. Dropout is known as a regularization technique in deep learning but can be also understood as an approximation to bagging. Dropout in a neural network randomly (with a prescribed probability) "turns off" neurons of the network for each mini-batch of training, essentially evaluating a bagged ensemble of exponentially many neural networks. That being said, training a neural network with dropout is not exactly the same as bagging. There are two notable differences. First, in the case of bagging, the models are independent, while when training with dropout, the models share parameters. Second, in bagging, the models are trained to convergence on their respective training set. However, when training with dropout, the ensemble member models would only be trained for a single training step because different nodes are dropped out in each iteration of the training loop.

Decreased model interpretability

Another point to keep in mind is model interpretability. Already in deep learning, effectively explaining why our model makes the predictions it does can be difficult. This problem is compounded with ensemble models. Consider, for example, decision trees versus the random forest. A decision tree ultimately learns boundary values for each feature that guide a single instance to the model's final prediction. As such, it is easy to explain why a decision tree makes the predictions it did. The random forest, being an ensemble of many decision trees, loses this level of local interpretability.

Choosing the right tool for the problem

It's also important to keep in mind the bias–variance trade-off. Some ensemble techniques are better at addressing bias or variance than others (Table 3-2). In particular, boosting is adapted for addressing high bias, while bagging is useful for correcting high variance. That being said, as we saw in the section on "Bagging" on page 100, combining two models with highly correlated errors will do nothing to help lower the variance. In short, using the wrong ensemble method for our problem won't necessarily improve performance; it will just add unnecessary overhead.

Table 3-2. A summary of the trade-off between bias and variance

Problem	Ensemble solution
High bias (underfitting)	Boosting
High variance (overfitting)	Bagging

Other ensemble methods

We've discussed some of the more common ensemble techniques in machine learning. The list discussed earlier is by no means exhaustive and there are different algorithms that fit with these broad categories. There are also other ensemble techniques, including many that incorporate a Bayesian approach or that combine neural architecture search and reinforcement learning, like Google's AdaNet or AutoML techniques. In short, the Ensemble design pattern encompasses techniques that combine multiple machine learning models to improve overall model performance and can be particularly useful when addressing common training issues like high bias or high variance.

Design Pattern 8: Cascade

The Cascade design pattern addresses situations where a machine learning problem can be profitably broken into a series of ML problems. Such a cascade often requires careful design of the ML experiment.

Problem

What happens if we need to predict a value during both usual and unusual activity? The model will learn to ignore the unusual activity because it is rare. If the unusual activity is also associated with abnormal values, then trainability suffers.

For example, suppose we are trying to train a model to predict the likelihood that a customer will return an item that they have purchased. If we train a single model, the resellers' return behavior will be lost because there are millions of retail buyers (and retail transactions) and only a few thousand resellers. We don't really know at the time that a purchase is being made whether this is a retail buyer or a reseller. However, by monitoring other marketplaces, we have identified when items bought from us are subsequently being resold, and so our training dataset has a label that identifies a purchase as having been done by a reseller.

One way to solve this problem is to overweight the reseller instances when training the model. This is suboptimal because we need to get the more common retail buyer use case as correct as possible. We do not want to trade off a lower accuracy on the retail buyer use case for a higher accuracy on the reseller use case. However, retail buyers and resellers behave very differently; for example, while retail buyers return items within a week or so, resellers return items only if they are unable to sell them, and so the returns may take place after several months. The business decision of stocking inventory is different for likely returns from retail buyers versus resellers. Therefore, it is necessary to get both types of returns as accurate as possible. Simply overweighting the reseller instances will not work.

An intuitive way to address this problem is by using the Cascade design pattern. We break the problem into four parts:

1. Predicting whether a specific transaction is by a reseller
2. Training one model on sales to retail buyers
3. Training the second model on sales to resellers
4. In production, combining the output of the three separate models to predict return likelihood for every item purchased and the probability that the transaction is by a reseller

This allows for the possibility of different decisions on items likely to be returned depending on the type of buyer and ensures that the models in steps 2 and 3 are as accurate as possible on their segment of the training data. Each of these models is relatively easy to train. The first is simply a classifier, and if the unusual activity is extremely rare, we can use the Rebalancing pattern to address it. The next two models are essentially classification models trained on different segments of the training data. The combination is deterministic since we choose which model to run based on whether the activity belonged to a reseller.

The problem comes during prediction. At prediction time, we don't have true labels, just the output of the first classification model. Based on the output of the first model, we will have to determine which of the two sales models we invoke. The problem is that we are training on labels, but at inference time, we will have to make decisions based on predictions. And predictions have errors. So, the second and third models will be required to make predictions on data that they might have never seen during training.

As an extreme example, assume that the address that resellers provide is always in an industrial area of the city, whereas retail buyers can live anywhere. If the first (classification) model makes a mistake and a retail buyer is wrongly identified as a reseller, the cancellation prediction model that is invoked will not have the neighborhood where the customer lives in its vocabulary.

How do we train a cascade of models where the output of one model is an input to the following model or determines the selection of subsequent models?

Solution

Any machine learning problem where the output of the one model is an input to the following model or determines the selection of subsequent models is called a *cascade*. Special care has to be taken when training a cascade of ML models.

For example, a machine learning problem that sometimes involves unusual circumstances can be solved by treating it as a cascade of four machine learning problems:

1. A classification model to identify the circumstance

2. One model trained on unusual circumstances

3. A separate model trained on typical circumstances

4. A model to combine the output of the two separate models, because the output is a probabilistic combination of the two outputs

This appears, at first glance, to be a specific case of the Ensemble design pattern, but is considered separately because of the special experiment design required when doing a cascade.

As an example, assume that, in order to estimate the cost of stocking bicycles at stations, we wish to predict the distance between rental and return stations for bicycles in San Francisco. The goal of the model, in other words, is to predict the distance we need to transport the bicycle back to the rental location given features such as the time of day the rental starts, where the bicycle is being rented from, whether the renter is a subscriber or not, etc. The problem is that rentals that are longer than four hours involve extremely different renter behavior than shorter rentals, and the stocking algorithm requires both outputs (the probability that the rental is longer than four hours and the likely distance the bicycle needs to be transported). However, only a very small fraction of rentals involve such abnormal trips.

One way to solve this problem is to train a classification model to first classify trips based on whether they are Long or Typical (the full code (*https://github.com/Google CloudPlatform/ml-design-patterns/blob/master/03_problem_representation/ cascade.ipynb*) is in the code repository of this book):

```
CREATE OR REPLACE MODEL mlpatterns.classify_trips
TRANSFORM(
  trip_type,
  EXTRACT (HOUR FROM start_date) AS start_hour,
  EXTRACT (DAYOFWEEK FROM start_date) AS day_of_week,
  start_station_name,
  subscriber_type,
  ...
)
OPTIONS(model_type='logistic_reg',
        auto_class_weights=True,
        input_label_cols=['trip_type']) AS

SELECT
  start_date, start_station_name, subscriber_type, ...
  IF(duration_sec > 3600*4, 'Long', 'Typical') AS trip_type
FROM `bigquery-public-data.san_francisco_bikeshare.bikeshare_trips`
```

It can be tempting to simply split the training dataset into two parts based on the actual duration of the rental and train the next two models, one on Long rentals and the other on Typical rentals. The problem is that the classification model just discussed will have errors. Indeed, evaluating the model on a held-out portion of the San Francisco bicycle data shows that the accuracy of the model is only around 75% (see Figure 3-16). Given this, training a model on a perfect split of the data will lead to tears.

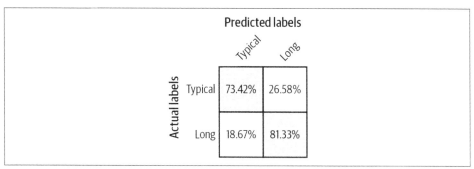

Figure 3-16. The accuracy of a classification model to predict atypical behavior is unlikely to be 100%.

Instead, after training this classification model, we need to use the predictions of this model to create the training dataset for the next set of models. For example, we could create the training dataset for the model to predict the distance of Typical rentals using:

```
CREATE OR REPLACE TABLE mlpatterns.Typical_trips AS
SELECT
  * EXCEPT(predicted_trip_type_probs, predicted_trip_type)
FROM
ML.PREDICT(MODEL mlpatterns.classify_trips,
  (SELECT
  start_date, start_station_name, subscriber_type, ...,
  ST_Distance(start_station_geom, end_station_geom) AS distance
  FROM `bigquery-public-data.san_francisco_bikeshare.bikeshare_trips`)
)
WHERE predicted_trip_type = 'Typical' AND distance IS NOT NULL
```

Then, we should use this dataset to train the model to predict distances:

```
CREATE OR REPLACE MODEL mlpatterns.predict_distance_Typical
TRANSFORM(
  distance,
  EXTRACT (HOUR FROM start_date) AS start_hour,
  EXTRACT (DAYOFWEEK FROM start_date) AS day_of_week,
  start_station_name,
  subscriber_type,
  ...
)
OPTIONS(model_type='linear_reg', input_label_cols=['distance']) AS

SELECT
  *
FROM
  mlpatterns.Typical_trips
```

Finally, our evaluation, prediction, etc. should take into account that we need to use three trained models, not just one. This is what we term the Cascade design pattern.

In practice, it can become hard to keep a Cascade workflow straight. Rather than train the models individually, it is better to automate the entire workflow using the Workflow Pipelines pattern (Chapter 6) as shown in Figure 3-17. The key is to ensure that training datasets for the two downstream models are created each time the experiment is run based on the predictions of upstream models.

Although we introduced the Cascade pattern as a way of predicting a value during both usual and unusual activity, the Cascade pattern's solution is capable of addressing a more general situation. The pipeline framework allows us to handle any situation where a machine learning problem can be profitably broken into a series (or cascade) of ML problems. Whenever the output of a machine learning model needs to be fed as the input to another model, the second model needs to be trained on the predictions of the first model. In all such situations, a formal pipeline experimentation framework will be helpful.

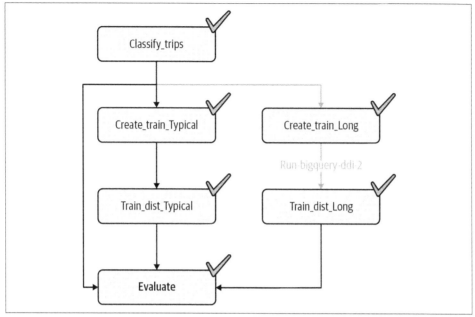

Figure 3-17. A pipeline to train the cascade of models as a single job.

Kubeflow Pipelines provides such a framework. Because it works with containers, the underlying machine learning models and glue code can be written in nearly any programming or scripting language. Here, we will wrap the BigQuery SQL models above into Python functions using the BigQuery client library. We could use TensorFlow or scikit-learn or even R to implement individual components.

The pipeline code using Kubeflow Pipelines can be expressed quite simply as the following (the full code (*https://github.com/GoogleCloudPlatform/ml-design-patterns/ blob/master/03_problem_representation/cascade.ipynb*) can be found in the code repository of this book):

```
@dsl.pipeline(
    name='Cascade pipeline on SF bikeshare',
    description='Cascade pipeline on SF bikeshare'
)
def cascade_pipeline(
    project_id = PROJECT_ID
):
    ddlop = comp.func_to_container_op(run_bigquery_ddl,
                packages_to_install=['google-cloud-bigquery'])

    c1 = train_classification_model(ddlop, PROJECT_ID)
```

```
c1_model_name = c1.outputs['created_table']

c2a_input = create_training_data(ddlop,
            PROJECT_ID, c1_model_name, 'Typical')
c2b_input = create_training_data(ddlop,
            PROJECT_ID, c1_model_name, 'Long')

c3a_model = train_distance_model(ddlop,
            PROJECT_ID, c2a_input.outputs['created_table'], 'Typical')
c3b_model = train_distance_model(ddlop,
            PROJECT_ID, c2b_input.outputs['created_table'], 'Long')

...
```

The entire pipeline can be submitted for running, and different runs of the experiment tracked using the Pipelines framework.

 If we are using TFX as our pipeline framework (we can run TFX on Kubeflow Pipelines), then it is not necessary to deploy the upstream models in order to use their output predictions in downstream models. Instead, we can use the TensorFlow Transform method `tft.apply_saved_model` as part of our preprocessing operations. The Transform design pattern is discussed in Chapter 6.

Use of a pipeline-experiment framework is strongly suggested whenever we will have chained ML models. Such a framework will ensure that downstream models are retrained whenever upstream models are revised and that we have a history of all the previous training runs.

Trade-Offs and Alternatives

Don't go overboard with the Cascade design pattern—unlike many of the design patterns we cover in this book, Cascade is not necessarily a best practice. It adds quite a bit of complexity to your machine learning workflows and may actually result in poorer performance. Note that a pipeline-experiment framework is definitely best practice, but as much as possible, try to limit a pipeline to a single machine learning problem (ingest, preprocessing, data validation, transformation, training, evaluation, and deployment). Avoid having, as in the Cascade pattern, multiple machine learning models in the same pipeline.

Deterministic inputs

Splitting an ML problem is usually a bad idea, since an ML model can/should learn combinations of multiple factors. For example:

- If a condition can be known deterministically from the input (holiday shopping versus weekday shopping), we should just add the condition as one more input to the model.
- If the condition involves an extrema in just one input (some customers who live nearby versus far away, with the meaning of near/far needing to be learned from the data), we can use Mixed Input Representation to handle it.

The Cascade design pattern addresses an unusual scenario for which we do not have a categorical input, and for which extreme values need to be learned from multiple inputs.

Single model

The Cascade design pattern should not be used for common scenarios where a single model will suffice. For example, suppose we are trying to learn a customer's propensity to buy. We may think we need to learn different models for people who have been comparison shopping versus those who aren't. We don't really know who has been comparison shopping, but we can make an educated guess based on the number of visits, how long the item has been in the cart, and so on. This problem does not need the Cascade design pattern because it is common enough (a large fraction of customers will be comparison shopping) that the machine learning model should be able to learn it implicitly in the course of training. For common scenarios, train a single model.

Internal consistency

The Cascade is needed when we need to maintain internal consistency amongst the predictions of multiple models. Note that we are trying to do more than just predict the unusual activity. We are trying to predict returns, considering that there will be some reseller activity also. If the task is only to predict whether or not a sale is by a reseller, we'd use the Rebalancing pattern. The reason to use Cascade is that the imbalanced label output is needed as an input to subsequent models and is useful in and of itself.

Similarly, suppose that the reason we are training the model to predict a customer's propensity to buy is to make a discounted offer. Whether or not we make the discounted offer, and the amount of discount, will very often depend on whether this customer is comparison shopping or not. Given this, we need internal consistency between the two models (the model for comparison shoppers and the model for propensity to buy). In this case, the Cascade design pattern might be needed.

Pre-trained models

The Cascade is also needed when we wish to reuse the output of a pre-trained model as an input into our model. For example, let's say we are building a model to detect authorized entrants to a building so that we can automatically open the gate. One of the inputs to our model might be the license plate of the vehicle. Instead of using the security photo directly in our model, we might find it simpler to use the output of an optical character recognition (OCR) model. It is critical that we recognize that OCR systems will have errors, and so we should not train our model with perfect license plate information. Instead, we should train the model on the actual output of the OCR system. Indeed, because different OCR models will behave differently and have different errors, it is necessary to retrain the model if we change the vendor of our OCR system.

 A common scenario of using a pre-trained model as the first step of a pipeline is using an object-detection model followed by a fine-grained image classification model. For example, the object-detection model might find all handbags in the image, an intermediate step might crop the image to the bounding boxes of the detected objects, and subsequent model might identify the type of handbag. We recommend using a Cascade so that the entire pipeline can be retrained whenever the object-detection model is updated (such as with a new version of the API).

Reframing instead of Cascade

Note that in our example problem, we were trying to predict the likelihood that an item would be returned, and so this was a classification problem. Suppose instead we wish to predict hourly sales amounts. Most of the time, we will serve just retail buyers, but once in a while (perhaps four or five times a year), we will have a wholesale buyer.

This is notionally a regression problem of predicting daily sales amounts where we have a confounding factor in the form of wholesale buyers. Reframing the regression problem to be a classification problem of different sales amounts might be a better approach. Although it will involve training a classification model for each sales amount bucket, it avoids the need to get the retail versus wholesale classification correct.

Regression in rare situations

The Cascade design pattern can be helpful when carrying out regression when some values are much more common than others. For example, we might want to predict the quantity of rainfall from a satellite image. It might be the case that on 99% of the pixels, it doesn't rain. In such a case, it can be helpful to create a stacked classification model followed by a regression model:

1. First, predict whether or not it is going to rain.
2. For pixels where the model predicts rain is not likely, predict a rainfall amount of zero.
3. Train a regression model to predict the rainfall amount on pixels where the model predicts that rain is likely.

It is critical to realize that the classification model is not perfect, and so the regression model has to be trained on the pixels that the classification model predicts as likely to be raining (and not just on pixels that correspond to rain in the labeled dataset). For complementary solutions to this problem, also see the discussions on "Design Pattern 10: Rebalancing " on page 122 and "Design Pattern 5: Reframing " on page 80.

Design Pattern 9: Neutral Class

In many classification situations, creating a neutral class can be helpful. For example, instead of training a binary classifier that outputs the probability of an event, train a three-class classifier that outputs disjoint probabilities for Yes, No, and Maybe. Disjoint here means that the classes do not overlap. A training pattern can belong to only one class, and so there is no overlap between Yes and Maybe, for example. The Maybe in this case is the neutral class.

Problem

Imagine that we are trying to create a model that provides guidance on pain relievers. There are two choices, ibuprofen and acetaminophen,[2] and it turns out in our historical dataset that acetaminophen tends to be prescribed preferentially to patients at risk of stomach problems, and ibuprofen tends to be prescribed preferentially to patients at risk of liver damage. Beyond that, things tend to be quite random; some physicians default to acetaminophen and others to ibuprofen.

Training a binary classifier on such a dataset will lead to poor accuracy because the model will need to get the essentially arbitrary cases correct.

---

2 This is just an example being used for illustrative purposes; please don't take this as medical advice!

Solution

Imagine a different scenario. Suppose the electronic record that captures the doctor's prescriptions also asks them whether the alternate pain medication would be acceptable. If the doctor prescribes acetaminophen, the application asks the doctor whether the patient can use ibuprofen if they already have it in their medicine cabinet.

Based on the answer to the second question, we have a neutral class. The prescription might still be written as "acetaminophen," but the record captures that the doctor was neutral for this patient. Note that this fundamentally requires us to design the data collection appropriately—we cannot manufacture a neutral class after the fact. We have to correctly design the machine learning problem. Correct design, in this case, starts with how we pose the problem in the first place.

If all we have is a historical dataset, we would need to get a labeling service (*https://oreil.ly/OSZsi*) involved. We could ask the human labelers to validate the doctor's original choice and answer the question of whether an alternate pain medication would be acceptable.

Why It Works

We can explore the mechanism by which this works by simulating the mechanism involved with a synthetic dataset. Then, we will show that something akin to this also happens in the real world with marginal cases.

Synthetic data

Let's create a synthetic dataset of length N where 10% of the data represents patients with a history of jaundice. Since they are at risk of liver damage, their correct prescription is ibuprofen (the full code (*https://github.com/GoogleCloudPlatform/ml-design-patterns/blob/master/03_problem_representation/neutral.ipynb*) is in GitHub):

```
jaundice[0:N//10] = True
prescription[0:N//10] = 'ibuprofen'
```

Another 10% of the data will represent patients with a history of stomach ulcers; since they are at risk of stomach damage, their correct prescription is acetaminophen:

```
ulcers[(9*N)//10:] = True
prescription[(9*N)//10:] = 'acetaminophen'
```

The remaining patients will be arbitrarily assigned to either medication. Naturally, this random assignment will cause the overall accuracy of a model trained on just two classes to be low. In fact, we can calculate the upper bound on the accuracy. Because 80% of the training examples have random labels, the best that the model can do is to guess half of them correctly. So, the accuracy on that subset of the training examples will be 40%. The remaining 20% of the training examples have systematic labels, and an ideal model will learn this, so we expect that overall accuracy can be at best 60%.

Indeed, training a model using scikit-learn as follows, we get an accuracy of 0.56:

```
ntrain = 8*len(df)//10 # 80% of data for training
lm = linear_model.LogisticRegression()
lm = lm.fit(df.loc[:ntrain-1, ['jaundice', 'ulcers']],
            df[label][:ntrain])
acc = lm.score(df.loc[ntrain:, ['jaundice', 'ulcers']],
            df[label][ntrain:])
```

If we create three classes, and put all the randomly assigned prescriptions into that class, we get, as expected, perfect (100%) accuracy. The purpose of the synthetic data was to illustrate that, provided there is random assignment at work, the Neutral Class design pattern can help us avoid losing model accuracy because of arbitrarily labeled data.

In the real world

In real-world situations, things may not be precisely random as in the synthetic dataset, but the arbitrary assignment paradigm still holds. For example, one minute after a baby is born, the baby is assigned an "Apgar score," a number between 1 and 10, with 10 being a baby that has come through the birthing process perfectly.

Consider a model that is trained to predict whether or not a baby will come through the birthing process healthily, or will require immediate attention (the full code (*https://github.com/GoogleCloudPlatform/ml-design-patterns/blob/master/03_prob lem_representation/neutral.ipynb*) is on GitHub):

```
CREATE OR REPLACE MODEL mlpatterns.neutral_2classes
OPTIONS(model_type='logistic_reg', input_label_cols=['health']) AS

SELECT
  IF(apgar_1min >= 9, 'Healthy', 'NeedsAttention') AS health,
  plurality,
  mother_age,
  gestation_weeks,
  ever_born
FROM `bigquery-public-data.samples.natality`
WHERE apgar_1min <= 10
```

We are thresholding the Apgar score at 9 and treating babies whose Apgar score is 9 or 10 as healthy, and babies whose Apgar score is 8 or lower as requiring attention. The accuracy of this binary classification model when trained on the natality dataset and evaluated on held-out data is 0.56.

Yet, assigning an Apgar score involves a number of relatively subjective assessments, and whether a baby is assigned 8 or 9 often reduces to matters of physician preference. Such babies are neither perfectly healthy, nor do they need serious medical intervention. What if we create a neutral class to hold these "marginal" scores? This requires creating three classes, with an Apgar score of 10 defined as healthy, scores of 8 to 9 defined as neutral, and lower scores defined as requiring attention:

```
CREATE OR REPLACE MODEL mlpatterns.neutral_3classes
OPTIONS(model_type='logistic_reg', input_label_cols=['health']) AS

SELECT
  IF(apgar_1min = 10, 'Healthy',
      IF(apgar_1min >= 8, 'Neutral', 'NeedsAttention')) AS health,
  plurality,
  mother_age,
  gestation_weeks,
  ever_born
FROM `bigquery-public-data.samples.natality`
WHERE apgar_1min <= 10
```

This model achieves an accuracy of 0.79 on a held-out evaluation dataset, much higher than the 0.56 that was achieved with two classes.

Trade-Offs and Alternatives

The Neutral Class design pattern is one to keep in mind at the beginning of a machine learning problem. Collect the right data, and we can avoid a lot of sticky problems down the line. Here are a few situations where having a neutral class can be helpful.

When human experts disagree

The neutral class is helpful in dealing with disagreements among human experts. Suppose we have human labelers to whom we show patient history and ask them what medication they would prescribe. We might have a clear signal for acetaminophen in some cases, a clear signal for ibuprofen in other cases, and a huge swath of cases for which human labelers disagree. The neutral class provides a way to deal with such cases.

In the case of human labeling (unlike with the historical dataset of actual doctor actions where a patient was seen by only one doctor), every pattern is labeled by multiple experts. Therefore, we know a priori which cases humans disagree about. It might seem far simpler to simply discard such cases, and simply train a binary classifier. After all, it doesn't matter what the model does on the neutral cases. This has two problems:

1. False confidence tends to affect the acceptance of the model by human experts. A model that outputs a neutral determination is often more acceptable to experts than a model that is wrongly confident in cases where the human expert would have chosen the alternative.

2. If we are training a cascade of models, then downstream models will be extremely sensitive to the neutral classes. If we continue to improve this model, downstream models could change dramatically from version to version.

Another alternative is to use the agreement among human labelers as the weight of a pattern during training. Thus, if 5 experts agree on a diagnosis, the training pattern gets a weight of 1, while if the experts are split 3 to 2, the weight of the pattern might be only 0.6. This allows us to train a binary classifier, but overweight the classifier toward the "sure" cases. The drawback to this approach is that when the probability output by the model is 0.5, it is unclear whether it is because this reflects a situation where there was insufficient training data, or whether it is a situation where human experts disagree. Using a neutral class to capture areas of disagreement allows us to disambiguate the two situations.

Customer satisfaction

The need for a neutral class also arises with models that attempt to predict customer satisfaction. If the training data consists of survey responses where customers grade their experience on a scale of 1 to 10, it might be helpful to bucket the ratings into three categories: 1 to 4 as bad, 8 to 10 as good, and 5 to 7 is neutral. If, instead, we attempt to train a binary classifier by thresholding at 6, the model will spend too much effort trying to get essentially neutral responses correct.

As a way to improve embeddings

Suppose we are creating a pricing model for flights and wish to predict whether or not a customer will buy a flight at a certain price. To do this, we can look at historical transactions of flight purchases and abandoned shopping carts. However, suppose many of our transactions also include purchases by consolidators and travel agents—these are people who have contracted fares, and so the fares for them were not actually set dynamically. In other words, they don't pay the currently displayed price.

We could throw away all the nondynamic purchases and train the model only on customers who made the decision to buy or not buy based on the price being displayed. However, such a model will miss all the information held in the destinations that the consolidator or travel agent was interested in at various times—this will affect things like how airports and hotels are embedded. One way to retain that information while not affecting the pricing decision is to use a neutral class for these transactions.

Reframing with neutral class

Suppose we are training an automated trading system that makes trades based on whether it expects a security to go up or down in price. Because of stock market volatility and the speed with which new information is reflected in stock prices, trying to trade on small predicted ups and downs is likely to lead to high trading costs and poor profits over time.

In such cases, it is helpful to consider what the end goal is. The end goal of the ML model is not to predict whether a stock will go up or down. We will be unable to buy every stock that we predict will go up, and unable to sell stocks that we don't hold.

The better strategy might be to buy call options[3] for the 10 stocks that are most likely to go up more than 5% over the next 6 months, and buy put options for stocks that are most likely to go down more than 5% over the next 6 months.

The solution, then, is to create a training dataset consisting of three classes:

- Stocks that went up more than 5%—call.
- Stocks that went down more than 5%—put.
- The remaining stocks are in the neutral category.

Rather than train a regression model on how much stocks will go up, we can now train a classification model with these three classes and pick the most confident predictions from our model.

Design Pattern 10: Rebalancing

The Rebalancing design pattern provides various approaches for handling datasets that are inherently imbalanced. By this we mean datasets where one label makes up the majority of the dataset, leaving far fewer examples of other labels.

This design pattern does *not* address scenarios where a dataset lacks representation for a specific population or real-world environment. Cases like this can often only be solved by additional data collection. The Rebalancing design pattern primarily addresses how to build models with datasets where few examples exist for a specific class or classes.

Problem

Machine learning models learn best when they are given a similar number of examples for each label class in a dataset. Many real-world problems, however, are not so neatly balanced. Take for example a fraud detection use case, where you are building a model to identify fraudulent credit card transactions. Fraudulent transactions are much rarer than regular transactions, and as such, there is less data on fraud cases available to train a model. The same is true for other problems like detecting whether someone will default on a loan, identifying defective products, predicting the

---

3 See *https://oreil.ly/kDndF* for a primer on call and put options.

presence of a disease given medical images, filtering spam emails, flagging error logs in a software application, and more.

Imbalanced datasets apply to many types of models, including binary classification, multiclass classification, multilabel classification, and regression. In regression cases, imbalanced datasets refer to data with outlier values that are either much higher or lower than the median in your dataset.

A common pitfall in training models with imbalanced label classes is relying on misleading accuracy values for model evaluation. If we train a fraud detection model and only 5% of our dataset contains fraudulent transactions, chances are our model will train to 95% accuracy without any modifications to the dataset or underlying model architecture. While this 95% accuracy number is *technically* correct, there's a good chance the model is guessing the majority class (in this case, nonfraud) for each example. As such, it's not learning anything about how to distinguish the minority class from other examples in our dataset.

To avoid leaning too much on this misleading accuracy value, it's worth looking at the model's confusion matrix to see accuracy for each class. The confusion matrix for a poorly performing model trained on an imbalanced dataset often looks something like Figure 3-18.

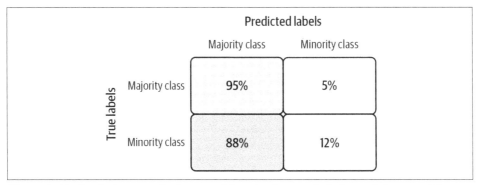

Figure 3-18. Confusion matrix for a model trained on an imbalanced dataset without dataset or model adjustments.

In this example, the model correctly guesses the majority class 95% of the time, but only guesses the minority class correctly 12% of the time. Typically, the confusion matrix for a high performing model has percentages close to 100 down the diagonal.

Solution

First, since accuracy can be misleading on imbalanced datasets, it's important to choose an appropriate evaluation metric when building our model. Then, there are various techniques we can employ for handling inherently imbalanced datasets at both the dataset and model level. *Downsampling* changes the balance of our underlying dataset, while *weighting* changes how our model handles certain classes. *Upsampling* duplicates examples from our minority class, and often involves applying augmentations to generate additional samples. We'll also look at approaches for *reframing* the problem: changing it to a regression task, analyzing our model's error values for each example, or clustering.

Choosing an evaluation metric

For imbalanced datasets like the one in our fraud detection example, it's best to use metrics like precision, recall, or F-measure to get a complete picture of how our model is performing. *Precision* measures the percentage of positive classifications that were correct out of all positive predictions made by the model. Conversely, *recall* measures the proportion of actual positive examples that were identified correctly by the model. The biggest difference between these two metrics is the denominator used to calculate them. For precision, the denominator is the total number of positive class predictions made by our model. For recall, it is the number of *actual* positive class examples present in our dataset.

A perfect model would have both precision and recall of 1.0, but in practice, these two measures are often at odds with each other. The *F-measure* is a metric that ranges from 0 to 1 and takes both precision and recall into account. It is calculated as:

```
2 * (precision * recall / (precision + recall))
```

Let's return to the fraud detection use case to see how each of these metrics plays out in practice. For this example, let's say our test set contains a total of 1,000 examples, 50 of which should be labeled as fraudulent transactions. For these examples, our model predicts 930/950 nonfraudulent examples correctly, and 15/50 fraudulent examples correctly. We can visualize these results in Figure 3-19.

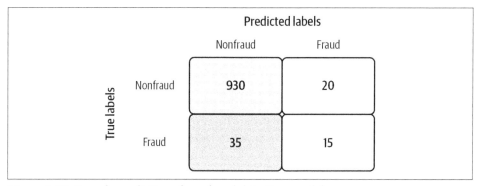

Figure 3-19. Sample predictions for a fraud detection model.

In this case, our model's precision is 15/35 (42%), recall is 15/50 (30%), and F-measure is 35%. These do a much better job capturing our model's inability to correctly identify fraudulent transactions compared to accuracy, which is 945/1000 (94.5%). Therefore, for models trained on imbalanced datasets, metrics other than accuracy are preferred. In fact, accuracy may even go down when optimizing for these metrics, but that is OK since precision, recall, and F-score are a better indication of model performance in this case.

Note that, when evaluating models trained on imbalanced datasets, we need to use *unsampled data* when calculating success metrics. This means that no matter how we modify our dataset for training per the solutions we'll outline below, we should leave our test set as is so that it provides an accurate representation of the original dataset. In other words, our test set should have roughly the same class balance as the original dataset. For the example above, that would be 5% fraud/95% nonfraud.

If we are looking for a metric that captures the performance of the model across all thresholds, average precision-recall is a more informative (*https://oreil.ly/5iJX2*) metric than area under the ROC curve (AUC) for model evaluation. This is because average precision-recall places more emphasis on how many predictions the model got right out of the *total* number it assigned to the positive class. This gives more weight to the positive class, which is important for imbalanced datasets. The AUC, on the other hand, treats both classes equally and is less sensitive to model improvements, which isn't optimal in situations with imbalanced data.

Downsampling

Downsampling is a solution for handling imbalanced datasets by changing the underlying dataset, rather than the model. With downsampling, we decrease the number of examples from the majority class used during model training. To see how this works, let's take a look at the synthetic fraud detection dataset on Kaggle (*https://oreil.ly/WqUM-*).[4] Each example in the dataset contains various information about the transaction, including the transaction type, the amount of the transaction, and the account balance both before and after the transaction took place. The dataset contains 6.3 million examples, only 8,000 of which are fraudulent transactions. That's a mere 0.1% of the entire dataset.

While a large dataset can often improve a model's ability to identify patterns, it's less helpful when the data is significantly imbalanced. If we train a model (*https://github.com/GoogleCloudPlatform/ml-design-patterns/blob/master/03_problem_repre sentation/rebalancing.ipynb*) on this entire dataset (6.3M rows) without any modifications, chances are we'll see a misleading accuracy of 99.9% as a result of the model

---

4 The dataset was generated based on the PaySim research proposed in this paper: EdgarLopez-Rojas , Ahmad Elmir, and Stefan Axelsson, "PaySim: A financial mobile money simulator for fraud detection," *28th European Modeling and Simulation Symposium,* EMSS, Larnaca, Cyprus (2016): 249–255.

randomly guessing the nonfraudulent class each time. We can solve for this by removing a large chunk of the majority class from the dataset.

We'll take all 8,000 of the fraudulent examples and set them aside to use when training the model. Then, we'll take a small, random sample of the nonfraudulent transactions. We'll then combine with our 8,000 fraudulent examples, reshuffle the data, and use this new, smaller dataset to train a model. Here's how we could implement this with pandas:

```
data = pd.read_csv('fraud_data.csv')

# Split into separate dataframes for fraud / not fraud
fraud = data[data['isFraud'] == 1]
not_fraud = data[data['isFraud'] == 0]

# Take a random sample of non fraud rows
not_fraud_sample = not_fraud.sample(random_state=2, frac=.005)

# Put it back together and shuffle
df = pd.concat([not_fraud_sample,fraud])
df = shuffle(df, random_state=2)
```

Following this, our dataset would contain 25% fraudulent transactions, much more balanced than the original dataset with only 0.1% in the minority class. It's worth experimenting with the exact balance used when downsampling. Here we used a 25/75 split, but different problems might require closer to a 50/50 split to achieve decent accuracy.

Downsampling is usually combined with the Ensemble pattern, following these steps:

1. Downsample the majority class and use all the instances of the minority class.
2. Train a model and add it to the ensemble.
3. Repeat.

During inference, take the median output of the ensemble models.

We discussed a classification example here, but downsampling can also be applied to regression models where we're predicting a numerical value. In this case, taking a random sample of majority class samples will be more nuanced since the majority "class" in our data includes a range of values rather than a single label.

Weighted classes

Another approach to handling imbalanced datasets is to change the *weight* our model gives to examples from each class. Note that this is a different use of the term "weight" than the weights (or parameters) learned by our model during training, which you cannot set manually. By weighting *classes*, we tell our model to treat specific label classes with more importance during training. We'll want our model to assign more weight to examples from the minority class. Exactly how much importance your model should give to certain examples is up to you, and is a parameter you can experiment with.

In Keras, we can pass a `class_weights` parameter to our model when we train it with `fit()`. The parameter `class_weights` is a dict, mapping each class to the weight Keras should assign to examples from that class. But how should we determine the exact weights for each class? The class weight values should relate to the balance of each class in our dataset. For example, if the minority class accounts for only 0.1% of the dataset, a reasonable conclusion is that our model should treat examples from that class with 1000× more weight than the majority class. In practice, it's common to divide this weight value by 2 for each class so that the average weight of an example is *1.0*. Therefore, given a dataset with 0.1% of values representing the minority class, we could calculate the class weights with the following code:

```
num_minority_examples = 1
num_majority_examples = 999
total_examples = num_minority_examples + num_majority_examples

minority_class_weight = 1/(num_minority_examples/total_examples)/2
majority_class_weight = 1/(num_majority_examples/total_examples)/2

# Pass the weights to Keras in a dict
# The key is the index of each class
keras_class_weights = {0: majority_class_weight, 1: minority_class_weight}
```

We'd then pass these weights to our model during training:

```
model.fit(
    train_data,
    train_labels,
    class_weight=keras_class_weights
)
```

In BigQuery ML, we can set `AUTO_CLASS_WEIGHTS = True` in the `OPTIONS` block when creating our model to have different classes weighted based on their frequency of occurrence in the training data.

While it can be helpful to follow a heuristic of class balance for setting class weights, the business application of a model might also dictate the class weights we choose to assign. For example, let's say we have a model classifying images of defective products. If the cost of shipping a defective product is 10 times that of incorrectly classifying a normal product, we would choose 10 as the weight for our minority class.

Output Layer Bias

In conjunction with assigning class weights, it is also helpful to initialize the model's output layer with a bias to account for dataset imbalance. Why would we want to manually set the initial bias for our output layer? When we have imbalanced datasets, setting the output bias will help our model converge faster. This is because the bias of the last (prediction) layer of a trained model will output, on average, the log of the ratio of minority to majority examples in the dataset. By setting the bias, the model already starts out at the "correct" value without having to discover it through gradient descent.

By default, Keras uses a bias of zero. This corresponds with the bias we'd want to use for a perfectly balanced dataset where log(1/1) = 0. To calculate the correct bias while taking our dataset balance into account, use:

```
bias = log(num_minority_examples / num_majority_examples)
```

Upsampling

Another common technique for handling imbalanced datasets is *upsampling*. With upsampling, we overrepresent our minority class by both replicating minority class examples and generating additional, synthetic examples. This is often done in combination with downsampling the majority class. This approach—combining downsampling and upsampling—was proposed in 2002 and referred to as Synthetic Minority Over-sampling Technique (SMOTE (*https://oreil.ly/CFJPz*)). SMOTE provides an algorithm that constructs these synthetic examples by analyzing the feature space of minority class examples in the dataset and then generates similar examples within this feature space using a nearest neighbors approach. Depending on how many similar data points we choose to consider at once (also referred to as the number of nearest neighbors), the SMOTE approach randomly generates a new minority class example between these points.

Let's look at the Pima Indian Diabetes Dataset (*https://oreil.ly/ljqnc*) to see how this works at a high level. 34% of this dataset contains examples of patients who had diabetes, so we'll consider this our minority class. Table 3-3 shows a subset of columns for two minority class examples.

Table 3-3. A subset of features for two training examples from the minority class (has diabetes) in the Pima Indian Diabetes Dataset

Glucose	BloodPressure	SkinThickness	BMI
148	72	35	33.6
183	64	0	23.3

A new, synthetic example based on these two actual examples from the dataset might look like Table 3-4, calculating by the midpoint between each of these column values.

Table 3-4. A synthetic example generated from the two minority training examples using the SMOTE approach

Glucose	BloodPressure	SkinThickness	BMI
165.5	68	17.5	28.4

The SMOTE technique refers primarily to tabular data, but similar logic can be applied to image datasets. For example, if we're building a model to distinguish between Bengal and Siamese cats and only 10% of our dataset contains images of Bengals, we can generate additional variations of the Bengal cats in our dataset through image augmentation using the Keras `ImageDataGenerator` class. With a few parameters, this class will generate multiple variations of the same image by rotating, cropping, adjusting brightness, and more.

Trade-Offs and Alternatives

There are a few other alternative solutions for building models with inherently imbalanced datasets, including reframing the problem and handling cases of anomaly detection. We'll also explore several important considerations for imbalanced datasets: overall dataset size, the optimal model architectures for different problem types, and explaining minority class prediction.

Reframing and Cascade

Reframing the problem is another approach for handling imbalanced datasets. First, we might consider switching the problem from classification to regression or vice versa utilizing the techniques described in the Reframing design pattern section and training a cascade of models. For example, let's say we have a regression problem where the majority of our training data falls within a certain range, with a few outliers. Assuming we care about predicting outlier values, we could convert this to a classification problem by bucketing the majority of the data in one bucket and the outliers in another.

Imagine we're building a model to predict baby weight using the BigQuery natality dataset. Using pandas, we can create a histogram of a sample of the baby weight data to see the weight distribution:

```
%%bigquerydf
SELECT
  weight_pounds
FROM
  `bigquery-public-data.samples.natality`
LIMIT 10000
```

```
df.plot(kind='hist')
```

Figure 3-20 shows the resulting histogram.

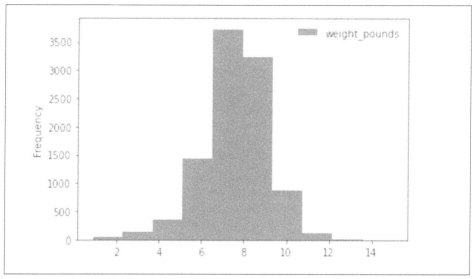

Figure 3-20. A histogram depicting the distribution of baby weight for 10,000 examples in the BigQuery natality dataset.

If we count the number of babies weighing 3 lbs in the entire dataset, there are approximately 96,000 (.06% of the data). Babies weighing 12 lbs make up only .05% of the dataset. To get good regression performance over the entire range, we can combine downsampling with the Reframing and Cascade design patterns. First, we'll split the data into three buckets: "underweight," "average," and "overweight." We can do that with the following query:

```
SELECT
  CASE
    WHEN weight_pounds < 5.5 THEN "underweight"
    WHEN weight_pounds > 9.5 THEN "overweight"
    ELSE
    "average"
  END
  AS weight,
  COUNT(*) AS num_examples,
  round(count(*) / sum(count(*)) over(), 4) as percent_of_dataset
FROM
  `bigquery-public-data.samples.natality`
GROUP BY
  1
```

Table 3-5 shows the results.

Table 3-5. The percentage of each weight class present in the natality dataset

weight	num_examples	percent_of_dataset
Average	123781044	0.8981
Underweight	9649724	0.07
Overweight	4395995	0.0319

For demo purposes, we'll take 100,000 examples from each class to train a model on an updated, balanced dataset:

```
SELECT
  is_male,
  gestation_weeks,
  mother_age,
  weight_pounds,
  weight
FROM (
  SELECT
    *,
    ROW_NUMBER() OVER (PARTITION BY weight ORDER BY RAND()) AS row_num
  FROM (
    SELECT
      is_male,
      gestation_weeks,
      mother_age,
      weight_pounds,
      CASE
        WHEN weight_pounds < 5.5 THEN "underweight"
        WHEN weight_pounds > 9.5 THEN "overweight"
      ELSE
      "average"
    END
      AS weight,
    FROM
      `bigquery-public-data.samples.natality`
    LIMIT
      4000000) )
WHERE
  row_num < 100000
```

We can save the results of that query to a table, and with a more balanced dataset, we can now train a classification model to label babies as "underweight," "average," or "overweight":

```
CREATE OR REPLACE MODEL
  `project.dataset.baby_weight_classification` OPTIONS(model_type='logistic_reg',
    input_label_cols=['weight']) AS
SELECT
  is_male,
  weight_pounds,
  mother_age,
```

```
  gestation_weeks,
  weight
FROM
  `project.dataset.baby_weight`
```

Another approach is to use the Cascade pattern, training three separate regression models for each class. Then, we can use our multidesign pattern solution by passing our initial classification model an example and using the result of that classification to decide which regression model to send the example to for numeric prediction.

Anomaly detection

There are two approaches to handling regression models for imbalanced datasets:

- Use the model's error on a prediction as a signal.
- Cluster incoming data and compare the distance of each new data point to existing clusters.

To better understand each solution, let's say we're training a model on data collected by a sensor to predict temperature in the future. In this case, we'd need the model output to be a numerical value.

For the first approach—using error as a signal—after training a model, we would then compare the model's predicted value with the actual value for the current point in time. If there was a significant difference between the predicted and actual current value, we could flag the incoming data point as an anomaly. Of course, this requires a model trained with good accuracy on enough historical data to rely on its quality for future predictions. The main caveat for this approach is that it requires us to have new data readily available, so that we can compare the incoming data with the model's prediction. As a result, it works best for problems involving streaming or time-series data.

In the second approach—clustering data—we start by building a model with a clustering algorithm, a modeling technique that organizes our data into clusters. Clustering is an *unsupervised learning* method, meaning it looks for patterns in the dataset without any knowledge of ground truth labels. A common clustering algorithm is k-means, which we can implement with BigQuery ML. The following shows how to train a k-means model on the BigQuery natality dataset using three features:

```
CREATE OR REPLACE MODEL
  `project-name.dataset-name.baby_weight` OPTIONS(model_type='kmeans',
    num_clusters=4) AS
SELECT
  weight_pounds,
  mother_age,
  gestation_weeks
FROM
```

```
      `bigquery-public-data.samples.natality`
   LIMIT 10000
```

The resulting model will cluster our data into four groups. Once the model has been created, we can then generate predictions on new data and look at that prediction's distance from existing clusters. If the distance is high, we can flag the data point as an anomaly. To generate a cluster prediction on our model, we can run the following query, passing it a made-up average example from the dataset:

```
SELECT
   *
FROM
   ML.PREDICT (MODEL `project-name.dataset-name.baby_weight`,
     (
     SELECT
        7.0 as weight_pounds,
        28 as mother_age,
        40 as gestation_weeks
     )
   )
```

The query results in Table 3-6 show us the distance between this data point and the model's generated clusters, called centroids.

Table 3-6. The distance between our average weight example data point and each of the clusters generated by our k-means model

CENTROID_ID	NEAREST_CENTROIDS_DISTANCE.CENTROID_ID	NEAREST_CENTROIDS_DISTANCE.DISTANCE
4	4	0.29998627812137374
1	1.2370167418282159	
2	1.376651161584178	
3	1.6853517159990536	

This example clearly fits into centroid 4, as seen by the small distance (.29).

We can compare this to the results we get if we send an outlier, underweight example to the model, as shown in Table 3-7.

Table 3-7. The distance between our underweight example data point and each of the clusters generated by our k-means model

CENTROID_ID	NEAREST_CENTROIDS_DISTANCE.CENTROID_ID	NEAREST_CENTROIDS_DISTANCE.DISTANCE
3	3	3.061985789261998
4	3.3124603501734966	
2	4.330205096751425	
1	4.658614918595627	

Here, the distance between this example and each centroid is quite large. We could then use these high-distance values to conclude that this data point might be an anomaly. This unsupervised clustering approach is especially useful if we don't know the labels for our data in advance. Once we've generated cluster predictions on enough examples, we could then build a supervised learning model using the predicted clusters as labels.

Number of minority class examples available

While the minority class in our first fraud detection example only made up 0.1% of the data, the dataset was large enough that we still had 8,000 fraudulent data points to work with. For datasets with even fewer examples of the minority class, downsampling may make the resulting dataset too small for a model to learn from. There isn't a hard-and-fast rule for determining how many examples is too few to use downsampling, since it largely depends on our problem and model architecture. A general rule of thumb is that if you only have hundreds of examples of the minority class, you might want to consider a solution other than downsampling for handling dataset imbalance.

It's also worth noting that the natural effect of removing a subset of our majority class is losing some information stored in those examples. This might slightly decrease our model's ability to identify the majority class, but often the benefits of downsampling still outweigh this.

Combining different techniques

The downsampling and class weight techniques described above can be combined for optimal results. To do this, we start by downsampling our data until we find a balance that works for our use case. Then, based on the label ratios for the rebalanced dataset, use the method described in the weighted classes section to pass new weights to our model. Combining these approaches can be especially useful when we have an anomaly detection problem and care most about predictions for our minority class. For example, if we're building a fraud detection model, we're likely much more concerned about the transactions our model flags as "fraud" rather than the ones it flags as "nonfraud." Additionally, as mentioned by SMOTE, the approach of generating synthetic examples from the minority class is often combined with removing a random sample of examples from the minority class.

Downsampling is also often combined with the Ensemble design pattern. Using this approach, instead of entirely removing a random sample of our majority class, we use different subsets of it to train multiple models and then ensemble those models. To illustrate this, let's say we have a dataset with 100 minority class examples and 1,000 majority examples. Rather than removing 900 examples from our majority class to perfectly balance the dataset, we'd randomly split the majority examples into 10 groups with 100 examples each. We'd then train 10 classifiers, each with the same 100 examples from our minority class and 100 different, randomly selected values from our majority class. The bagging technique illustrated in Figure 3-11 would work well for this approach.

In addition to combining these data-centric approaches, we can also adjust the threshold for our classifier to optimize for precision or recall depending on our use case. If we care more that our model is correct whenever it makes a positive class prediction, we'd optimize our prediction threshold for recall. This can apply in any situation where we want to avoid false positives. Alternatively, if it is more costly to *miss* a potential positive classification even when we might get it wrong, we optimize our model for recall.

Choosing a model architecture

Depending on our prediction task, there are different model architectures to consider when solving problems with the Rebalancing design pattern. If we're working with tabular data and building a classification model for anomaly detection, research (*https://oreil.ly/EnAab*) has shown that decision tree models perform well on these types of tasks. Tree-based models also work well on problems involving small and imbalanced datasets. XGBoost, scikit-learn, and TensorFlow all have methods for implementing decision tree models.

We can implement a binary classifier in XGBoost with the following code:

```
# Build the model
model = xgb.XGBClassifier(
    objective='binary:logistic'
)

# Train the model
model.fit(
    train_data,
    train_labels
)
```

We can use downsampling and class weights in each of these frameworks to further optimize our model using the Rebalancing design pattern. For example, to add weighted classes to our XGBClassifier above, we'd add a scale_pos_weight parameter, calculated based on the balance of classes in our dataset.

If we're detecting anomalies in time-series data, long short-term memory (LSTM) models work well for identifying patterns present in sequences. Clustering models are also an option for tabular data with imbalanced classes. For imbalanced datasets with image input, use deep learning architectures with downsampling, weighted classes, upsampling, or a combination of these techniques. For text data, however, generating synthetic data is less straightforward (*https://oreil.ly/2ai2k*), and it's best to rely on downsampling and weighted classes.

Regardless of the data modality we're working with, it's useful to experiment with different model architectures to see which performs best on our imbalanced data.

Importance of explainability

When building models for flagging rare occurrences in data such as anomalies, it's especially important to understand how our model is making predictions. This can both verify that the model is picking up on the correct signals to make its predictions and help explain the model's behavior to end users. There are a few tools available to help us interpret models and explain predictions, including the open source framework SHAP (*https://github.com/slundberg/shap*), the What-If Tool (*https://oreil.ly/Vf3D-*), and Explainable AI on Google Cloud (*https://oreil.ly/lDocn*).

Model explanations can take many forms, one of which is called *attribution values*. Attribution values tell us how much each feature in our model influenced the model's prediction. Positive attribution values mean a particular feature pushed our model's prediction up, and negative attribution values mean the feature pushed our model's prediction down. The higher the absolute value of an attribution, the bigger impact it had on our model's prediction. In image and text models, attributions can show you the pixels or words that signaled your model's prediction most. For tabular models, attributions provide numerical values for each feature, indicating its overall effect on the model's prediction.

After training a TensorFlow model on the synthetic fraud detection dataset from Kaggle and deploying it to Explainable AI on Google Cloud, let's take a look at some examples of instance-level attributions. In Figure 3-21, we see two example transactions that our model correctly identified as fraud, along with their feature attributions.

In the first example where the model predicted a 99% chance of fraud, the old balance at the origin account before the transaction was made was the biggest indicator of fraud. In the second example, our model was 89% confident in its prediction of fraud with the amount of the transaction identified as the biggest signal of fraud. However, the balance at the origin account made our model *less confident* in its prediction of fraud and explains *why* the prediction confidence is slightly *lower* by 10 percentage points.

Explanations are important for any type of machine learning model, but we can see how they are especially useful for models following the Rebalancing design pattern. When dealing with imbalanced data, it's important to look beyond our model's accuracy and error metrics to verify that it's picking up on meaningful signals in our data.

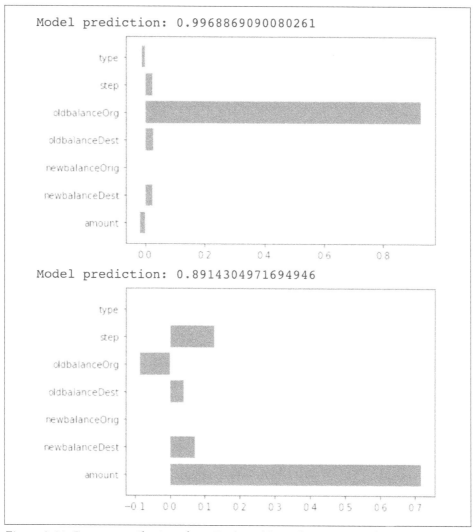

Figure 3-21. Feature attributions from Explainable AI for two correctly classified fraudulent transactions.

Summary

This chapter looked at different ways to represent a prediction task through the lens of model architecture and model output. Thinking about how you'll apply your model can guide your decision on the type of model to build, and how to format your output for prediction. With this in mind, we started with the *Reframing* design pattern, which explores changing your problem from a regression task to a classification task (or vice versa) to improve the quality of your model. You can do this by reformatting the label column in your data. Next we explored the *Multilabel* design pattern, which addresses cases where an input to your model can be associated with more than one label. To handle this case, use the sigmoid activation function on your output layer with binary cross entropy loss.

Whereas the Reframing and Multilabel patterns focus on formatting model *output*, the *Ensemble* design pattern addresses model *architecture* and includes various methods for combining multiple models to improve upon machine learning results from a single model. Specifically, the Ensemble pattern includes bagging, boosting, and stacking—all different techniques for aggregating multiple models into one ML system. The *Cascade* design pattern is also a model-level approach, and involves breaking a machine learning problem into several smaller problems. Unlike ensemble models, the Cascade pattern requires outputs from an initial model to be inputs into downstream models. Because of the complexity cascade models can create, you should only use them when you have a scenario where the initial classification labels are disparate and equally important.

Next, we looked at the *Neutral Class* design pattern, which addresses problem representation at the output level. This pattern improves a binary classifier by adding a third "neutral" class. This is useful in cases where you want to capture arbitrary or less-polarizing classifications that don't fall into either of the distinct binary categories. Finally, the *Rebalancing* design pattern provides solutions for cases where you have an inherently imbalanced dataset. This pattern proposes using downsampling, weighted classes, or specific reframing techniques to solve for datasets with imbalanced label classes.

Chapters 2 and 3 focused on the initial steps for structuring your machine learning problem, specifically formatting input data, model architecture options, and model output representation. In the next chapter, we'll navigate the next step in the machine learning workflow—design patterns for training models.

## CHAPTER 4

Model Training Patterns

Machine learning models are usually trained iteratively, and this iterative process is informally called the *training loop*. In this chapter, we discuss what the typical training loop looks like, and catalog a number of situations in which you might want to do something different.

Typical Training Loop

Machine learning models can be trained using different types of optimization. Decision trees are often built node by node based on an information gain measure. In genetic algorithms, the model parameters are represented as genes, and the optimization method involves techniques that are based on evolutionary theory. However, the most common approach to determining the parameters of machine learning models is *gradient descent*.

Stochastic Gradient Descent

On large datasets, gradient descent is applied to mini-batches of the input data to train everything from linear models and boosted trees to deep neural networks (DNNs) and support vector machines (SVMs). This is called *stochastic gradient descent (SGD)*, and extensions of SGD (such as Adam and Adagrad) are the de facto optimizers used in modern-day machine learning frameworks.

Because SGD requires training to take place iteratively on small batches of the training dataset, training a machine learning model happens in a loop. SGD finds a minimum, but is not a closed-form solution, and so we have to detect whether the model convergence has happened. Because of this, the error (called the *loss*) on the training dataset has to be monitored. Overfitting can happen if the model complexity is higher than can be afforded by the size and coverage of the dataset. Unfortunately, you cannot know whether the model complexity is too high for a particular dataset until you actually train that model on that dataset. Therefore, evaluation needs to be done within the training loop, and *error metrics* on a withheld split of the training data, called the *validation dataset*, have to be monitored as well. Because the training and validation datasets have been used in the training loop, it is necessary to withhold yet another split of the training dataset, called the *testing dataset*, to report the actual error metrics that could be expected on new and unseen data. This evaluation is carried out at the end.

Keras Training Loop

The typical training loop in Keras looks like this:

```
model = keras.Model(...)
model.compile(optimizer=keras.optimizers.Adam(),
              loss=keras.losses.categorical_crossentropy(),
              metrics=['accuracy'])

history = model.fit(x_train, y_train,
                    batch_size=64,
                    epochs=3,
                    validation_data=(x_val, y_val))
results = model.evaluate(x_test, y_test, batch_size=128))
model.save(...)
```

Here, the model uses the Adam optimizer to carry out SGD on the cross entropy over the training dataset and reports out the final accuracy obtained on the testing dataset. The model fitting loops over the training dataset three times (each traversal over the training dataset is termed an *epoch*) with the model seeing batches consisting of 64 training examples at a time. At the end of every epoch, the error metrics are calculated on the validation dataset and added to the history. At the end of the fitting loop, the model is evaluated on the testing dataset, saved, and potentially deployed for serving, as shown in Figure 4-1.

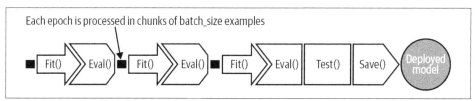

Figure 4-1. A typical training loop consisting of three epochs. Each epoch is processed in chunks of batch_size examples. At the end of the third epoch, the model is evaluated on the testing dataset, and saved for potential deployment as a web service.

Instead of using the prebuilt `fit()` function, we could also write a custom training loop that iterates over the batches explicitly, but we will not need to do this for any of the design patterns discussed in this chapter.

Training Design Patterns

The design patterns covered in this chapter all have to do with modifying the typical training loop in some way. In *Useful Overfitting*, we forgo the use of a validation or testing dataset because we want to intentionally overfit on the training dataset. In *Checkpoints*, we store the full state of the model periodically, so that we have access to partially trained models. When we use checkpoints, we usually also use *virtual epochs*, wherein we decide to carry out the inner loop of the fit() function, not on the full training dataset but on a fixed number of training examples. In *Transfer Learning*, we take part of a previously trained model, freeze the weights, and incorporate these nontrainable layers into a new model that solves the same problem, but on a smaller dataset. In *Distribution Strategy,* the training loop is carried out at scale over multiple workers, often with caching, hardware acceleration, and parallelization. Finally, in *Hyperparameter Tuning*, the training loop is itself inserted into an optimization method to find the optimal set of model hyperparameters.

Design Pattern 11: Useful Overfitting

Useful Overfitting is a design pattern where we forgo the use of generalization mechanisms because we want to intentionally overfit on the training dataset. In situations where overfitting can be beneficial, this design pattern recommends that we carry out machine learning without regularization, dropout, or a validation dataset for early stopping.

Problem

The goal of a machine learning model is to generalize and make reliable predictions on new, unseen data. If your model *overfits* the training data (for example, it continues to decrease the training error beyond the point at which validation error starts to increase), then its ability to generalize suffers and so do your future predictions. Introductory machine learning textbooks advise avoiding overfitting by using early stopping and regularization techniques.

Consider, however, a situation of simulating the behavior of physical or dynamical systems like those found in climate science, computational biology, or computational finance. In such systems, the time dependence of observations can be described by a mathematical function or set of partial differential equations (PDEs). Although the equations that govern many of these systems can be formally expressed, they don't have a closed-form solution. Instead, classical numerical methods have been developed to approximate solutions to these systems. Unfortunately, for many real-world applications, these methods can be too slow to be used in practice.

Consider the situation shown in Figure 4-2. Observations collected from the physical environment are used as inputs (or initial starting conditions) for a physics-based model that carries out iterative, numerical calculations to calculate the precise state of the system. Suppose all the observations have a finite number of possibilities (for example, temperature will be between 60°C and 80°C in increments of 0.01°C). It is then possible to create a training dataset for the machine learning system consisting of the complete input space and calculate the labels using the physical model.

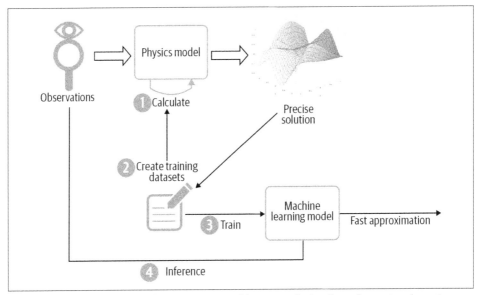

Figure 4-2. One situation when it is acceptable to overfit is when the entire domain space of observations can be tabulated and a physical model capable of computing the precise solution is available.

The ML model needs to learn this precisely calculated and nonoverlapping lookup table of inputs to outputs. Splitting such a dataset into a training dataset and an evaluation dataset is counterproductive because we would then be expecting the model to learn parts of the input space it will not have seen in the training dataset.

Solution

In this scenario, there is no "unseen" data that needs to be generalized to, since all possible inputs have been tabulated. When building a machine learning model to learn such a physics model or dynamical system, there is no such thing as overfitting. The basic machine learning training paradigm is slightly different. Here, there is some physical phenomenon that you are trying to learn that is governed by an underlying PDE or system of PDEs. Machine learning merely provides a data-driven approach to approximate the precise solution, and concepts like overfitting must be reevaluated.

For example, a ray-tracing approach is used to simulate the satellite imagery that would result from the output of numerical weather prediction models. This involves calculating how much of a solar ray gets absorbed by the predicted hydrometeors (rain, snow, hail, ice pellets, and so on) at each atmospheric level. There is a finite number of possible hydrometeor types and a finite number of heights that the numerical model predicts. So the ray-tracing model has to apply optical equations to a large but finite set of inputs.

The equations of radiative transfer govern the complex dynamical system of how electromagnetic radiation propagates in the atmosphere, and forward radiative transfer models are an effective means of inferring the future state of satellite images. However, classical numerical methods to compute the solutions to these equations can take tremendous computational effort and are too slow to use in practice.

Enter machine learning. It is possible to use machine learning to build a model that approximates solutions (*https://oreil.ly/IkYKm*) to the forward radiative transfer model (see Figure 4-3). This ML approximation can be made close enough to the solution of the model that was originally achieved by using more classical methods. The advantage is that inference using the learned ML approximation (which needs to just calculate a closed formula) takes only a fraction of the time required to carry out ray tracing (which would require numerical methods). At the same time, the training dataset is too large (multiple terabytes) and too unwieldy to use as a lookup table in production.

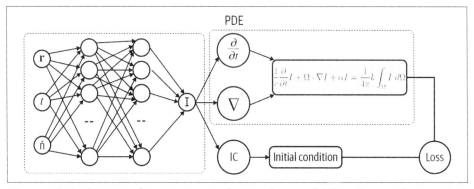

Figure 4-3. Architecture for using a neural network to model the solution of a partial differential equation to solve for I(r,t,n).

There is an important difference between training an ML model to approximate the solution to a dynamical system like this and training an ML model to predict baby weight based on natality data collected over the years. Namely, the dynamical system is a set of equations governed by the laws of electromagnetic radiation—there is no unobserved variable, no noise, and no statistical variability. For a given set of inputs, there is only one precisely calculable output. There is no overlap between different examples in the training dataset. For this reason, we can toss out concerns about generalization. We *want* our ML model to fit the training data as perfectly as possible, to "overfit."

This is counter to the typical approach of training an ML model where considerations of bias, variance, and generalization error play an important role. Traditional training says that it is possible for a model to learn the training data "too well," and that training your model so that the train loss function is equal to zero is more of a red flag than cause for celebration. Overfitting of the training dataset in this way causes the model to give misguided predictions on new, unseen data points. The difference here is that we know in advance there won't be unseen data, thus the model is approximating a solution to a PDE over the full input spectrum. If your neural network is able to learn a set of parameters where the loss function is zero, then that parameter set determines the actual solution of the PDE in question.

Why It Works

If all possible inputs can be tabulated, then as shown by the dotted curve in Figure 4-4, an overfit model will still make the same predictions as the "true" model if all possible input points are trained for. So overfitting is not a concern. We have to take care that inferences are made on rounded-off values of the inputs, with the rounding determined by the resolution with which the input space was gridded.

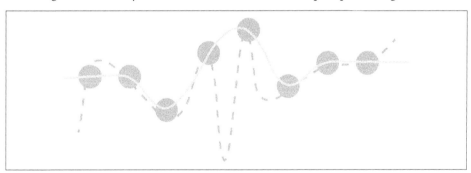

Figure 4-4. Overfitting is not a concern if all possible input points are trained for because predictions are the same with both curves.

Is it possible to find a model function that gets arbitrarily close to the true labels? One bit of intuition as to why this works comes from the Uniform Approximation Theorem of deep learning, which, loosely put, states that any function (and its derivatives) can be approximated by a neural network with at least one hidden layer and any "squashing" activation function, like sigmoid. This means that no matter what function we are given, so long as it's relatively well behaved, there exists a neural network with just one hidden layer that approximates that function as closely as we want.[1]

Deep learning approaches to solving differential equations or complex dynamical systems aim to represent a function defined implicitly by a differential equation, or system of equations, using a neural network.

Overfitting is useful when the following two conditions are met:

- There is no noise, so the labels are accurate for all instances.
- You have the complete dataset at your disposal (you have all the examples there are). In this case, overfitting becomes interpolating the dataset.

Trade-Offs and Alternatives

We introduced overfitting as being useful when the set of inputs can be exhaustively listed and the accurate label for each set of inputs can be calculated. If the full input space can be tabulated, overfitting is not a concern because there is no unseen data. However, the Useful Overfitting design pattern is useful beyond this narrow use case. In many real-world situations, even if one or more of these conditions has to be relaxed, the concept that overfitting can be useful remains valid.

Interpolation and chaos theory

The machine learning model essentially functions as an approximation to a lookup table of inputs to outputs. If the lookup table is small, just use it as a lookup table! There is no need to approximate it by a machine learning model. An ML approximation is useful in situations where the lookup table will be too large to effectively use. It is when the lookup table is too unwieldy that it becomes better to treat it as the training dataset for a machine learning model that approximates the lookup table.

Note that we assumed that the observations would have a finite number of possibilities. For example, we posited that temperature would be measured in 0.01°C increments and lie between 60°C and 80°C. This will be the case if the observations are made by digital instruments. If this is not the case, the ML model is needed to interpolate between entries in the lookup table.

Machine learning models interpolate by weighting unseen values by the distance of these unseen values from training examples. Such interpolation works only if the underlying system is not chaotic. In chaotic systems, even if the system is deterministic, small differences in initial conditions can lead to dramatically different outcomes.

---

1 It may, of course, not be the case that we can learn the network using gradient descent just because there exists such a neural network (this is why changing the model architecture by adding layers helps—it makes the loss landscape more amenable to SGD).

Nevertheless, in practice, each specific chaotic phenomenon has a specific resolution threshold (*https://oreil.ly/F-drU*) beyond which it is possible for models to forecast it over short time periods. Therefore, provided the lookup table is fine-grained enough and the limits of resolvability are understood, useful approximations can result.

Monte Carlo methods

In reality, tabulating all possible inputs might not be possible, and you might take a Monte Carlo approach (*https://oreil.ly/pTgS9*) of sampling the input space to create the set of inputs, especially where not all possible combinations of inputs are physically possible.

In such cases, overfitting is technically possible (see Figure 4-5, where the unfilled circles are approximated by wrong estimates shown by crossed circles).

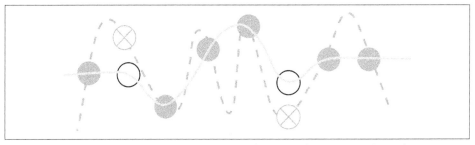

Figure 4-5. If the input space is sampled, not tabulated, then you need to take care to limit model complexity.

However, even here, you can see that the ML model will be interpolating between known answers. The calculation is always deterministic, and it is only the input points that are subject to random selection. Therefore, these known answers do not contain noise, and because there are no unobserved variables, errors at unsampled points will be strictly bounded by the model complexity. Here, the overfitting danger comes from model complexity and not from fitting to noise. Overfitting is not as much of a concern when the size of the dataset is larger than the number of free parameters. Therefore, using a combination of low-complexity models and mild regularization provides a practical way to avoid unacceptable overfitting in the case of Monte Carlo selection of the input space.

Data-driven discretizations

Although deriving a closed-form solution is possible for some PDEs, determining solutions using numerical methods is more common. Numerical methods of PDEs are already a deep field of research, and there are many books (*https://oreil.ly/RJWVQ*), courses (*https://oreil.ly/wcl_n*), and journals (*https://msp.org/apde*) devoted to the subject. One common approach is to use finite difference methods, similar to

Runge-Kutta methods, for solving ordinary differential equations. This is typically done by discretizing the differential operator of the PDE and finding a solution to the discrete problem on a spatio-temporal grid of the original domain. However, when the dimension of the problem becomes large, this mesh-based approach fails dramatically due to the curse of dimensionality because the mesh spacing of the grid must be small enough (*https://oreil.ly/TxHD-*) to capture the smallest feature size of the solution. So, to achieve 10× higher resolution of an image requires 10,000× more compute power, because the mesh grid must be scaled in four dimensions accounting for space and time.

However, it is possible to use machine learning (rather than Monte Carlo methods) to select the sampling points to create data-driven discretizations of PDEs. In the paper "Learning data-driven discretizations for PDEs (*https://oreil.ly/djDkK*)," Bar-Sinai et al. demonstrate the effectiveness of this approach. The authors use a low-resolution grid of fixed points to approximate a solution via a piecewise polynomial interpolation using standard finite-difference methods as well as one obtained from a neural network. The solution obtained from the neural network vastly outperforms the numeric simulation in minimizing the absolute error, in some places achieving a 10^2 order of magnitude improvement. While increasing the resolution requires substantially more compute power using finite-difference methods, the neural network is able to maintain high performance with only marginal additional cost. Techniques like the Deep Galerkin Method can then use deep learning to provide a mesh-free approximation of the solution to the given PDE. In this way, solving the PDE is reduced to a chained optimization problem (see "Design Pattern 8: Cascade " on page 108).

Deep Galerkin Method

The Deep Galerkin Method (*https://oreil.ly/rQy4d*) is a deep learning algorithm for solving partial differential equations. The algorithm is similar in spirit to Galerkin methods used in the field of numeric analysis, where the solution is approximated using a neural network instead of a linear combination of basis functions.

Unbounded domains

The Monte Carlo and data-driven discretization methods both assume that sampling the entire input space, even if imperfectly, is possible. That's why the ML model was treated as an interpolation between known points.

Generalization and the concern of overfitting become difficult to ignore whenever we are unable to sample points in the full domain of the function—for example, for functions with unbounded domains or projections along a time axis into the future. In these settings, it is important to consider overfitting, underfitting, and generalization error. In fact, it's been shown that although techniques like the Deep Galerkin Method do well on regions that are well sampled, a function that is learned this way does not generalize well on regions outside the domain that were not sampled in the training phase. This can be problematic for using ML to solve PDEs that are defined on unbounded domains, since it would be impossible to capture a representative sample for training.

Distilling knowledge of neural network

Another situation where overfitting is warranted is in distilling, or transferring knowledge, from a large machine learning model into a smaller one. Knowledge distillation is useful when the learning capacity of the large model is not fully utilized. If that is the case, the computational complexity of the large model may not be necessary. However, it is also the case that training smaller models is harder. While the smaller model has enough capacity to represent the knowledge, it may not have enough capacity to learn the knowledge efficiently.

The solution is to train the smaller model on a large amount of generated data that is labeled by the larger model. The smaller model learns the soft output of the larger model, instead of actual labels on real data. This is a simpler problem that can be learned by the smaller model. As with approximating a numerical function by a machine learning model, the aim is for the smaller model to faithfully represent the predictions of the larger machine learning model. This second training step can employ Useful Overfitting.

Overfitting a batch

In practice, training neural networks requires a lot of experimentation, and a practitioner must make many choices, from the size and architecture of the network to the choice of the learning rate, weight initializations, or other hyperparameters.

Overfitting on a small batch is a good sanity check (*https://oreil.ly/AcLtu*) both for the model code as well as the data input pipeline. Just because the model compiles and the code runs without errors doesn't mean you've computed what you think you have or that the training objective is configured correctly. A complex enough model *should* be able to overfit on a small enough batch of data, assuming everything is set up correctly. So, if you're not able to overfit a small batch with any model, it's worth rechecking your model code, input pipeline, and loss function for any errors or simple bugs. Overfitting on a batch is a useful technique when training and troubleshooting neural networks.

Overfitting goes beyond just a batch. From a more holistic perspective, overfitting follows the general advice commonly given with regards to deep learning and regularization. The best fitting model is a large model that has been properly regularized (*https://oreil.ly/A7DFC*). In short, if your deep neural network isn't capable of overfitting your training dataset, you should be using a bigger one. Then, once you have a large model that overfits the training set, you can apply regularization to improve the validation accuracy, even though training accuracy may decrease.

You can test your Keras model code in this way using the `tf.data.Dataset` you've written for your input pipeline. For example, if your training data input pipeline is called `trainds`, we'll use `batch()` to pull a single batch of data. You can find the full code for this example (*https://github.com/GoogleCloudPlatform/ml-design-patterns/blob/master/04_hacking_training_loop/distribution_strategies.ipynb*) in the repository accompanying this book:

```
BATCH_SIZE = 256
single_batch = trainds.batch(BATCH_SIZE).take(1)
```

Then, when training the model, instead of calling the full `trainds` dataset inside the `fit()` method, use the single batch that we created:

```
model.fit(single_batch.repeat(),
          validation_data=evalds,
          …)
```

Note that we apply `repeat()` so that we won't run out of data when training on that single batch. This ensures that we take the one batch over and over again while training. Everything else (the validation dataset, model code, engineered features, and so on) remains the same.

Rather than choose an arbitrary sample of the training dataset, we recommend that you overfit on a small dataset, each of whose examples has been carefully verified to have correct labels. Design your neural network architecture such that it is able to learn this batch of data precisely and get to zero loss. Then take the same network and train it on the full training dataset.

Design Pattern 12: Checkpoints

In Checkpoints, we store the full state of the model periodically so that we have partially trained models available. These partially trained models can serve as the final model (in the case of early stopping) or as the starting points for continued training (in the cases of machine failure and fine-tuning).

Problem

The more complex a model is (for example, the more layers and nodes a neural network has), the larger the dataset that is needed to train it effectively. This is because more complex models tend to have more tunable parameters. As model sizes increase, the time it takes to fit one batch of examples also increases. As the data size increases (and assuming batch sizes are fixed), the number of batches also increases. Therefore, in terms of computational complexity, this double whammy means that training will take a long time.

At the time of writing, training an English-to-German translation model on a state-of-the-art tensor processing unit (TPU) pod on a relatively small dataset takes about two hours (*https://oreil.ly/vDRve*). On real datasets of the sort used to train smart devices, the training can take several days.

When we have training that takes this long, the chances of machine failure are uncomfortably high. If there is a problem, we'd like to be able to resume from an intermediate point, instead of from the very beginning.

Solution

At the end of every epoch, we can save the model state. Then, if the training loop is interrupted for any reason, we can go back to the saved model state and restart. However, when doing this, we have to make sure to save the intermediate model state, not just the model. What does that mean?

Once training is complete, we save or *export* the model so that we can deploy it for inference. An exported model does not contain the entire model state, just the information necessary to create the prediction function. For a decision tree, for example, this would be the final rules for each intermediate node and the predicted value for each of the leaf nodes. For a linear model, this would be the final values of the weights and biases. For a fully connected neural network, we'd also need to add the activation functions and the weights of the hidden connections.

What data on model state do we need when restoring from a checkpoint that an exported model does not contain? An exported model does not contain which epoch and batch number the model is currently processing, which is obviously important in order to resume training. But there is more information that a model training loop can contain. In order to carry out gradient descent effectively, the optimizer might be changing the learning rate on a schedule. This learning rate state is not present in an exported model. Additionally, there might be stochastic behavior in the model, such as dropout. This is not captured in the exported model state either. Models like recurrent neural networks incorporate history of previous input values. In general, the full model state can be many times the size of the exported model.

Saving the full model state so that model training can resume from a point is called *checkpointing*, and the saved model files are called *checkpoints*. How often should we checkpoint? The model state changes after every batch because of gradient descent. So, technically, if we don't want to lose any work, we should checkpoint after every batch. However, checkpoints are huge and this I/O would add considerable overhead. Instead, model frameworks typically provide the option to checkpoint at the end of every epoch. This is a reasonable tradeoff between never checkpointing and checkpointing after every batch.

To checkpoint a model in Keras, provide a callback to the `fit()` method:

```
checkpoint_path = '{}/checkpoints/taxi'.format(OUTDIR)
cp_callback = tf.keras.callbacks.ModelCheckpoint(checkpoint_path,
                                                 save_weights_only=False,
                                                 verbose=1)
history = model.fit(x_train, y_train,
                    batch_size=64,
                    epochs=3,
                    validation_data=(x_val, y_val),
                    verbose=2,
                    callbacks=[cp_callback])
```

With checkpointing added, the training looping becomes what is shown in Figure 4-6.

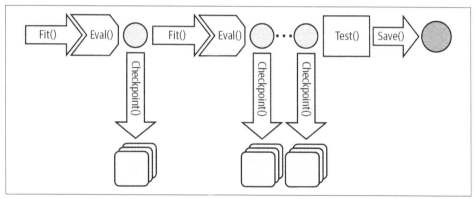

Figure 4-6. Checkpointing saves the full model state at the end of every epoch.

Checkpoints in PyTorch

At the time of writing, PyTorch doesn't support checkpoints directly. However, it does support externalizing the state of most objects. To implement checkpoints in PyTorch, ask for the epoch, model state, optimizer state, and any other information needed to resume training to be serialized along with the model:

```
torch.save({
            'epoch': epoch,
            'model_state_dict': model.state_dict(),
            'optimizer_state_dict': optimizer.state_dict(),
            'loss': loss,

            ...
            }, PATH)
```

When loading from a checkpoint, you need to create the necessary classes and then load them from the checkpoint:

```
model = ...
optimizer = ...
checkpoint = torch.load(PATH)
model.load_state_dict(checkpoint['model_state_dict'])
optimizer.load_state_dict(checkpoint['optimizer_state_dict'])
epoch = checkpoint['epoch']
loss = checkpoint['loss']
```

This is lower level than TensorFlow but provides the flexibility of storing multiple models in a checkpoint and choosing which parts of the model state to load or not load.

Why It Works

TensorFlow and Keras automatically resume training from a checkpoint if checkpoints are found in the output path. To start training from scratch, therefore, you have to start from a new output directory (or delete previous checkpoints from the output directory). This works because enterprise-grade machine learning frameworks honor the presence of checkpoint files.

Even though checkpoints are designed primarily to support resilience, the availability of partially trained models opens up a number of other use cases. This is because the partially trained models are usually more generalizable than the models created in later iterations. A good intuition of why this occurs can be obtained from the Tensor-Flow playground (*https://oreil.ly/sRjkN*), as shown in Figure 4-7.

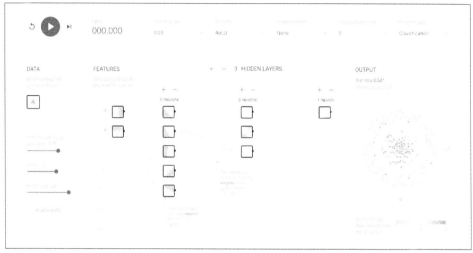

Figure 4-7. Starting point of the spiral classification problem. You can get to this setup by opening up this link (https://oreil.ly/ISg9X) in a web browser.

In the playground, we are trying to build a classifier to distinguish between blue dots and orange dots (if you are reading this in the print book, please do follow along by navigating to the link in a web browser). The two input features are x_1 and x_2, which are the coordinates of the points. Based on these features, the model needs to output the probability that the point is blue. The model starts with random weights and the background of the dots shows the model prediction for each coordinate point. As you can see, because the weights are random, the probability tends to hover near the center value for all the pixels.

Starting the training by clicking on the arrow at the top left of the image, we see the model slowly start to learn with successive epochs, as shown in Figure 4-8.

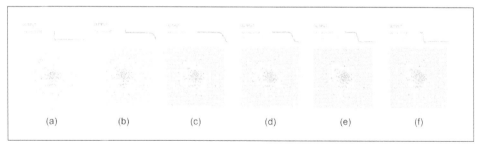

(a)   (b)   (c)   (d)   (e)   (f)

Figure 4-8. What the model learns as training progresses. The graphs at the top are the training loss and validation error, while the images show how the model at that stage would predict the color of a point at each coordinate in the grid.

We see the first hint of learning in Figure 4-8(b), and see that the model has learned the high-level view of the data by Figure 4-8(c). From then on, the model is adjusting the boundaries to get more and more of the blue points into the center region while keeping the orange points out. This helps, but only up to point. By the time we get to Figure 4-8(e), the adjustment of weights is starting to reflect random perturbations in the training data, and these are counterproductive on the validation dataset.

We can therefore break the training into three phases. In the first phase, between stages (a) and (c), the model is learning high-level organization of the data. In the second phase, between stages and (c) and (e), the model is learning the details. By the time we get to the third phase, stage (f), the model is overfitting. A partially trained model from the end of phase 1 or from phase 2 has some advantages precisely because it has learned the high-level organization but is not caught up in the details.

Trade-Offs and Alternatives

Besides providing resilience, saving intermediate checkpoints also allows us to implement early stopping and fine-tuning capabilities.

Early stopping

In general, the longer you train, the lower the loss on the training dataset. However, at some point, the error on the validation dataset might stop decreasing. If you are starting to overfit to the training dataset, the validation error might even start to increase, as shown in Figure 4-9.

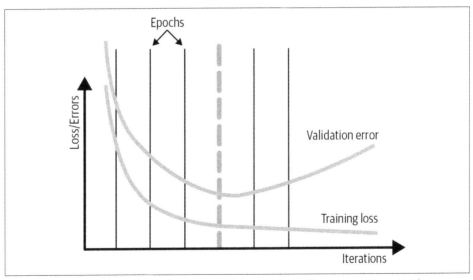

Figure 4-9. Typically, the training loss continues to drop the longer you train, but once overfitting starts, the validation error on a withheld dataset starts to go up.

In such cases, it can be helpful to look at the validation error at the end of every epoch and stop the training process when the validation error is more than that of the previous epoch. In Figure 4-9, this will be at the end of the fourth epoch, shown by the thick dashed line. This is called *early stopping*.

Had we been checkpointing at the end of every batch, we might have been able to capture the true minimum, which might have been a bit before or after the epoch boundary. See the discussion on virtual epochs in this section for a more frequent way to checkpoint.

If we are checkpointing much more frequently, it can be helpful if early stopping isn't overly sensitive to small perturbations in the validation error. Instead, we can apply early stopping only after the validation error doesn't improve for more than N checkpoints.

Checkpoint selection.   While early stopping can be implemented by stopping the training as soon as the validation error starts to increase, we recommend training longer and choosing the optimal run as a postprocessing step. The reason we suggest training well into phase 3 (see the preceding "Why It Works" section for an explanation of the three phases of the training loop) is that it is not uncommon for the validation error to increase for a bit and then start to drop again. This is usually because the training initially focuses on more common scenarios (phase 1), then starts to home in on the rarer situations (phase 2). Because rare situations may be imperfectly sampled between the training and validation datasets, occasional increases in the validation error during the training run are to be expected in phase 2. In addition, there are situations endemic to big models where deep double descent (*https://oreil.ly/Kya8h*) is expected, and so it is essential to train a bit longer just in case.

In our example, instead of exporting the model at the end of the training run, we will load up the fourth checkpoint and export our final model from there instead. This is called *checkpoint selection*, and in TensorFlow, it can be achieved using BestExporter (*https://oreil.ly/UpN1a*).

Regularization.   Instead of using early stopping or checkpoint selection, it can be helpful to try to add L2 regularization to your model so that the validation error does not increase and the model never gets into phase 3. Instead, both the training loss and the validation error should plateau, as shown in Figure 4-10. We term such a training loop (where both training and validation metrics reach a plateau) a *well-behaved* training loop.

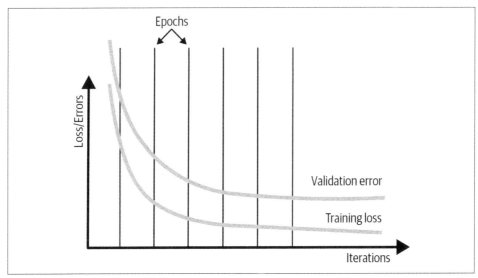

Figure 4-10. In the ideal situation, validation error does not increase. Instead, both the training loss and validation error plateau.

If early stopping is not carried out, and only the training loss is used to decide convergence, then we can avoid having to set aside a separate testing dataset. Even if we are not doing early stopping, displaying the progress of the model training can be helpful, particularly if the model takes a long time to train. Although the performance and progress of the model training is normally monitored on the validation dataset during the training loop, it is for visualization purposes only. Since we don't have to take any action based on metrics being displayed, we can carry out visualization on the test dataset.

The reason that using regularization might be better than early stopping is that regularization allows you to use the entire dataset to change the weights of the model, whereas early stopping requires you to waste 10% to 20% of your dataset purely to decide when to stop training. Other methods to limit overfitting (such as dropout and using models with lower complexity) are also good alternatives to early stopping. In addition, recent research (*https://oreil.ly/FJ_iy*) indicates that double descent happens in a variety of machine learning problems, and therefore it is better to train longer rather than risk a suboptimal solution by stopping early.

Two splits. Isn't the advice in the regularization section in conflict with the advice in the previous sections on early stopping or checkpoint selection? Not really.

We recommend that you split your data into two parts: a training dataset and an evaluation dataset. The evaluation dataset plays the part of the test dataset during experimentation (where there is no validation dataset) and plays the part of the validation dataset in production (where there is no test dataset).

The larger your training dataset, the more complex a model you can use, and the more accurate a model you can get. Using regularization rather than early stopping or checkpoint selection allows you to use a larger training dataset. In the experimentation phase (when you are exploring different model architectures, training techniques, and hyperparameters), we recommend that you turn off early stopping and train with larger models (see also "Design Pattern 11: Useful Overfitting" on page 141). This is to ensure that the model has enough capacity to learn the predictive patterns. During this process, monitor error convergence on the training split. At the end of experimentation, you can use the evaluation dataset to diagnose how well your model does on data it has not encountered during training.

When training the model to deploy in production, you will need to prepare to be able to do continuous evaluation and model retraining. Turn on early stopping or checkpoint selection and monitor the error metric on the evaluation dataset. Choose between early stopping and checkpoint selection depending on whether you need to control cost (in which case, you would choose early stopping) or want to prioritize model accuracy (in which case, you would choose checkpoint selection).

Fine-tuning

In a well-behaved training loop, gradient descent behaves such that you get to the neighborhood of the optimal error quickly on the basis of the majority of your data, then slowly converge toward the lowest error by optimizing on the corner cases.

Now, imagine that you need to periodically retrain the model on fresh data. You typically want to emphasize the fresh data, not the corner cases from last month. You are often better off resuming your training, not from the last checkpoint, but the checkpoint marked by the blue line in Figure 4-11. This corresponds to the start of phase 2 in our discussion of the phases of model training described earlier in "Why It Works" on page 169. This helps ensure that you have a general method that you are able to then fine-tune for a few epochs on just the fresh data.

When you resume from the checkpoint marked by the thick dashed vertical line, you will be on the fourth epoch, and so the learning rate will be quite low. Therefore, the fresh data will not dramatically change the model. However, the model will behave optimally (in the context of the larger model) on the fresh data because you will have sharpened it on this smaller dataset. This is called *fine-tuning*. Fine-tuning is also discussed in "Design Pattern 13: Transfer Learning" on page 161.

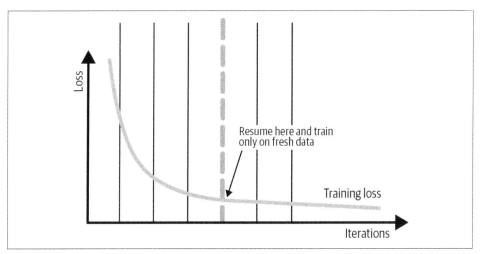

Figure 4-11. Resume from a checkpoint from before the training loss starts to plateau. Train only on fresh data for subsequent iterations.

 Fine-tuning only works as long as you are not changing the model architecture.

It is not necessary to always start from an earlier checkpoint. In some cases, the final checkpoint (that is used to serve the model) can be used as a warm start for another model training iteration. Still, starting from an earlier checkpoint tends to provide better generalization.

Redefining an epoch

Machine learning tutorials often have code like this:

```
model.fit(X_train, y_train,
        batch_size=100,
        epochs=15)
```

This code assumes that you have a dataset that fits in memory, and consequently that your model can iterate through 15 epochs without running the risk of machine failure. Both these assumptions are unreasonable—ML datasets range into terabytes, and when training can last hours, the chances of machine failure are high.

To make the preceding code more resilient, supply a TensorFlow dataset (*https://oreil.ly/EKJ4V*) (not just a NumPy array) because the TensorFlow dataset is an out-of-memory dataset. It provides iteration capability and lazy loading. The code is now as follows:

```
cp_callback = tf.keras.callbacks.ModelCheckpoint(...)
history = model.fit(trainds,
                    validation_data=evalds,
                    epochs=15,
                    batch_size=128,
                    callbacks=[cp_callback])
```

However, using epochs on large datasets remains a bad idea. Epochs may be easy to understand, but the use of epochs leads to bad effects in real-world ML models. To see why, imagine that you have a training dataset with one million examples. It can be tempting to simply go through this dataset 15 times (for example) by setting the number of epochs to 15. There are several problems with this:

- The number of epochs is an integer, but the difference in training time between processing the dataset 14.3 times and 15 times can be hours. If the model has converged after having seen 14.3 million examples, you might want to exit and not waste the computational resources necessary to process 0.7 million more examples.

- You checkpoint once per epoch, and waiting one million examples between checkpoints might be way too long. For resilience, you might want to checkpoint more often.

- Datasets grow over time. If you get 100,000 more examples and you train the model and get a higher error, is it because you need to do an early stop, or is the new data corrupt in some way? You can't tell because the prior training was on 15 million examples and the new one is on 16.5 million examples.

- In distributed, parameter-server training (see "Design Pattern 14: Distribution Strategy" on page 175) with data parallelism and proper shuffling, the concept of an epoch is not clear anymore. Because of potentially straggling workers, you can only instruct the system to train on some number of mini-batches.

Steps per epoch. Instead of training for 15 epochs, we might decide to train for 143,000 steps where the batch_size is 100:

```
NUM_STEPS = 143000
BATCH_SIZE = 100
NUM_CHECKPOINTS = 15
cp_callback = tf.keras.callbacks.ModelCheckpoint(...)
history = model.fit(trainds,
                    validation_data=evalds,
                    epochs=NUM_CHECKPOINTS,
                    steps_per_epoch=NUM_STEPS // NUM_CHECKPOINTS,
                    batch_size=BATCH_SIZE,
                    callbacks=[cp_callback])
```

Each step involves weight updates based on a single mini-batch of data, and this allows us to stop at 14.3 epochs. This gives us much more granularity, but we have to define an "epoch" as 1/15th of the total number of steps:

```
steps_per_epoch=NUM_STEPS // NUM_CHECKPOINTS,
```

This is so that we get the right number of checkpoints. It works as long as we make sure to repeat the `trainds` infinitely:

```
trainds = trainds.repeat()
```

The `repeat()` is needed because we no longer set `num_epochs`, so the number of epochs defaults to one. Without the `repeat()`, the model will exit once the training patterns are exhausted after reading the dataset once.

Retraining with more data. What happens when we get 100,000 more examples? Easy! We add it to our data warehouse but do not update the code. Our code will still want to process 143,000 steps, and it will get to process that much data, except that 10% of the examples it sees are newer. If the model converges, great. If it doesn't, we know that these new data points are the issue because we are not training longer than we were before. By keeping the number of steps constant, we have been able to separate out the effects of new data from training on more data.

Once we have trained for 143,000 steps, we restart the training and run it a bit longer (say, 10,000 steps), and as long as the model continues to converge, we keep training it longer. Then, we update the number 143,000 in the code above (in reality, it will be a parameter to the code) to reflect the new number of steps.

This all works fine, until you want to do hyperparameter tuning. When you do hyperparameter tuning, you will want to want to change the batch size. Unfortunately, if you change the batch size to 50, you will find yourself training for half the time because we are training for 143,000 steps, and each step is only half as long as before. Obviously, this is no good.

Virtual epochs. The answer is to keep the total number of training examples shown to the model (not number of steps; see Figure 4-12) constant:

```
NUM_TRAINING_EXAMPLES = 1000 * 1000
STOP_POINT = 14.3
TOTAL_TRAINING_EXAMPLES = int(STOP_POINT * NUM_TRAINING_EXAMPLES)
BATCH_SIZE = 100
NUM_CHECKPOINTS = 15
steps_per_epoch = (TOTAL_TRAINING_EXAMPLES //
                   (BATCH_SIZE*NUM_CHECKPOINTS))
cp_callback = tf.keras.callbacks.ModelCheckpoint(...)
history = model.fit(trainds,
                    validation_data=evalds,
                    epochs=NUM_CHECKPOINTS,
                    steps_per_epoch=steps_per_epoch,
```

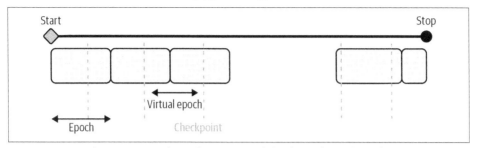

```
                    batch_size=BATCH_SIZE,
                    callbacks=[cp_callback])
```

Figure 4-12. Defining a virtual epoch in terms of the desired number of steps between checkpoints.

When you get more data, first train it with the old settings, then increase the number of examples to reflect the new data, and finally change the STOP_POINT to reflect the number of times you have to traverse the data to attain convergence.

This is now safe even with hyperparameter tuning (discussed later in this chapter) and retains all the advantages of keeping the number of steps constant.

Design Pattern 13: Transfer Learning

In Transfer Learning, we take part of a previously trained model, freeze the weights, and incorporate these nontrainable layers into a new model that solves a similar problem, but on a smaller dataset.

Problem

Training custom ML models on unstructured data requires extremely large datasets, which are not always readily available. Consider the case of a model identifying whether an x-ray of an arm contains a broken bone. To achieve high accuracy, you'll need hundreds of thousands of images, if not more. Before your model learns what a broken bone looks like, it needs to first learn to make sense of the pixels, edges, and shapes that are part of the images in your dataset. The same is true for models trained on text data. Let's say we're building a model that takes descriptions of patient symptoms and predicts the possible conditions associated with those symptoms. In addition to learning which words differentiate a cold from pneumonia, the model also needs to learn basic language semantics and how the sequence of words creates meaning. For example, the model would need to not only learn to detect the presence of the word *fever*, but that the sequence *no fever* carries a very different meaning than *high fever*.

To see just how much data is required to train high-accuracy models, we can look at ImageNet (*https://oreil.ly/t6583*), a database of over 14 million labeled images. Image-Net is frequently used as a benchmark for evaluating machine learning frameworks on various hardware. As an example, the MLPerf benchmark suite (*https://oreil.ly/hDPiJ*) uses ImageNet to compare the time it took for various ML frameworks running on different hardware to reach 75.9% classification accuracy. In the v0.7 MLPerf Training results, a TensorFlow model running on a Google TPU v3 took around 30 seconds to reach this target accuracy.[2] With more training time, models can reach even higher accuracy on ImageNet. However, this is largely due to ImageNet's size. Most organizations with specialized prediction problems don't have nearly as much data available.

Because use cases like the image and text examples described above involve particularly specialized data domains, it's also not possible to use a general-purpose model to successfully identify bone fractures or diagnose diseases. A model that is trained on ImageNet might be able to label an x-ray image as *x-ray* or *medical imaging* but is unlikely to be able to label it as a *broken femur*. Because such models are often trained on a wide variety of high-level label categories, we wouldn't expect them to understand conditions present in the images that are specific to our dataset. To handle this, we need a solution that allows us to build a custom model using only the data we have available and with the labels that we care about.

Solution

With the Transfer Learning design pattern, we can take a model that has been trained on the same type of data for a similar task and apply it to a specialized task using our own custom data. By "same type of data," we mean the same data modality—images, text, and so forth. Beyond just the broad category like images, it is also ideal to use a model that has been pre-trained on the same types of images. For example, use a model that has been pre-trained on photographs if you are going to use it for photograph classification and a model that has been pre-trained on remotely sensed imagery if you are going to use it to classify satellite images. By *similar task*, we're referring to the problem being solved. To do transfer learning for image classification, for example, it is better to start with a model that has been trained for image classification, rather than object detection.

Continuing with the example, let's say we're building a binary classifier to determine whether an image of an x-ray contains a broken bone. We only have 200 images of each class: *broken* and *not broken*. This isn't enough to train a high-quality model from scratch, but it is sufficient for transfer learning. To solve this with transfer

---

2 MLPerf v0.7 Training Closed ResNet. Retrieved from www.mlperf.org 23 September 2020, entry 0.7-67. MLPerf name and logo are trademarks. See www.mlperf.org for more information.

---

learning, we'll need to find a model that has already been trained on a large dataset to do image classification. We'll then remove the last layer from that model, freeze the weights of that model, and continue training using our 400 x-ray images. We'd ideally find a model trained on a dataset with similar images to our x-rays, like images taken in a lab or another controlled condition. However, we can still utilize transfer learning if the datasets are different, so long as the prediction task is the same. In this case we're doing image classification.

You can use transfer learning for many prediction tasks in addition to image classification, so long as there is an existing pre-trained model that matches the task you'd like to perform on your dataset. For example, transfer learning is also frequently applied in image object detection, image style transfer, image generation, text classification, machine translation, and more.

 Transfer learning works because it lets us stand on the shoulders of giants, utilizing models that have already been trained on extremely large, labeled datasets. We're able to use transfer learning thanks to years of research and work others have put into creating these datasets for us, which has advanced the state-of-the-art in transfer learning. One example of such a dataset is the ImageNet project, started in 2006 by Fei-Fei Li and published in 2009. ImageNet[3] has been essential to the development of transfer learning and paved the way for other large datasets like COCO (*https://oreil.ly/mXt77*) and Open Images (*https://oreil.ly/QN9KU*).

The idea behind transfer learning is that you can utilize the weights and layers from a model trained in the same domain as your prediction task. In most deep learning models, the final layer contains the classification label or output specific to your prediction task. With transfer learning, we remove this layer, freeze the model's trained weights, and replace the final layer with the output for our specialized prediction task before continuing to train. We can see how this works in Figure 4-13.

Typically, the penultimate layer of the model (the layer before the model's output layer) is chosen as the *bottleneck layer*. Next, we'll explain the bottleneck layer, along with different ways to implement transfer learning in TensorFlow.

---

3 Jia Deng et al.,"ImageNet: A Large-Scale Hierarchical Image Database," (*https://oreil.ly/Wio_D*) IEEE Computer Society Conference on Computer Vision and Pattern Recognition (CVPR) (2009): 248–255.

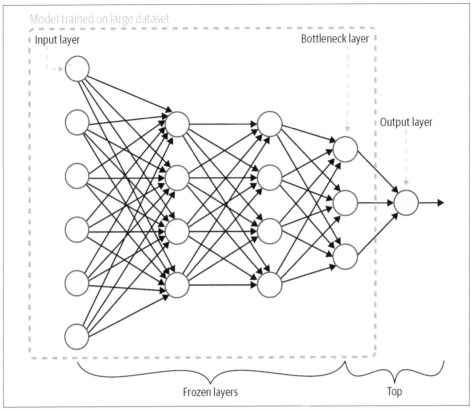

Figure 4-13. Transfer learning involves training a model on a large dataset. The "top" of the model (typically, just the output layer) is removed and the remaining layers have their weights frozen. The last layer of the remaining model is called the bottleneck layer.

Bottleneck layer

In relation to an entire model, the bottleneck layer represents the input (typically an image or text document) in the lowest-dimensionality space. More specifically, when we feed data into our model, the first layers see this data nearly in its original form. To see how this works, let's continue with a medical imaging example, but this time we'll build a model (*https://oreil.ly/QfOU_*) with a colorectal histology dataset (*https://oreil.ly/r4HHq*) to classify the histology images into one of eight categories.

To explore the model we are going to use for transfer learning, let's load the VGG model architecture pre-trained on the ImageNet dataset:

```
vgg_model_withtop = tf.keras.applications.VGG19(
    include_top=True,
    weights='imagenet',
)
```

Notice that we've set `include_top=True`, which means we're loading the full VGG model, including the output layer. For ImageNet, the model classifies images into 1,000 different classes, so the output layer is a 1,000-element array. Let's look at the output of `model.summary()` to understand which layer will be used as the bottleneck. For brevity, we've left out some of the middle layers here:

```
Model: "vgg19"
_____
Layer (type)                 Output Shape              Param #
=================================================================
input_3 (InputLayer)         [(None, 224, 224, 3)]     0

block1_conv1 (Conv2D)        (None, 224, 224, 64)      1792
...more layers here...

block5_conv3 (Conv2D)        (None, 14, 14, 512)       2359808

block5_conv4 (Conv2D)        (None, 14, 14, 512)       2359808

block5_pool (MaxPooling2D)   (None, 7, 7, 512)         0

flatten (Flatten)            (None, 25088)             0

fc1 (Dense)                  (None, 4096)              102764544

fc2 (Dense)                  (None, 4096)              16781312

predictions (Dense)          (None, 1000)              4097000
=================================================================
Total params: 143,667,240
Trainable params: 143,667,240
Non-trainable params: 0
_____
```

As you can see, the VGG model accepts images as a 224×224×3-pixel array. This 128-element array is then passed through successive layers (each of which may change the dimensionality of the array) until it is flattened into a 25,088×1-dimensional array in the layer called `flatten`. Finally, it is fed into the output layer, which returns a 1,000-element array (for each class in ImageNet). In this example, we'll choose the `block5_pool` layer as the bottleneck layer when we adapt this model to be trained on our medical histology images. The bottleneck layer produces a 7×7×512-dimensional array, which is a low-dimensional representation of the input image. It has retained enough of the information from the input image to be able to classify it. When we apply this model to our medical image classification task, we hope that the information distillation will be sufficient to successfully carry out classification on our dataset.

The histology dataset comes with images as (150,150,3) dimensional arrays. This 150×150×3 representation is the *highest* dimensionality. To use the VGG model with our image data, we can load it with the following:

```
vgg_model = tf.keras.applications.VGG19(
    include_top=False,
    weights='imagenet',
    input_shape=((150,150,3))
)

vgg_model.trainable = False
```

By setting `include_top=False`, we're specifying that the last layer of VGG we want to load is the bottleneck layer. The `input_shape` we passed in matches the input shape of our histology images. A summary of the last few layers of this updated VGG model looks like the following:

```
block5_conv3 (Conv2D)       (None, 9, 9, 512)        2359808
_____
block5_conv4 (Conv2D)       (None, 9, 9, 512)        2359808
_____
block5_pool (MaxPooling2D)  (None, 4, 4, 512)        0
============================================================
Total params: 20,024,384
Trainable params: 0
Non-trainable params: 20,024,384
_____
```

The last layer is now our bottleneck layer. You may notice that the size of `block5_pool` is (4,4,512), whereas before, it was (7,7,512). This is because we instantiated VGG with an `input_shape` parameter to account for the size of the images in our dataset. It's also worth noting that setting `include_top=False` is hardcoded to use `block5_pool` as the bottleneck layer, but if you want to customize this, you can load the full model and delete any additional layers you don't want to use.

Before this model is ready to be trained, we'll need to add a few layers on top, specific to our data and classification task. It's also important to note that because we've set `trainable=False`, there are 0 trainable parameters in the current model.

 As a general rule of thumb, the bottleneck layer is typically the last, lowest-dimensionality, flattened layer before a flattening operation.

Because they both represent features in reduced dimensionality, bottleneck layers are conceptually similar to embeddings. For example, in an autoencoder model with an encoder-decoder architecture, the bottleneck layer *is* an embedding. In this case, the bottleneck serves as the middle layer of the model, mapping the original input data to a lower-dimensionality representation, which the decoder (the second half of the network) uses to map the input back to its original, higher-dimensional representation. To see a diagram of the bottleneck layer in an autoencoder, refer to Figure 2-13 in Chapter 2.

An embedding layer is essentially a lookup table of weights, mapping a particular feature to some dimension in vector space. The main difference is that the weights in an embedding layer can be trained, whereas all the layers leading up to and including the bottleneck layer have their weights frozen. In other words, the entire network up to and including the bottleneck layer is nontrainable, and the weights in the layers after the bottleneck are the only trainable layers in the model.

 It's also worth noting that pre-trained embeddings can be used in the Transfer Learning design pattern. When you build a model that includes an embedding layer, you can either utilize an existing (pre-trained) embedding lookup, or train your own embedding layer from scratch.

To summarize, transfer learning is a solution you can employ to solve a similar problem on a smaller dataset. Transfer learning always makes use of a bottleneck layer with nontrainable, frozen weights. Embeddings are a type of data representation. Ultimately, it comes down to purpose. If the purpose is to train a similar model, you would use transfer learning. Consequently, if the purpose is to represent an input image more concisely, you would use an embedding. The code might be exactly the same.

Implementing transfer learning

You can implement transfer learning in Keras using one of these two methods:

- Loading a pre-trained model on your own, removing the layers after the bottleneck, and adding a new final layer with your own data and labels
- Using a pre-trained TensorFlow Hub (*https://tfhub.dev*) module as the foundation for your transfer learning task

Let's start by looking at how to load and use a pre-trained model on your own. For this, we'll build on the VGG model example we introduced earlier. Note that VGG is a model architecture, whereas ImageNet is the data it was trained on. Together, these make up the pre-trained model we'll be using for transfer learning. Here, we're using transfer learning to classify colorectal histology images. Whereas the original

ImageNet dataset contains 1,000 labels, our resulting model will *only* return 8 possible classes that we'll specify, as opposed to the thousands of labels present in ImageNet.

 Loading a pre-trained model and using it to get classifications on the *original labels* that model was trained on is not transfer learning. Transfer learning is going one step further, replacing the final layers of the model with your own prediction task.

The VGG model we've loaded will be our base model. We'll need to add a few layers to flatten the output of our bottleneck layer and feed this flattened output into an 8-element softmax array:

```
global_avg_layer = tf.keras.layers.GlobalAveragePooling2D()
feature_batch_avg = global_avg_layer(feature_batch)

prediction_layer = tf.keras.layers.Dense(8, activation='softmax')
prediction_batch = prediction_layer(feature_batch_avg)
```

Finally, we can use the `Sequential,` API to create our new transfer learning model as a stack of layers:

```
histology_model = keras.Sequential([
  vgg_model,
  global_avg_layer,
  prediction_layer
])
```

Let's take note of the output of `model.summary()` on our transfer learning model:

Layer (type)	Output Shape	Param #
vgg19 (Model)	(None, 4, 4, 512)	20024384
global_average_pooling2d (Gl	(None, 512)	0
dense (Dense)	(None, 8)	4104

```
Total params: 20,028,488
Trainable params: 4,104
Non-trainable params: 20,024,384
```

The important piece here is that the only trainable parameters are the ones *after* our bottleneck layer. In this example, the bottleneck layer is the feature vectors from the VGG model. After compiling this model, we can train it using our dataset of histology images.

Pre-trained embeddings

While we can load a pre-trained model on our own, we can also implement transfer learning by making use of the many pre-trained models available in TF Hub, a library of pre-trained models (called modules). These modules span a variety of data domains and use cases, including classification, object detection, machine translation, and more. In TensorFlow, you can load these modules as a layer, then add your own classification layer on top.

To see how TF Hub works, let's build a model that classifies movie reviews as either *positive* or *negative*. First, we'll load a pre-trained embedding model trained on a large corpus of news articles. We can instantiate this model as a `hub.KerasLayer`:

```
hub_layer = hub.KerasLayer(
    "https://tfhub.dev/google/tf2-preview/gnews-swivel-20dim/1",
    input_shape=[], dtype=tf.string, trainable=True)
```

We can stack additional layers on top of this to build our classifier:

```
model = keras.Sequential([
  hub_layer,
  keras.layers.Dense(32, activation='relu'),
  keras.layers.Dense(1, activation='sigmoid')
])
```

We can now train this model, passing it our own text dataset as input. The resulting prediction will be a 1-element array indicating whether our model thinks the given text is positive or negative.

Why It Works

To understand why transfer learning works, let's first look at an analogy. When children are learning their first language, they are exposed to many examples and corrected if they misidentify something. For example, the first time they learn to identify a cat, they'll see their parents point to the cat and say the word *cat,* and this repetition strengthens pathways in their brain. Similarly, they are corrected when they say *cat* referring to an animal that is not a cat. When the child then learns how to identify a dog, they don't need to start from scratch. They can use a similar recognition process to the one they used for the cat but apply it to a slightly different task. In this way, the child has built a foundation for learning. In addition to learning new things, they have also learned *how* to learn new things. Applying these learning methods to different domains is roughly how transfer learning works, too.

How does this play out in neural networks? In a typical convolutional neural network (CNN), the learning is hierarchical. The first layers learn to recognize edges and shapes present in an image. In the cat example, this might mean that the model can identify areas in an image where the edge of the cat's body meets the background. The next layers in the model begin to understand groups of edges—perhaps that there are two edges that meet toward the top-left corner of the image. A CNN's final layers can then piece together these groups of edges, developing an understanding of different features in the image. In the cat example, the model might be able to identify two triangular shapes toward the top of the image and two oval shapes below them. As humans, we know that these triangular shapes are ears and the oval shapes are eyes.

We can visualize this process in Figure 4-14, from research by Zeiler and Fergus (*https://oreil.ly/VzRV_*) on deconstructing CNNs to understand the different features that were activated throughout each layer of the model. For each layer in a five-layer CNN, this shows an image's feature map for a given layer alongside the actual image. This lets us see how the model's perception of an image progresses as it moves throughout the network. Layers 1 and 2 recognize only edges, layer 3 begins to recognize objects, and layers 4 and 5 can understand focal points within the entire image.

Remember, though, that to our model, these are simply groupings of pixel values. It doesn't know that the triangular and oval shapes are ears and eyes—it only knows to associate specific groupings of features with the labels it has been trained on. In this way, the model's process of learning what groupings of features make up a cat isn't *much* different from learning the groups of features that are part of other objects, like a table, a mountain, or even a celebrity. To a model, these are all just different combinations of pixel values, edges, and shapes.

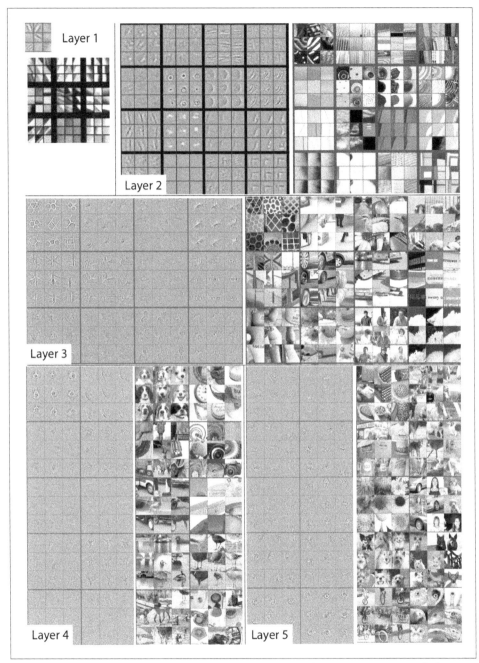

Figure 4-14. Research from Zeiler and Fergus (2013) in deconstructing CNNs helps us visualize how a CNN sees images at each layer of the network.

Trade-Offs and Alternatives

So far, we haven't discussed methods of modifying the weights of our original model when implementing transfer learning. Here, we'll examine two approaches for this: feature extraction and fine-tuning. We'll also discuss why transfer learning is primarily focused on image and text models and look at the relationship between text sentence embeddings and transfer learning.

Fine-tuning versus feature extraction

Feature extraction describes an approach to transfer learning where you freeze the weights of all layers before the bottleneck layer and train the following layers on your own data and labels. Another option is instead *fine-tuning* the weights of the pre-trained model's layers. With fine-tuning, you can either update the weights of each layer in the pre-trained model, or just a few of the layers right before the bottleneck. Training a transfer learning model using fine-tuning typically takes longer than feature extraction. You'll notice in our text classification example above, we set `trainable=True` when initializing our TF Hub layer. This is an example of fine-tuning.

When fine-tuning, it's common to leave the weights of the model's initial layers frozen since these layers have been trained to recognize basic features that are often common across many types of images. To fine-tune a MobileNet model, for example, we'd set `trainable=False` only for a subset of layers in the model, rather than making every layer non-trainable. For example, to fine-tune after the 100th layer, we could run:

```
base_model = tf.keras.applications.MobileNetV2(input_shape=(160,160,3),
                                               include_top=False,
                                               weights='imagenet')

for layer in base_model.layers[:100]:
    layer.trainable = False
```

One recommended approach to determining how many layers to freeze is known as *progressive fine-tuning* (*https://oreil.ly/fAv1S*), and it involves iteratively unfreezing layers after every training run to find the ideal number of layers to fine-tune. This works best and is most efficient if you keep your learning rate low (0.001 is common) and the number of training iterations relatively small. To implement progressive fine-tuning, start by unfreezing only the last layer of your transferred model (the layer closest to the output) and calculate your model's loss after training. Then, one by one, unfreeze more layers until you reach the Input layer or until the loss starts to plateau. Use this to inform the number of layers to fine-tune.

How should you determine whether to fine-tune or freeze all layers of your pre-trained model? Typically, when you've got a small dataset, it's best to use the pre-trained model as a feature extractor rather than fine-tuning. If you're retraining the weights of a model that was likely trained on thousands or millions of examples, fine-tuning can cause the updated model to overfit to your small dataset and lose the more general information learned from those millions of examples. Although it depends on your data and prediction task, when we say "small dataset" here, we're referring to datasets with hundreds or a few thousand training examples.

Another factor to take into account when deciding whether to fine-tune is how similar your prediction task is to that of the original pre-trained model you're using. When the prediction task is similar or a continuation of the previous training, as it was in our movie review sentiment analysis model, fine-tuning can produce higher-accuracy results. When the task is different or the datasets are significantly different, it's best to freeze all the layers of the pre-trained model instead of fine-tuning them. Table 4-1 summarizes the key points.[4]

Table 4-1. Criteria to help choose between feature extraction and fine-tuning

Criterion	Feature extraction	Fine-tuning
How large is the dataset?	Small	Large
Is your prediction task the same as that of the pre-trained model?	Different tasks	Same task, or similar task with same class distribution of labels
Budget for training time and computational cost	Low	High

In our text example, the pre-trained model was trained on a corpus of news text but our use case was sentiment analysis. Because these tasks are different, we should use the original model as a feature extractor rather than fine-tune it. An example of different prediction tasks in an image domain might be using our MobileNet model trained on ImageNet as a basis for doing transfer learning on a dataset of medical images. Although both tasks involve image classification, the nature of the images in each dataset are very different.

Focus on image and text models

You may have noticed that all of the examples in this section focused on image and text data. This is because transfer learning is primarily for cases where you can apply a similar task to the same data domain. Models trained with tabular data, however, cover a potentially infinite number of possible prediction tasks and data types. You could train a model on tabular data to predict how you should price tickets to your event, whether or not someone is likely to default on loan, your company's revenue next quarter, the duration of a taxi trip, and so forth. The specific data for these tasks

---

4 For more information, see "CS231n Convolutional Neural Networks for Visual Recognition (*https://oreil.ly/w109T*)."

is also incredibly varied, with the ticket problem depending on information about artists and venues, the loan problem on personal income, and the taxi duration on urban traffic patterns. For these reasons, there are inherent challenges in transferring the learnings from one tabular model to another.

Although transfer learning is not yet as common on tabular data as it is for image and text domains, a new model architecture called TabNet (*https://oreil.ly/HI5Xl*) presents novel research in this area. Most tabular models require significant feature engineering when compared with image and text models. TabNet employs a technique that first uses unsupervised learning to learn representations for tabular features, and then fine-tunes these learned representations to produce predictions. In this way, TabNet automates feature engineering for tabular models.

Embeddings of words versus sentences

In our discussion of text embeddings so far, we've referred mostly to *word* embeddings. Another type of text embedding is *sentence* embeddings. Where word embeddings represent individual words in a vector space, sentence embeddings represent entire sentences. Consequently, word embeddings are context agnostic. Let's see how this plays out with the following sentence:

"I've left you fresh baked cookies on the left side of the kitchen counter."

Notice that the word *left* appears twice in that sentence, first as a verb and then as an adjective. If we were to generate word embeddings for this sentence, we'd get a separate array for each word. With word embeddings, the array for both instances of the word *left* would be the same. Using sentence-level embeddings, however, we'd get a single vector to represent the entire sentence. There are several approaches for generating sentence embeddings—from averaging a sentence's word embeddings to training a supervised learning model on a large corpus of text to generate the embeddings.

How does this relate to transfer learning? The latter method—training a supervised learning model to generate sentence-level embeddings—is actually a form of transfer learning. This is the approach used by Google's Universal Sentence Encoder (*https://oreil.ly/Y0Ry9*) (available in TF Hub) and BERT (*https://oreil.ly/l_gQf*). These methods differ from word embeddings in that they go beyond simply providing a weight lookup for individual words. Instead, they have been built by training a model on a large dataset of varied text to understand the meaning conveyed by *sequences* of words. In this way, they are designed to be transferred to different natural language tasks and can thus be used to build models that implement transfer learning.

Design Pattern 14: Distribution Strategy

In Distribution Strategy, the training loop is carried out at scale over multiple workers, often with caching, hardware acceleration, and parallelization.

Problem

These days, it's common for large neural networks to have millions of parameters and be trained on massive amounts of data. In fact, it's been shown that increasing the scale of deep learning, with respect to the number of training examples, the number of model parameters, or both, drastically improves model performance. However, as the size of models and data increases, the computation and memory demands increase proportionally, making the time it takes to train these models one of the biggest problems of deep learning.

GPUs provide a substantial computational boost and bring the training time of modestly sized deep neural networks within reach. However, for very large models trained on massive amounts of data, individual GPUs aren't enough to make the training time tractible. For example, at the time of writing, training ResNet-50 on the benchmark ImageNet dataset for 90 epochs on a single NVIDIA M40 GPU requires 10^{18} single precision operations and takes 14 days. As AI is being used more and more to solve problems within complex domains, and open source libraries like Tensorflow and PyTorch make building deep learning models more accessible, large neural networks comparable to ResNet-50 have become the norm.

This is a problem. If it takes two weeks to train your neural network, then you have to wait two weeks before you can iterate on new ideas or experiment with tweaking the settings. Furthermore, for some complex problems like medical imaging, autonomous driving, or language translation, it's not always feasible to break the problem down into smaller components or work with only a subset of the data. It's only with the full scale of the data that you can assess whether things work or not.

Training time translates quite literally to money. In the world of serverless machine learning, rather than buying your own expensive GPU, it is possible to submit training jobs via a cloud service where you are charged for training time. The cost of training a model, whether it is to pay for a GPU or to pay for a serverless training service, quickly adds up.

Is there a way to speed up the training of these large neural networks?

Solution

One way to accelerate training is through distribution strategies in the training loop. There are different distribution techniques, but the common idea is to split the effort of training the model across multiple machines. There are two ways this can be done:

data parallelism and *model parallelism*. In data parallelism, computation is split across different machines and different workers train on different subsets of the training data. In model parallelism, the model is split and different workers carry out the computation for different parts of the model. In this section, we'll focus on data parallelism and show implementations in TensorFlow using the `tf.distrib ute.Strategy` library. We'll discuss model parallelism in "Trade-Offs and Alternatives" on page 183.

To implement data parallelism, there must be a method in place for different workers to compute gradients and share that information to make updates to the model parameters. This ensures that all workers are consistent and each gradient step works to train the model. Broadly speaking, data parallelism can be carried out either synchronously or asynchronously.

Synchronous training

In synchronous training, the workers train on different slices of input data in parallel and the gradient values are aggregated at the end of each training step. This is performed via an *all-reduce* algorithm. This means that each worker, typically a GPU, has a copy of the model on device and, for a single stochastic gradient descent (SGD) step, a mini-batch of data is split among each of the separate workers. Each device performs a forward pass with their portion of the mini-batch and computes gradients for each parameter of the model. These locally computed gradients are then collected from each device and aggregated (for example, averaged) to produce a single gradient update for each parameter. A central server holds the most current copy of the model parameters and performs the gradient step according to the gradients received from the multiple workers. Once the model parameters are updated according to this aggregated gradient step, the new model is sent back to the workers along with another split of the next mini-batch, and the process repeats. Figure 4-15 shows a typical all-reduce architecture for synchronous data distribution.

As with any parallelism strategy, this introduces additional overhead to manage timing and communication between workers. Large models could cause I/O bottlenecks as data is passed from the CPU to the GPU during training, and slow networks could also cause delays.

In TensorFlow, `tf.distribute.MirroredStrategy` supports synchronous distributed training across multiple GPUs on the same machine. Each model parameter is mirrored across all workers and stored as a single conceptual variable called `MirroredVariable`. During the all-reduce step, all gradient tensors are made available on each device. This helps to significantly reduce the overhead of synchronization. There are also various other implementations for the all-reduce algorithm available, many of which use NVIDIA NCCL (*https://oreil.ly/HX4NE*).

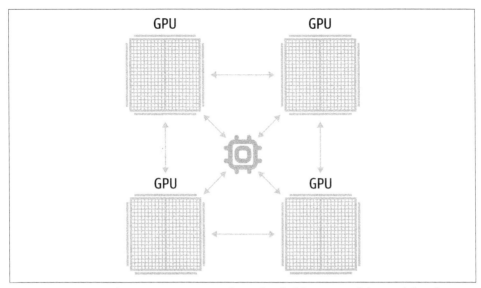

Figure 4-15. In synchronous training, each worker holds a copy of the model and computes gradients using a slice of the training data mini-batch.

To implement this mirrored strategy in Keras, you first create an instance of the mirrored distribution strategy, then move the creation and compiling of the model inside the scope of that instance. The following code shows how to use `MirroredStrategy` when training a three-layer neural network:

```
mirrored_strategy = tf.distribute.MirroredStrategy()
with mirrored_strategy.scope():
    model = tf.keras.Sequential([tf.keras.layers.Dense(32, input_shape=(5,)),
                                 tf.keras.layers.Dense(16, activation='relu'),
                                 tf.keras.layers.Dense(1)])
    model.compile(loss='mse', optimizer='sgd')
```

By creating the model inside this scope, the parameters of the model are created as mirrored variables instead of regular variables. When it comes to fitting the model on the dataset, everything is performed exactly the same as before. The model code stays the same! Wrapping the model code in the distribution strategy scope is all you need to do to enable distributed training. The `MirroredStrategy` handles replicating the model parameters on the available GPUs, aggregating gradients, and more. To train or evaluate the model, we just call `fit()` or `evaluate()` as usual:

```
model.fit(train_dataset, epochs=2)
model.evaluate(train_dataset)
```

During training, each batch of the input data is divided equally among the multiple workers. For example, if you are using two GPUs, then a batch size of 10 will be split among the 2 GPUs, with each receiving 5 training examples each step. There are also other synchronous distribution strategies within Keras, such as `CentralStorageStrategy` and `MultiWorkerMirroredStrategy`. `MultiWorkerMirroredStrategy` enables the distribution to be spread not just on GPUs on a single machine, but on multiple machines. In `CentralStorageStrategy`, the model variables are not mirrored; instead, they are placed on the CPU and operations are replicated across all local GPUs. So the variable updates only happen in one place.

When choosing between different distribution strategies, the best option depends on your computer topology and how fast the CPUs and GPUs can communicate with one another. Table 4-2 summarizes how the different strategies described here compare on these criteria.

Table 4-2. Choosing between distribution strategies depends on your computer topology and how fast the CPUs and GPUs can communicate with one another

	Faster CPU-GPU connection	Faster GPU-GPU connection
One machine with multiple GPUs	`CentralStorageStrategy`	`MirroredStrategy`
Multiple machines with multiple GPUs	`MultiWorkerMirroredStrategy`	`MultiWorkerMirroredStrategy`

Distributed Data Parallelism in PyTorch

In PyTorch, the code always uses `DistributedDataParallel` whether you have one GPU or multiple GPUs and whether the model is run on one machine or multiple machines. Instead, how and where you start the processes and how you wire up sampling, data loading, and so on determines the distribution strategy.

First, we initialize the process and wait for other processes to start and set up communication using:

```
torch.distributed.init_process_group(backend="nccl")
```

Second, specify the device number by obtaining a rank from the command line. Rank = 0 is the master process, and 1,2,3,... are the workers:

```
device = torch.device("cuda:{}".format(local_rank))
```

The model is created as normal in each of the processes, but is sent to this device. A distributed version of the model that will process its shard of batch is created using `DistributedDataParallel`:

```
model = model.to(device)
ddp_model = DistributedDataParallel(model, device_ids=[local_rank],
                                    output_device=local_rank)
```

The data itself is sharded using a `DistributedSampler`, and each batch of data is also sent to the device:

```
        sampler = DistributedSampler(dataset=trainds)
        loader = DataLoader(dataset=trainds, batch_size=batch_size,
                            sampler=sampler, num_workers=4)
        ...
        for data in train_loader:
            features, labels = data[0].to(device), data[1].to(device)
```

When a PyTorch trainer is launched, it is told the total number of nodes and its own rank:

```
python -m torch.distributed.launch --nproc_per_node=4 \
       --nnodes=16 --node_rank=3 --master_addr="192.168.0.1" \
       --master_port=1234 my_pytorch.py
```

If the number of nodes is one, we have the equivalent of TensorFlow's `MirroredStrategy`, and if the number of nodes is more than one, we have the equivalent of TensorFlow's `MultiWorkerMirroredStrategy`. If the number of processes per node and number of nodes are both one, then we have a `OneDeviceStrategy`. Optimized communication for all these cases is provided if supported by the backend (NCCL, in this case) passed into `init_process_group`.

Asynchronous training

In asynchronous training, the workers train on different slices of the input data independently, and the model weights and parameters are updated asynchronously, typically through a parameter server architecture (*https://oreil.ly/Wkk5B*). This means that no one worker waits for updates to the model from any of the other workers. In the parameter-server architecture, there is a single parameter server that manages the current values of the model weights, as in Figure 4-16.

As with synchronous training, a mini-batch of data is split among each of the separate workers for each SGD step. Each device performs a forward pass with their portion of the mini-batch and computes gradients for each parameter of the model. Those gradients are sent to the parameter server, which performs the parameter update and then sends the new model parameters back to the worker with another split of the next mini-batch.

The key difference between synchronous and asynchronous training is that the parameter server does not do an *all*-reduce. Instead, it computes the new model parameters periodically based on whichever gradient updates it received since the last computation. Typically, asynchronous distribution achieves higher throughput than synchronous training because a slow worker doesn't block the progression of training steps. If a single worker fails, the training continues as planned with the other workers while that worker reboots. As a result, some splits of the mini-batch may be lost during training, making it too difficult to accurately keep track of how many epochs of data have been processed. This is another reason why we typically specify virtual epochs when training large distributed jobs instead of epochs; see "Design Pattern 12: Checkpoints" on page 149 for a discussion of virtual epochs.

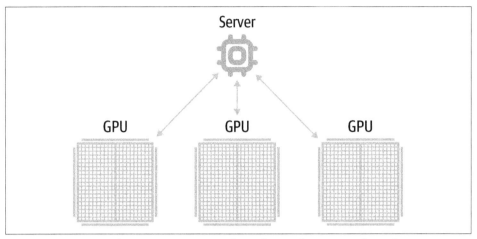

Figure 4-16. In asynchronous training, each worker performs a gradient descent step with a split of the mini-batch. No one worker waits for updates to the model from any of the other workers.

In addition, since there is no synchronization between the weight updates, it is possible that one worker updates the model weights based on stale model state. However, in practice, this doesn't seem to be a problem. Typically, large neural networks are trained for multiple epochs, and these small discrepancies become negligible in the end.

In Keras, `ParameterServerStrategy` implements asynchronous parameter server training on multiple machines. When using this distribution, some machines are designated as workers and some are held as parameter servers. The parameter servers hold each variable of the model, and computation is performed on the workers, typically GPUs.

The implementation is similar to that of other distribution strategies in Keras. For example, in your code, you would just replace `MirroredStrategy()` with `ParameterServerStrategy()`.

 Another distribution strategy supported in Keras worth mentioning is `OneDeviceStrategy`. This strategy will place any variables created in its scope on the specified device. This strategy is particularly useful as a way to test your code before switching to other strategies that actually distribute to multiple devices/machines.

Synchronous and asynchronous training each have their advantages, and disadvantages and choosing between the two often comes down to hardware and network limitations.

Synchronous training is particularly vulnerable to slow devices or poor network connection because training will stall waiting for updates from all workers. This means synchronous distribution is preferable when all devices are on a single host and there are fast devices (for example, TPUs or GPUs) with strong links. On the other hand, asynchronous distribution is preferable if there are many low-power or unreliable workers. If a single worker fails or stalls in returning a gradient update, it won't stall the training loop. The only limitation is I/O constraints.

Why It Works

Large, complex neural networks require massive amounts of training data to be effective. Distributed training schemes drastically increase the throughput of data processed by these models and can effectively decrease training times from weeks to hours. Sharing resources between workers and parameter server tasks leads to a dramatic increase in data throughput. Figure 4-17 compares the throughput of training data, in this case images, with different distribution setups.[5] Most notable is that throughput increases with the number of worker nodes and, even though parameter servers perform tasks not related to the computation done on the GPU's workers, splitting the workload among more machines is the most advantageous strategy.

In addition, data parallelization decreases time to convergence during training. In a similar study, it was shown that increasing workers leads to minimum loss much faster.[6] Figure 4-18 compares the time to minimum for different distribution strategies. As the number of workers increases, the time to minimum training loss dramatically decreases, showing nearly a 5× speed up with 8 workers as opposed to just 1.

---

5 Victor Campos et al., "Distributed training strategies for a computer vision deep learning algorithm on a distributed GPU cluster," *International Conference on Computational Science, ICCS 2017*, June 12–14, 2017.

6 Ibid.

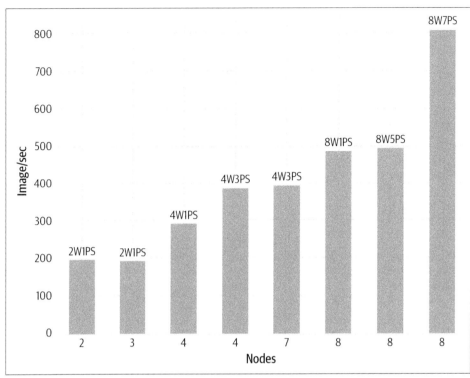

Figure 4-17. Comparison of throughput between different distribution setups. Here, 2W1PS indicates two workers and one parameter server.

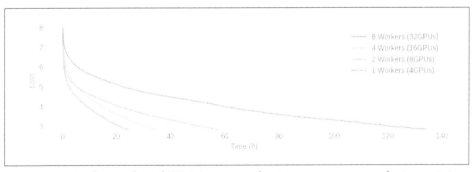

Figure 4-18. As the number of GPUs increases, the time to convergence during training decreases.

Trade-Offs and Alternatives

In addition to data parallelism, there are other aspects of distribution to consider, such as model parallelism, other training accelerators—(such as TPUs) and other considerations (such as I/O limitations and batch size).

Model parallelism

In some cases, the neural network is so large it cannot fit in the memory of a single device; for example, Google's Neural Machine Translation (*https://oreil.ly/xL4Cu*) has billions of parameters. In order to train models this big, they must be split up over multiple devices,[7] as shown in Figure 4-19. This is called *model parallelism*. By partitioning parts of a network and their associated computations across multiple cores, the computation and memory workload is distributed across multiple devices. Each device operates over the same mini-batch of data during training, but carries out computations related only to their separate components of the model.

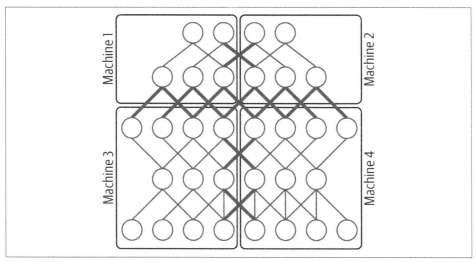

Figure 4-19. Model parallelism partitions the model over multiple devices.

---

7 Jeffrey Dean et al. "Large Scale Distributed Deep Networks," *NIPS Proceedings* (2012).

Model Parallelism or Data Parallelism?

A priori, neither scheme is better than the other. Each has its own benefits. Typically, the model architecture determines whether it is better to use data parallelism or model parallelism.

In particular, model parallelism improves efficiency when the amount of computation per neuron activity is high, such as in wide models with many fully connected layers. This is because it is the neuron value that is being communicated between different components of the model. Outside of the training paradigm, model parallelism provides an added benefit for serving very large models where low latency is needed. Distributing the computation of a large model across multiple devices can vastly reduce the overall computation time when making online predictions.

On the other hand, data parallelism is more efficient when the amount of computation per weight is high, such as when there are convolutional layers involved. This is because it is the model weights (and their gradient updates) that are being passed between different workers.

Depending on the scale of your model and problem, it may be necessary to exploit both. Mesh TensorFlow (*https://oreil.ly/svS4q*) is a library (*https://github.com/tensor flow/mesh*) optimized for distributed deep learning that combines synchronous data parallelism with model parallelism. It is implemented as a layer over TensorFlow and allows tensors to be easily split across different dimensions. Splitting across the batch layer is synonymous with data parallelism, while splitting over any other dimension —for example, a dimension representing the size of a hidden layer—achieves model parallelism.

ASICs for better performance at lower cost

Another way to speed up the training process is by accelerating the underlying hardware, such as by using application-specific integrated circuits (ASICs). In machine learning, this refers to hardware components designed specifically to optimize performance on the types of large matrix computations at the heart of the training loop. TPUs in Google Cloud are ASICs that can be used for both model training and making predictions. Similarly, Microsoft Azure offers the Azure FPGA (field-programmable gate array), which is also a custom machine learning chip like the ASIC except that it can be reconfigured over time. These chips are able to vastly minimize the time to accuracy when training large, complex neural network models. A model that takes two weeks to train on GPUs can converge in hours on TPUs.

There are other advantages to using custom machine learning chips. For example, as accelerators (GPUs, FPGAs, TPUs, and so on) have gotten faster, I/O has become a significant bottleneck in ML training. Many training processes waste cycles waiting to read and move data to the accelerator and waiting for gradient updates to carry out all-reduce. TPU pods have high-speed interconnect, so we tend to not worry about communication overhead within a pod (a pod consists of thousands of TPUs). In addition, there is lots of memory available on-disk, which means that it is possible to preemptively fetch data and make less-frequent calls to the CPU. As a result, you should use much higher batch sizes to take full advantage of high-memory, high-interconnect chips like TPUs.

In terms of distributed training, `TPUStrategy` allows you to run distributed training jobs on TPUs. Under the hood, `TPUStrategy` is the same as `MirroredStrategy` although TPUs have their own implementation of the all-reduce algorithm.

Using `TPUStrategy` is similar to using the other distribution strategies in TensorFlow. One difference is you must first set up a `TPUClusterResolver`, which points to the location of the TPUs. TPUs are currently available to use for free in Google Colab, and there you don't need to specify any arguments for `tpu_address`:

```
cluster_resolver = tf.distribute.cluster_resolver.TPUClusterResolver(
    tpu=tpu_address)
tf.config.experimental_connect_to_cluster(cluster_resolver)
tf.tpu.experimental.initialize_tpu_system(cluster_resolver)
tpu_strategy = tf.distribute.experimental.TPUStrategy(cluster_resolver)
```

Choosing a batch size

Another important factor to consider is batch size. Particular to synchronous data parallelism, when the model is particularly large, it's better to decrease the total number of training iterations because each training step requires the updated model to be shared among different workers, causing a slowdown for transfer time. Thus, it's important to increase the mini-batch size as much as possible so that the same performance can be met with fewer steps.

However, it has been shown (*https://oreil.ly/FOtIX*) that very large batch sizes adversely affect the rate at which stochastic gradient descent converges as well as the quality of the final solution.[8] Figure 4-20 shows that increasing the batch size alone ultimately causes the top-1 validation error to increase. In fact, they argue that linearly scaling the learning rate as a function of the large batch size is necessary to maintain a low validation error while decreasing the time of distributed training.

---

8 Priya Goyal et al., "Accurate, Large Minibatch SGD: Training ImageNet in 1 Hour" (2017), arXiv: 1706.02677v2 [cs.CV].

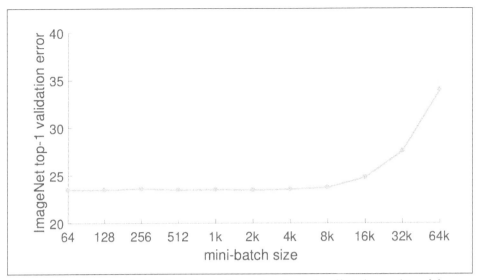

Figure 4-20. Large batch sizes have been shown to adversely affect the quality of the final trained model.

Thus, setting the mini-batch size in the context of distributed training is a complex optimization space of its own, as it affects both statistical accuracy (generalization) and hardware efficiency (utilization) of the model. Related work (*https://oreil.ly/ yeALI*), focusing on this optimization, introduces a layerwise adaptive large batch optimization technique called LAMB, which has been able to reduce BERT training time from 3 days to just 76 minutes.

Minimizing I/O waits

GPUs and TPUs can process data much faster than CPUs, and when using distributed strategies with multiple accelerators, I/O pipelines can struggle to keep up, creating a bottleneck to more efficient training. Specifically, before a training step finishes, the data for the next step is not available for processing. This is shown in Figure 4-21. The CPU handles the input pipeline: reading data from storage, preprocessing, and sending to the accelerator for computation. As distributed strategies speed up training, more than ever it becomes necessary to have efficient input pipelines to fully utilize the computing power available.

This can be achieved in a number of ways, including using optimized file formats like TFRecords and building data pipelines using the TensorFlow `tf.data` API. The `tf.data` API makes it possible to handle large amounts of data and has built-in transformations useful for creating flexible, efficient pipelines. For example, `tf.data.Data set.prefetch` overlaps the preprocessing and model execution of a training step so that while the model is executing training step N, the input pipeline is reading and preparing data for training step $N + 1$, as shown in Figure 4-22.

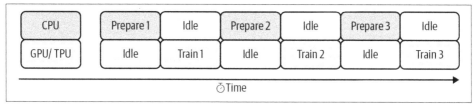

Figure 4-21. With distributed training on multiple GPU/TPUs available, it is necessary to have efficient input pipelines.

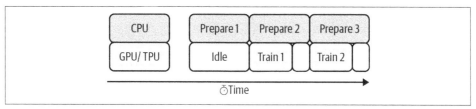

Figure 4-22. Prefetching overlaps preprocessing and model execution, so that while the model is executing one training step, the input pipeline is reading and preparing data for the next.

Design Pattern 15: Hyperparameter Tuning

In Hyperparameter Tuning, the training loop is itself inserted into an optimization method to find the optimal set of model hyperparameters.

Problem

In machine learning, model training involves finding the optimal set of breakpoints (in the case of decision trees), weights (in the case of neural networks), or support vectors (in the case of support vector machines). We term these *model* parameters. However, in order to carry out model training and find the optimal model parameters, we often have to hardcode a variety of things. For example, we might decide that the maximum depth of a tree will be 5 (in the case of decision trees), or that the activation function will be ReLU (for neural networks) or choose the set of kernels that we will employ (in SVMs). These parameters are called *hyperparameters*.

Model parameters refer to the weights and biases learned by your model. You do not have direct control over model parameters, since they are largely a function of your training data, model architecture, and many other factors. In other words, you cannot manually set model parameters. Your model's weights are initialized with random values and then optimized by your model as it goes through training iterations. Hyperparameters, on the other hand, refer to any parameters that you, as a model builder, can control. They include values like learning rate, number of epochs, number of layers in your model, and more.

Manual tuning

Because you can manually select the values for different hyperparameters, your first instinct might be a trial-and-error approach to finding the optimal combination of hyperparameter values. This might work for models that train in seconds or minutes, but it can quickly get expensive on larger models that require significant training time and infrastructure. Imagine you are training an image classification model that takes hours to train on GPUs. You settle on a few hyperparameter values to try and then wait for the results of the first training run. Based on these results, you tweak the hyperparameters, train the model again, compare the results with the first run, and then settle on the best hyperparameter values by looking at the training run with the best metrics.

There are a few problems with this approach. First, you've spent nearly a day and many compute hours on this task. Second, there's no way of knowing if you've arrived at the optimal combination of hyperparameter values. You've only tried two different combinations, and because you changed multiple values at once, you don't know which parameter had the biggest influence on performance. Even with additional trials, using this approach will quickly use up your time and compute resources and may not yield the most optimal hyperparameter values.

 We're using the term *trial* here to refer to a single training run with a set of hyperparameter values.

Grid search and combinatorial explosion

A more structured version of the trial-and-error approach described earlier is known as *grid search*. When implementing hyperparameter tuning with grid search, we choose a list of possible values we'd like to try for each hyperparameter we want to optimize. For example, in scikit-learn's `RandomForestRegressor()` model, let's say we want to try the following combination of values for the model's `max_depth` and `n_estimators` hyperparameters:

```
grid_values = {
  'max_depth': [5, 10, 100],
  'n_estimators': [100, 150, 200]
}
```

Using grid search, we'd try every combination of the specified values, then use the combination that yielded the best evaluation metric on our model. Let's see how this works on a random forest model trained on the Boston housing dataset, which comes pre-installed with scikit-learn. The model will predict the price of a house based on a number of factors. We can run grid search by creating an instance of the GridSearchCV class, and training the model passing it the values we defined earlier:

```
from sklearn.ensemble import RandomForestRegressor
from sklearn.datasets import load_boston

X, y = load_boston(return_X_y=True)
housing_model = RandomForestRegressor()

grid_search_housing = GridSearchCV(
    housing_model, param_grid=grid_vals, scoring='max_error')
grid_search_housing.fit(X, y)
```

Note that the scoring parameter here is the metric we want to optimize. In the case of this regression model, we want to use the combination of hyperparameters that results in the lowest error for our model. To get the best combination of values from the grid search, we can run grid_search_housing.best_params_. This returns the following:

```
{'max_depth': 100, 'n_estimators': 150}
```

We'd want to compare this to the error we'd get training a random forest regressor model *without* hyperparameter tuning, using scikit-learn's default values for these parameters. This grid search approach works OK on the small example we've defined above, but with more complex models, we'd likely want to optimize more than two hyperparameters, each with a wide range of possible values. Eventually, grid search will lead to *combinatorial explosion*—as we add additional hyperparameters and values to our grid of options, the number of possible combinations we need to try and the time required to try them all increases significantly.

Another problem with this approach is that no logic is being applied when choosing different combinations. Grid search is essentially a brute force solution, trying every possible combination of values. Let's say that after a certain max_depth value, our model's error increases. The grid search algorithm doesn't learn from previous trials, so it wouldn't know to stop trying max_depth values after a certain threshold. It will simply try every value you provide no matter the results.

scikit-learn supports an alternative to grid search called RandomizedSearchCV that implements *random search*. Instead of trying every possible combination of hyperparameters from a set, you determine the number of times you'd like to randomly sample values for each hyperparameter. To implement random search in scikit-learn, we'd create an instance of RandomizedSearchCV and pass it a dict similar to grid_values above, specifying *ranges* instead of specific values. Random search runs faster than grid search since it doesn't try every combination in your set of possible values, but it is very likely that the optimal set of hyperparameters will not be among the ones randomly selected.

For robust hyperparameter tuning, we need a solution that scales and learns from previous trials to find an optimal combination of hyperparameter values.

Solution

The keras-tuner library implements Bayesian optimization to do hyperparameter search directly in Keras. To use keras-tuner, we define our model inside a function that takes a hyperparameter argument, here called hp. We can then use hp throughout the function wherever we want to include a hyperparameter, specifying the hyperparameter's name, data type, the value range we'd like to search, and how much to increment it each time we try a new one.

Instead of hardcoding the hyperparameter value when we define a layer in our Keras model, we define it using a hyperparameter variable. Here, we want to tune the number of neurons in the first hidden layer of our neural network:

```
keras.layers.Dense(hp.Int('first_hidden', 32, 256, step=32), activation='relu')
```

first_hidden is the name we've given this hyperparameter, 32 is the minimum value we've defined for it, 256 is the maximum, and 32 is the amount we should increment this value by within the range we've defined. If we were building an MNIST classification model, the full function that we'd pass to keras-tuner might look like the following:

```
def build_model(hp):
  model = keras.Sequential([
    keras.layers.Flatten(input_shape=(28, 28)),
    keras.layers.Dense(
      hp.Int('first_hidden', 32, 256, step=32), activation='relu'),
    keras.layers.Dense(
      hp.Int('second_hidden', 32, 256, step=32), activation='relu'),
    keras.layers.Dense(10, activation='softmax')
  ])

  model.compile(
    optimizer=tf.keras.optimizers.Adam(
```

```
        hp.Float('learning_rate', .005, .01, sampling='log')),
    loss='sparse_categorical_crossentropy',
    metrics=['accuracy'])

    return model
```

The `keras-tuner` library supports many different optimization algorithms. Here, we'll instantiate our tuner with Bayesian optimization and optimize for validation accuracy:

```
import kerastuner as kt

tuner = kt.BayesianOptimization(
    build_model,
    objective='val_accuracy',
    max_trials=10
)
```

The code to run the tuning job looks similar to training our model with `fit()`. As this runs, we'll be able to see the values for the three hyperparameters that were selected for each trial. When the job completes, we can see the hyperparameter combination that resulted in the best trial. In Figure 4-23, we can see the example output for a single trial run using `keras-tuner`.

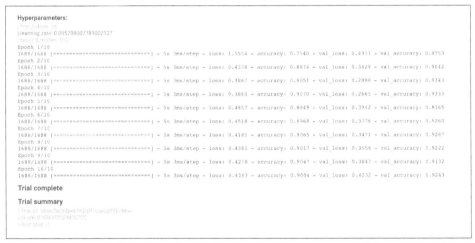

Figure 4-23. Output for one trial run of hyperparameter tuning with keras-tuner. At the top, we can see the hyperparameters selected by the tuner, and in the summary section, we see the resulting optimization metric.

In addition to the examples shown here, there is additional functionality provided by `keras-tuner` that we haven't covered. You can use it to experiment with different numbers of layers for your model by defining an `hp.Int()` parameter within a loop, and you can also provide a fixed set of values for a hyperparameter instead of a range.

For more complex models, this `hp.Choice()` parameter could be used to experiment with different types of layers, like `BasicLSTMCell` and `BasicRNNCell`. `keras-tuner` runs in any environment where you can train a Keras model.

Why It Works

Although grid and random search are more efficient than a trial-and-error approach to hyperparameter tuning, they quickly become expensive for models requiring significant training time or having a large hyperparameter search space.

Since both machine learning models themselves and the process of hyperparameter search are optimization problems, it would follow that we would be able to use an approach that *learns* to find the optimal hyperparameter combination within a given range of possible values just like our models learn from training data.

We can think of hyperparameter tuning as an outer optimization loop (see Figure 4-24) where the inner loop consists of typical model training. Even though we depict neural networks as the model whose parameters are being optimized, this solution is applicable to other types of machine learning models. Also, although the more common use case is to choose a single best model from all potential hyperparameters, in some cases, the hyperparameter framework can be used to generate a family of models that can act as an ensemble (see the discussion of the Ensembles pattern in Chapter 3).

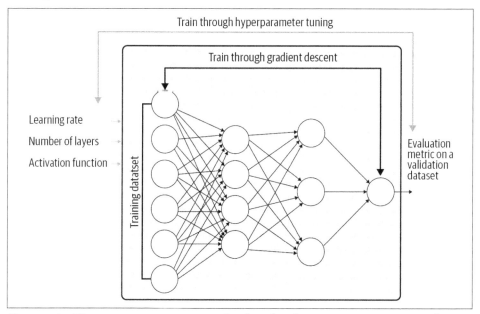

Figure 4-24. Hyperparameter tuning can be thought of as an outer optimization loop.

Nonlinear optimization

The hyperparameters that need to be tuned fall into two groups: those related to model *architecture* and those related to model *training*. Model architecture hyperparameters, like the number of layers in your model or the number of neurons per layer, control the mathematical function that underlies the machine learning model. Parameters related to model training, like the number of epochs, learning rate, and batch size, control the training loop and often have to do with the way that the gradient descent optimizer works. Taking both these types of parameters into consideration, it is clear that the overall model function with respect to these hyperparameters is, in general, not differentiable.

The inner training loop is differentiable, and the search for optimal parameters can be carried out through stochastic gradient descent. A single step of a machine learning model trained through stochastic gradient might take only a few milliseconds. On the other hand, a single trial in the hyperparameter tuning problem involves training a complete model on the training dataset and might take several hours. Moreover, the optimization problem for the hyperparameters will have to be solved through nonlinear optimization methods that apply to nondifferentiable problems.

Once we decide that we are going to use nonlinear optimization methods, our choice of metric becomes wider. This metric will be evaluated on the validation dataset and does not have to be the same as the training loss. For a classification model, your optimization metric might be accuracy, and you'd therefore want to find the combination of hyperparameters that leads to the highest model accuracy even if the loss is binary cross entropy. For a regression model, you might want to optimize median absolute error even if the loss is squared error. In that case, you'd want to find the hyperparameters that yield the *lowest* mean squared error. This metric can even be chosen based on business goals. For example, we might choose to maximize expected revenue or minimize losses due to fraud.

Bayesian optimization

Bayesian optimization is a technique for optimizing black-box functions, originally developed in the 1970s by Jonas Mockus (*https://oreil.ly/Ak24H*). The technique has been applied to many domains and was first applied to hyperparameter tuning in 2012 (*https://oreil.ly/KkGlG*). Here, we'll focus on Bayesian optimization as it relates to hyperparameter tuning. In this context, a machine learning model is our *black-box function*, since ML models produce a set of outputs from inputs we provide without requiring us to know the internal details of the model itself. The process of training our ML model is referred to as calling the *objective function*.

The goal of Bayesian optimization is to directly train our model as few times as possible since doing so is costly. Remember that each time we try a new combination of hyperparameters on our model, we need to run through our model's entire training cycle. This might seem trivial with a small model like the scikit-learn one we trained above, but for many production models, the training process requires significant infrastructure and time.

Instead of training our model each time we try a new combination of hyperparameters, Bayesian optimization defines a new function that emulates our model but is much cheaper to run. This is referred to as the *surrogate function*—the inputs to this function are your hyperparameter values and the output is your optimization metric. The surrogate function is called much more frequently than the objective function, with the goal of finding an optimal combination of hyperparameters *before* completing a training run on your model. With this approach, more compute time is spent choosing the hyperparameters for each trial as compared with grid search. However, because this is significantly cheaper than running our objective function each time we try different hyperparameters, the Bayesian approach of using a surrogate function is preferred. Common approaches to generate the surrogate function include a Gaussian process (*https://oreil.ly/-Srjj*) or a tree-structured Parzen estimator (*https://oreil.ly/UqxDd*).

So far, we've touched on the different pieces of Bayesian optimization, but how do they work together? First, we must choose the hyperparameters we want to optimize and define a range of values for each hyperparameter. This part of the process is manual and will define the space in which our algorithm will search for optimal values. We'll also need to define our objective function, which is the code that calls our model training process. From there, Bayesian optimization develops a surrogate function to simulate our model training process and uses that function to determine the best combination of hyperparameters to run on our model. It is only once this surrogate arrives at what it thinks is a good combination of hyperparameters that we do a full training run (trial) on our model. The results of this are then fed back to the surrogate function and the process is repeated for the number of trials we've specified.

Trade-Offs and Alternatives

Genetic algorithms are an alternative to Bayesian methods for hyperparameter tuning, but they tend to require many more model training runs than Bayesian methods. We'll also show you how to use a managed service for hyperparameter tuning optimization on models built with a variety of ML frameworks.

Fully managed hyperparameter tuning

The `keras-tuner` approach may not scale to large machine learning problems because we'd like the trials to happen in parallel, and the likelihood of machine error and other failure increases as the time for model training stretches into the hours. Hence, a fully managed, resilient approach that provides black-box optimization is useful for hyperparameter tuning. An example of a managed service that implements Bayesian optimization is the hyperparameter tuning service (*https://oreil.ly/MO8FZ*) provided by Google Cloud AI Platform. This service is based on Vizier (*https://oreil.ly/tScQa*), the black-box optimization tool used internally at Google.

The underlying concepts of the Cloud service work similarly to `keras-tuner`: you specify each hyperparameter's name, type, range, and scale, and these values are referenced in your model training code. We'll show you how to run hyperparameter tuning in AI Platform using a PyTorch model trained on the BigQuery natality dataset to predict a baby's birth weight.

The first step is to create a *config.yaml* file specifying the hyperparameters you want the job to optimize, along with some other metadata on your job. One benefit of using the Cloud service is that you can scale your tuning job by running it on GPUs or TPUs and spreading it across multiple parameter servers. In this config file, you also specify the total number of hyperparameter trials you want to run and how many of these trials you want to run in parallel. The more you run in parallel, the faster your job will run. However, the benefit of running fewer trials in parallel is that the service will be able to learn from the results of each completed trial to optimize the next ones.

For our model, a sample config file that makes use of GPUs might look like the following. In this example, we'll tune three hyperparameters—our model's learning rate, the optimizer's momentum value (*https://oreil.ly/8mHPQ*), and the number of neurons in our model's hidden layer. We also specify our optimization metric. In this example, our goal will be to *minimize* our model's loss on our validation set:

```
trainingInput:
  scaleTier: BASIC_GPU
  parameterServerType: large_model
  workerCount: 9
  parameterServerCount: 3
  hyperparameters:
    goal: MINIMIZE
    maxTrials: 10
    maxParallelTrials: 5
    hyperparameterMetricTag: val_error
    enableTrialEarlyStopping: TRUE
    params:
    - parameterName: lr
      type: DOUBLE
      minValue: 0.0001
      maxValue: 0.1
      scaleType: UNIT_LINEAR_SCALE
    - parameterName: momentum
      type: DOUBLE
      minValue: 0.0
      maxValue: 1.0
      scaleType: UNIT_LINEAR_SCALE
```

```
    - parameterName: hidden-layer-size
      type: INTEGER
      minValue: 8
      maxValue: 32
      scaleType: UNIT_LINEAR_SCALE
```

 Instead of using a config file to define these values, you can also do this using the AI Platform Python API.

In order to do this, we'll need to add an argument parser to our code that will specify the arguments we defined in the file above, then refer to these hyperparameters where they appear throughout our model code.

Next, we'll build our model using PyTorch's `nn.Sequential` API with the SGD optimizer. Since our model predicts baby weight as a float, this will be a regression model. We specify each of our hyperparameters using the `args` variable, which contains the variables defined in our argument parser:

```
import torch.nn as nn

model = nn.Sequential(nn.Linear(num_features, args.hidden_layer_size),
                      nn.ReLU(),
                      nn.Linear(args.hidden_layer_size, 1))

optimizer = torch.optim.SGD(model.parameters(), lr=args.lr,
                            momentum=args.momentum)
```

At the end of our model training code, we'll create an instance of `HyperTune()`, and tell it the metric we're trying to optimize. This will report the resulting value of our optimization metric after each training run. It's important that whichever optimization metric we choose is calculated on our test or validation datasets, and not our training dataset:

```
import hypertune

hpt = hypertune.HyperTune()

val_mse = 0
num_batches = 0

criterion = nn.MSELoss()

with torch.no_grad():
    for i, (data, label) in enumerate(validation_dataloader):
        num_batches += 1
        y_pred = model(data)
        mse = criterion(y_pred, label.view(-1,1))
```

```
        val_mse += mse.item()

    avg_val_mse = (val_mse / num_batches)

hpt.report_hyperparameter_tuning_metric(
    hyperparameter_metric_tag='val_mse',
    metric_value=avg_val_mse,
    global_step=epochs
)
```

Once we've submitted our training job to AI Platform, we can monitor logs in the Cloud console. After each trial completes, you'll be able to see the values chosen for each hyperparameter and the resulting value of your optimization metric, as seen in Figure 4-25.

HyperTune trials

	Trial ID	avg_val_mse ↑	Training step	lr	momentum	hidden-layer-size	
○ ⊘	6	1.58062	10	0.04151	0.37651	31	⋮
○ ⊘	7	1.58216	10	0.00821	0.97651	17	⋮
○ ⊘	1	1.58262	10	0.00542	0.90981	35	⋮
○ ⊘	10	1.58374	10	0.07829	0.69754	24	⋮
○ ⊘	4	1.58463	10	0.90905	0.4507	20	⋮
○ ⊘	2	1.59563	10	0.07565	0.0407	14	⋮
○ ⊘	3	1.60248	10	0.04735	0.6407	26	⋮
○ ⊘	9	1.60602	10	0.05561	0.62768	19	⋮
○ ⊘	8	1.61204	10	0.56797	0.85666	22	⋮
○ ⊘	5	1.80907	10	0.0484	0.24004	27	⋮

Figure 4-25. A sample of the HyperTune summary in the AI Platform console. This is for a PyTorch model optimizing three model parameters, with the goal of minimizing mean squared error on the validation dataset.

By default, AI Platform Training will use Bayesian optimization for your tuning job, but you can also specify if you'd like to use grid or random search algorithms instead. The Cloud service also optimizes your hyperparameter search *across* training jobs. If we run another training job similar to the one above, but with a few tweaks to our hyperparameters and search space, it'll use the results of our last job to efficiently choose values for the next set of trials.

We've shown a PyTorch example here, but you can use AI Platform Training for hyperparameter tuning in any machine learning framework by packaging your training code and providing a *setup.py* file that installs any library dependencies.

Genetic algorithms

We've explored various algorithms for hyperparameter optimization: manual search, grid search, random search, and Bayesian optimization. Another less-common alternative is a genetic algorithm, which is roughly based on Charles Darwin's evolutionary theory of natural selection. This theory, also known as "survival of the fittest," posits that the highest-performing ("fittest") members of a population will survive and pass their genes to future generations, while less-fit members will not. Genetic algorithms have been applied to different types of optimization problems, including hyperparameter tuning.

As it relates to hyperparameter search, a genetic approach works by first defining a *fitness function*. This function measures the quality of a particular trial, and can typically be defined by your model's optimization metric (accuracy, error, and so on). After defining your fitness function, you randomly select a few combinations of hyperparameters from your search space and run a trial for each of those combinations. You then take the hyperparameters from the trials that performed best, and use those values to define your new search space. This search space becomes your new "population," and you use it to generate new combinations of values to use in your next set of trials. You continue this process, narrowing down the number of trials you run until you've arrived at a result that satisfies your requirements.

Because they use the results of previous trials to improve, genetic algorithms are "smarter" than manual, grid, and random search. However, when the hyperparameter search space is large, the complexity of genetic algorithms increases. Rather than using a surrogate function as a proxy for model training like in Bayesian optimization, genetic algorithms require training your model for each possible combination of hyperparameter values. Additionally, at the time of writing, genetic algorithms are less common and there are fewer ML frameworks that support them out of the box for hyperparameter tuning.

Summary

This chapter focused on design patterns that modify the typical SGD training loop of machine learning. We started with looking at the *Useful Overfitting* pattern, which covered situations where overfitting is beneficial. For example, when using data-driven methods like machine learning to approximate solutions to complex dynamical systems or PDEs where the full input space can be covered, overfitting on the training set is the goal. Overfitting is also useful as a technique when developing and debugging ML model architectures. Next, we covered model *Checkpoints* and how to use them when training ML models. In this design pattern, we save the full state of the model periodically during training. These checkpoints can be used as the final model, as in the case of early stopping, or used as the starting points in the case of training failures or fine-tuning.

The *Transfer Learning* design pattern covered reusing parts of a previously trained model. Transfer learning is a useful way to leverage the learned feature extraction layers of the pre-trained model when your own dataset is limited. It can also be used to fine-tune a pre-trained model that was trained on a large generic dataset to your more specialized dataset. We then discussed the *Distribution Strategy* design pattern. Training large, complex neural networks can take a considerable amount of time. Distribution strategies offer various ways in which the training loop can be modified to be carried out at scale over multiple workers, using parallelization and hardware accelerators.

Lastly, the *Hyperparameter Tuning* design pattern discussed how the SGD training loop itself can be optimized with respect to model hyperparameters. We saw some useful libraries that can be used to implement hyperparameter tuning for models created with Keras and PyTorch.

The next chapter looks at design patterns related to *resilience* (to large numbers of requests, spiky traffic, or change management) when placing models into production.

## CHAPTER 5

Design Patterns for Resilient Serving

The purpose of a machine learning model is to use it to make inferences on data it hasn't seen during training. Therefore, once a model has been trained, it is typically deployed into a production environment and used to make predictions in response to incoming requests. Software that is deployed into production environments is expected to be resilient and require little in the way of human intervention to keep it running. The design patterns in this chapter solve problems associated with resilience under different circumstances as it relates to production ML models.

The *Stateless Serving Function* design pattern allows the serving infrastructure to scale and handle thousands or even millions of prediction requests per second. The *Batch Serving* design pattern allows the serving infrastructure to asynchronously handle occasional or periodic requests for millions to billions of predictions. These patterns are useful beyond resilience in that they reduce coupling between creators and users of machine learning models.

The *Continued Model Evaluation* design pattern handles the common problem of detecting when a deployed model is no longer fit-for-purpose. The *Two-Phase Predictions* design pattern provides a way to address the problem of keeping models sophisticated and performant when they have to be deployed onto distributed devices. The *Keyed Predictions* design pattern is a necessity to scalably implement several of the design patterns discussed in this chapter.

Design Pattern 16: Stateless Serving Function

The Stateless Serving Function design pattern makes it possible for a production ML system to synchronously handle thousands to millions of prediction requests per second. The production ML system is designed around a stateless function that captures the architecture and weights of a trained model.

Stateless Functions

A stateless function is a function whose outputs are determined purely by its inputs. This function, for example, is stateless:

```python
def stateless_fn(x):
    return 3*x + 15
```

Another way to think of a stateless function is as an immutable object, where the weights and biases are stored as constants:

```python
class Stateless:
    def __init__(self):
        self.weight = 3
        self.bias = 15
    def __call__(self, x):
        return self.weight*x + self.bias
```

A function that maintains a counter of the number of times it has been invoked and returns a different value depending on whether the counter is odd or even is an example of a function that is stateful, not stateless:

```python
class State:
    def __init__(self):
        self.counter = 0
    def __call__(self, x):
        self.counter += 1
        if self.counter % 2 == 0:
            return 3*x + 15
        else:
            return 3*x - 15
```

Invoking `stateless_fn(3)` or `Stateless()(3)` always returns 24, whereas

```python
a = State()
```

and then invoking

```python
a(3)
```

returns a value that rocks between −6 and 24. The counter in this case is the state of the function, and the output depends on both the input (`x`) and the state (`counter`). The state is typically maintained using class variables (as in our example) or using global variables.

Because stateless components don't have any state, they can be shared by multiple clients. Servers typically create an instance pool of stateless components and use them to service client requests as they come in. On the other hand, stateful components will need to represent each client's conversational state. The life cycle of stateless components needs to be managed by the server. For example, they need to be initialized on the first request and destroyed when the client terminates or times out. Because of these factors, stateless components are highly scalable, whereas stateful components are expensive and difficult to manage. When designing enterprise applications, architects are careful to minimize the number of stateful components. Web applications, for example, are often designed to work based on REST APIs, and these involve transfer of state from the client to the server with each call.

In a machine learning model, there is a lot of state captured during training. Things like the epoch number and learning rate are part of a model's state and have to be remembered because typically, the learning rate is decayed with each successive epoch. By saying that the model has to be exported as a stateless function, we are requiring the model framework creators to keep track of these stateful variables and not include them in the exported file.

When stateless functions are used, it simplifies the server code and makes it more scalable but can make client code more complicated. For example, some model functions are inherently stateful. A spelling correction model that takes a word and returns the corrected form will need to be stateful because it has to know the previous few words in order to correct a word like "there" to "their" depending on the context. Models that operate on sequences maintain history using special structures like recurrent neural network units. In such cases, needing to export the model as a stateless function requires changing the input from a single word to, for example, a sentence. This means clients of a spelling correction model will need to manage the state (to collect a sequence of words and break them up into sentences) and send it along with every request. The resulting client-side complexity is most visible when the spell-checking client has to go back and change a previous word because of context that gets added later.

Problem

Let's take a text classification model that uses, as its training data, movie reviews from the Internet Movie Database (IMDb). For the initial layer of the model, we will use a pre-trained embedding that maps text to 20-dimensional embedding vectors (for the full code, see the *serving_function.ipynb* notebook (*https://github.com/GoogleCloud Platform/ml-design-patterns/blob/master/05_resilience/serving_function.ipynb*) in the GitHub repository for this book):

```
model = tf.keras.Sequential()
embedding = (
        "https://tfhub.dev/google/tf2-preview/gnews-swivel-20dim-with-oov/1")
hub_layer = hub.KerasLayer(embedding, input_shape=[],
                           dtype=tf.string, trainable=True, name='full_text')
model.add(hub_layer)
model.add(tf.keras.layers.Dense(16, activation='relu', name='h1_dense'))
model.add(tf.keras.layers.Dense(1, name='positive_review_logits'))
```

The embedding layer is obtained from TensorFlow Hub and marked as being trainable so that we can carry out fine-tuning (see "Design Pattern 13: Transfer Learning" on page 161 in Chapter 4) on the vocabulary found in IMDb reviews. The subsequent layers are that of a simple neural network with one hidden layer and an output logits layer. This model can then be trained on the dataset of movie reviews to learn to predict whether or not a review is positive or negative.

Once the model has been trained, we can use it to carry out inferences on how positive a review is:

```
review1 = 'The film is based on a prize-winning novel.'
review2 = 'The film is fast moving and has several great action scenes.'
review3 = 'The film was very boring. I walked out half-way.'
logits = model.predict(x=tf.constant([review1, review2, review3]))
```

The result is a 2D array that might be something like:

```
[[ 0.6965847]
 [ 1.61773  ]
 [-0.7543597]]
```

There are several problems with carrying out inferences by calling model.predict() on an in-memory object (or a trainable object loaded into memory) as described in the preceding code snippet:

- We have to load the entire Keras model into memory. The text embedding layer, which was set up to be trainable, can be quite large because it needs to store embeddings for the full vocabulary of English words. Deep learning models with many layers can also be quite large.

- The preceding architecture imposes limits on the latency that can be achieved because calls to the predict() method have to be sent one by one.

- Even though the data scientist's programming language of choice is Python, model inference is likely to be invoked by programs written by developers who prefer other languages, or on mobile platforms like Android or iOS that require different languages.

- The model input and output that is most effective for training may not be user friendly. In our example, the model output was logits (*https://oreil.ly/qCWdH*) because it is better for gradient descent. This is why the second number in the output array is greater than 1. What clients will typically want is the sigmoid of this so that the output range is 0 to 1 and can be interpreted in a more user-friendly format as a probability. We will want to carry out this postprocessing on the server so that the client code is as simple as possible. Similarly, the model may have been trained from compressed, binary records, whereas during production, we might want to be able to handle self-descriptive input formats like JSON.

Solution

The solution consists of the following steps:

1. Export the model into a format that captures the mathematical core of the model and is programming language agnostic.
2. In the production system, the formula consisting of the "forward" calculations of the model is restored as a stateless function.
3. The stateless function is deployed into a framework that provides a REST endpoint.

Model export

The first step of the solution is to export the model into a format (TensorFlow uses SavedModel (*https://oreil.ly/9TjS3*), but ONNX (*https://onnx.ai*) is another choice) that captures the mathematical core of the model. The entire model state (learning rate, dropout, short-circuit, etc.) doesn't need to be saved—just the mathematical formula required to compute the output from the inputs. Typically, the trained weight values are constants in the mathematical formula.

In Keras, this is accomplished by:

```
model.save('export/mymodel')
```

The SavedModel format relies on protocol buffers (*https://oreil.ly/g3Vjc*) for a platform-neutral, efficient restoration mechanism. In other words, the `model.save()` method writes the model as a protocol buffer (with the extension `.pb`) and externalizes the trained weights, vocabularies, and so on into other files in a standard directory structure:

```
export/.../variables/variables.data-00000-of-00001
export/.../assets/tokens.txt
export/.../saved_model.pb
```

Inference in Python

In a production system, the model's formula is restored from the protocol buffer and other associated files as a stateless function that conforms to a specific model signature with input and output variable names and data types.

We can use the TensorFlow `saved_model_cli` tool to examine the exported files to determine the signature of the stateless function that we can use in serving:

```
saved_model_cli show --dir ${export_path} \
    --tag_set serve --signature_def serving_default
```

This outputs:

```
The given SavedModel SignatureDef contains the following input(s):
  inputs['full_text_input'] tensor_info:
      dtype: DT_STRING
      shape: (-1)
      name: serving_default_full_text_input:0
The given SavedModel SignatureDef contains the following output(s):
  outputs['positive_review_logits'] tensor_info:
      dtype: DT_FLOAT
      shape: (-1, 1)
      name: StatefulPartitionedCall_2:0
Method name is: tensorflow/serving/predict
```

The signature specifies that the prediction method takes a one-element array as input (called `full_text_input`) that is a string, and outputs one floating point number whose name is `positive_review_logits`. These names come from the names that we assigned to the Keras layers:

```
hub_layer = hub.KerasLayer(..., name='full_text')
...
model.add(tf.keras.layers.Dense(1, name='positive_review_logits'))
```

Here is how we can obtain the serving function and use it for inference:

```
serving_fn = tf.keras.models.load_model(export_path). \
                 signatures['serving_default']
outputs = serving_fn(full_text_input=
                 tf.constant([review1, review2, review3]))
logit = outputs['positive_review_logits']
```

Note how we are using the input and output names from the serving function in the code.

Create web endpoint

The code above can be put into a web application or serverless framework such as Google App Engine, Heroku, AWS Lambda, Azure Functions, Google Cloud Functions, Cloud Run, and so on. What all these frameworks have in common is that they allow the developer to specify a function that needs to be executed. The frameworks take care of autoscaling the infrastructure so as to handle large numbers of prediction requests per second at low latency.

For example, we can invoke the serving function from within Cloud Functions as follows:

```
serving_fn = None
def handler(request):
    global serving_fn
    if serving_fn is None:
        serving_fn = (tf.keras.models.load_model(export_path)
                          .signatures['serving_default'])
    request_json = request.get_json(silent=True)
```

```
if request_json and 'review' in request_json:
    review = request_json['review']
    outputs = serving_fn(full_text_input=tf.constant([review]))
    return outputs['positive_review_logits']
```

Note that we should be careful to define the serving function as a global variable (or a singleton class) so that it isn't reloaded in response to every request. In practice, the serving function will be reloaded from the export path (on Google Cloud Storage) only in the case of cold starts.

Why It Works

The approach of exporting a model to a stateless function and deploying the stateless function in a web application framework works because web application frameworks offer autoscaling, can be fully managed, and are language neutral. They are also familiar to software and business development teams who may not have experience with machine learning. This also has benefits for agile development—an ML engineer or data scientist can independently change the model, and all the application developer needs to do is change the endpoint they are accessing.

Autoscaling

Scaling web endpoints to millions of requests per second is a well-understood engineering problem. Rather than building services unique to machine learning, we can rely on the decades of engineering work that has gone into building resilient web applications and web servers. Cloud providers know how to autoscale web endpoints efficiently, with minimal warmup times.

We don't even need to write the serving system ourselves. Most modern enterprise machine learning frameworks come with a serving subsystem. For example, TensorFlow provides TensorFlow Serving and PyTorch provides TorchServe. If we use these serving subsystems, we can simply provide the exported file and the software takes care of creating a web endpoint.

Fully managed

Cloud platforms abstract away the managing and installation of components like TensorFlow Serving as well. Thus, on Google Cloud, deploying the serving function as a REST API is as simple as running this command-line program providing the location of the SavedModel output:

```
gcloud ai-platform versions create ${MODEL_VERSION} \
        --model ${MODEL_NAME} --origin ${MODEL_LOCATION} \
        --runtime-version $TFVERSION
```

In Amazon's SageMaker, deployment of a TensorFlow SavedModel is similarly simple, and achieved using:

```
model = Model(model_data=MODEL_LOCATION, role='SomeRole')
predictor = model.deploy(initial_instance_count=1,
                         instance_type='ml.c5.xlarge')
```

With a REST endpoint in place, we can send a prediction request as a JSON with the form:

```
{"instances":
  [
      {"reviews": "The film is based on a prize-winning novel."},
      {"reviews": "The film is fast moving and has several great action scenes."},
      {"reviews": "The film was very boring. I walked out half-way."}
  ]
}
```

We get back the predicted values also wrapped in a JSON structure:

```
{"predictions": [{ "positive_review_logits": [0.6965846419334412]},
                 {"positive_review_logits": [1.6177300214767456]},
                 {"positive_review_logits": [-0.754359781742096]}]}
```

 By allowing clients to send JSON requests with multiple instances in the request, called *batching*, we are allowing clients to trade off the higher throughput associated with fewer network calls against the increased parallelization if they send more requests with fewer instances per request.

Besides batching, there are other knobs and levers to improve performance or lower cost. Using a machine with more powerful GPUs, for example, typically helps to improve the performance of deep learning models. Choosing a machine with multiple accelerators and/or threads helps improve the number of requests per second. Using an autoscaling cluster of machines can help lower cost on spiky workloads. These kinds of tweaks are often done by the ML/DevOps team; some are ML-specific, some are not.

Language-neutral

Every modern programming language can speak REST, and a discovery service is provided to autogenerate the necessary HTTP stubs. Thus, Python clients can invoke the REST API as follows. Note that there is nothing framework specific in the code below. Because the cloud service abstracts the specifics of our ML model, we don't need to provide any references to Keras or TensorFlow:

```
credentials = GoogleCredentials.get_application_default()
api = discovery.build("ml", "v1", credentials = credentials,
          discoveryServiceUrl = "https://storage.googleapis.com/cloud-
ml/discovery/ml_v1_discovery.json")

request_data = {"instances":
  [
```

```
    {"reviews": "The film is based on a prize-winning novel."},
    {"reviews": "The film is fast moving and has several great action scenes."},
    {"reviews": "The film was very boring. I walked out half-way."}
  ]
}

parent = "projects/{}/models/imdb".format("PROJECT", "v1")
response = api.projects().predict(body = request_data,
                                  name = parent).execute()
```

The equivalent of the above code can be written in many languages (we show Python because we assume you are somewhat familiar with it). At the time that this book is being written, developers can access the Discovery API (*https://oreil.ly/zCZir*) from Java, PHP, .NET, JavaScript, Objective-C, Dart, Ruby, Node.js, and Go.

Powerful ecosystem

Because web application frameworks are so widely used, there is a lot of tooling available to measure, monitor, and manage web applications. If we deploy the ML model to a web application framework, the model can be monitored and throttled using tools that software reliability engineers (SREs), IT administrators, and DevOps personnel are familiar with. They do not have to know anything about machine learning.

Similarly, your business development colleagues know how to meter and monetize web applications using API gateways. They can carry over that knowledge and apply it to metering and monetizing machine learning models.

Trade-Offs and Alternatives

As the joke by David Wheeler (*https://oreil.ly/uskud*) goes, the solution to any problem in computer science is to add an extra level of indirection. Introduction of an exported stateless function specification provides that extra level of indirection. The Stateless Serving Function design pattern allows us to change the serving signature to provide extra functionality, like additional pre- and postprocessing, beyond what the ML model does. In fact, it is possible to use this design pattern to provide multiple endpoints for a model. This design pattern can also help with creating low-latency, online prediction for models that are trained on systems, such as data warehouses, that are typically associated with long-running queries.

Custom serving function

The output layer of our text classification model is a Dense layer whose output is in the range $(-\infty, \infty)$:

```
model.add(tf.keras.layers.Dense(1, name='positive_review_logits'))
```

Our loss function takes this into account:

```
model.compile(optimizer='adam',
              loss=tf.keras.losses.BinaryCrossentropy(
                  from_logits=True),
              metrics=['accuracy'])
```

When we use the model for prediction, the model naturally returns what it was trained to predict and outputs the logits. What clients expect, however, is the probability that the review is positive. To solve this, we need to return the sigmoid output of the model.

We can do this by writing a custom serving function and exporting it instead. Here is a custom serving function in Keras that adds a probability and returns a dictionary that contains both the logits and the probabilities for each of the reviews provided as input:

```
@tf.function(input_signature=[tf.TensorSpec([None],
                                            dtype=tf.string)])
def add_prob(reviews):
    logits = model(reviews, training=False) # call model
    probs = tf.sigmoid(logits)
    return {
        'positive_review_logits' : logits,
        'positive_review_probability' : probs
    }
```

We can then export the above function as the serving default:

```
model.save(export_path,
           signatures={'serving_default': add_prob})
```

The add_prob method definition is saved in the export_path and will be invoked in response to a client request.

The serving signature of the exported model reflects the new input name (note the name of the input parameter to add_prob) and the output dictionary keys and data types:

```
The given SavedModel SignatureDef contains the following input(s):
  inputs['reviews'] tensor_info:
      dtype: DT_STRING
      shape: (-1)
      name: serving_default_reviews:0
The given SavedModel SignatureDef contains the following output(s):
  outputs['positive_review_logits'] tensor_info:
      dtype: DT_FLOAT
      shape: (-1, 1)
      name: StatefulPartitionedCall_2:0
  outputs['positive_review_probability'] tensor_info:
      dtype: DT_FLOAT
      shape: (-1, 1)
      name: StatefulPartitionedCall_2:1
Method name is: tensorflow/serving/predict
```

When this model is deployed and used for inference, the output JSON contains both the logits and the probability:

```
{'predictions': [
    {'positive_review_probability': [0.6674301028251648],
     'positive_review_logits': [0.6965846419334412]},
    {'positive_review_probability': [0.8344818353652954],
     'positive_review_logits': [1.6177300214767456]},
    {'positive_review_probability': [0.31987208127975464],
     'positive_review_logits': [-0.754359781742096]}
]}
```

Note that `add_prob` is a function that we write. In this case, we did a bit of postprocessing of the output. However, we could have done pretty much any (stateless) thing that we wanted inside that function.

Multiple signatures

It is quite common for models to support multiple objectives or clients who have different needs. While outputting a dictionary can allow different clients to pull out whatever they want, this may not be ideal in some cases. For example, the function we had to invoke to get a probability from the logits was simply `tf.sigmoid()`. This is pretty inexpensive, and there is no problem with computing it even for clients who will discard it. On the other hand, if the function had been expensive, computing it for clients who don't need the value can add considerable overhead.

If a small number of clients require a very expensive operation, it is helpful to provide multiple serving signatures and have the client inform the serving framework which signature to invoke. This is done by specifying a name other than `serving_default` when the model is exported. For example, we might write out two signatures using:

```
model.save(export_path, signatures={
        'serving_default': func1,
        'expensive_result': func2,
    })
```

Then, the input JSON request includes the signature name to choose which serving endpoint of the model is desired:

```
{
  "signature_name": "expensive_result",
    {"instances": …}
}
```

Online prediction

Because the exported serving function is ultimately just a file format, it can be used to provide online prediction capabilities when the original machine learning training framework does not natively support online predictions.

For example, we can train a model to infer whether or not a baby will require attention by training a logistic regression model on the natality dataset:

```
CREATE OR REPLACE MODEL
 mlpatterns.neutral_3classes OPTIONS(model_type='logistic_reg',
  input_label_cols=['health']) AS
SELECT
IF
 (apgar_1min = 10,
  'Healthy',
 IF
  (apgar_1min >= 8,
   'Neutral',
   'NeedsAttention')) AS health,
 plurality,
 mother_age,
 gestation_weeks,
 ever_born
FROM
 `bigquery-public-data.samples.natality`
WHERE
 apgar_1min <= 10
```

Once the model is trained, we can carry out prediction using SQL:

```
SELECT * FROM ML.PREDICT(MODEL mlpatterns.neutral_3classes,
     (SELECT
       2 AS plurality,
       32 AS mother_age,
       41 AS gestation_weeks,
       1 AS ever_born
       )
  )
```

However, BigQuery is primarily for distributed data processing. While it was great for training the ML model on gigabytes of data, using such a system to carry out inference on a single row is not the best fit—latencies can be as high as a second or two. Rather, the ML.PREDICT functionality is more appropriate for batch serving.

In order to carry out online prediction, we can ask BigQuery to export the model as a TensorFlow SavedModel:

```
bq extract -m --destination_format=ML_TF_SAVED_MODEL \
    mlpatterns.neutral_3classes  gs://${BUCKET}/export/baby_health
```

Now, we can deploy the SavedModel into a serving framework like Cloud AI Platform that supports SavedModel to get the benefits of low-latency, autoscaled ML model serving. See the notebook in GitHub (*https://github.com/GoogleCloudPlatform/ml-design-patterns/blob/master/05_resilience/serving_function.ipynb*) for the complete code.

Even if this ability to export the model as a SavedModel did not exist, we could have extracted the weights, written a mathematical model to carry out the linear model, containerized it, and deployed the container image into a serving platform.

Prediction library

Instead of deploying the serving function as a microservice that can be invoked via a REST API, it is possible to implement the prediction code as a library function. The library function would load the exported model the first time it is called, invoke `model.predict()` with the provided input, and return the result. Application developers who need to predict with the library can then include the library with their applications.

A library function is a better alternative than a microservice if the model cannot be called over a network either because of physical reasons (there is no network connectivity) or because of performance constraints. The library function approach also places the computational burden on the client, and this might be preferable from a budgetary standpoint. Using the library approach with `TensorFlow.js` can avoid cross-site problems when there is a desire to have the model running in a browser.

The main drawback of the library approach is that maintenance and updates of the model are difficult—all the client code that uses the model will have to be updated to use the new version of the library. The more commonly a model is updated, the more attractive a microservices approach becomes. A secondary drawback is that the library approach is restricted to programming languages for which libraries are written, whereas the REST API approach opens up the model to applications written in pretty much any modern programming language.

The library developer should take care to employ a threadpool and use parallelization to support the necessary throughput. However, there is usually a limit to the scalability achievable with this approach.

Design Pattern 17: Batch Serving

The Batch Serving design pattern uses software infrastructure commonly used for distributed data processing to carry out inference on a large number of instances all at once.

Problem

Commonly, predictions are carried one at a time and on demand. Whether or not a credit card transaction is fraudulent is determined at the time a payment is being processed. Whether or not a baby requires intensive care is determined when the baby is examined immediately after birth. Therefore, when you deploy a model into an ML serving framework, it is set up to process one instance, or at most a few thousands of instances, embedded in a single request.

The serving framework is architected to process an individual request synchronously and as quickly as possible, as discussed in "Design Pattern 16: Stateless Serving Function" on page 201. The serving infrastructure is usually designed as a microservice that offloads the heavy computation (such as with deep convolutional neural networks) to high-performance hardware such as tensor processing units (TPUs) or graphics processing units (GPUs) and minimizes the inefficiency associated with multiple software layers.

However, there are circumstances where predictions need to be carried out asynchronously over large volumes of data. For example, determining whether to reorder a stock-keeping unit (SKU) might be an operation that is carried out hourly, not every time the SKU is bought at the cash register. Music services might create personalized daily playlists for every one of their users and push them out to those users. The personalized playlist is not created on-demand in response to every interaction that the user makes with the music software. Because of this, the ML model needs to make predictions for millions of instances at a time, not one instance at a time.

Attempting to take a software endpoint that is designed to handle one request at a time and sending it millions of SKUs or billions of users will overwhelm the ML model.

Solution

The Batch Serving design pattern uses a distributed data processing infrastructure (MapReduce, Apache Spark, BigQuery, Apache Beam, and so on) to carry out ML inference on a large number of instances asynchronously.

In the discussion on the Stateless Serving Function design pattern, we trained a text classification model to output whether a review was positive or negative. Let's say that we want to apply this model to every complaint that has ever been made to the United States Consumer Finance Protection Bureau (CFPB).

We can load the Keras model into BigQuery as follows (complete code is available in a notebook in GitHub (*https://github.com/GoogleCloudPlatform/ml-design-patterns/blob/master/05_resilience/batch_serving.ipynb*)):

```
CREATE OR REPLACE MODEL mlpatterns.imdb_sentiment
OPTIONS(model_type='tensorflow', model_path='gs://.../*')
```

Where normally, one would train a model using data in BigQuery, here we are simply loading an externally trained model. Having done that, though, it is possible to use BigQuery to carry out ML predictions. For example, the SQL query.

```
SELECT * FROM ML.PREDICT(MODEL mlpatterns.imdb_sentiment,
  (SELECT 'This was very well done.' AS reviews)
)
```

returns a `positive_review_probability` of 0.82.

Using a distributed data processing system like BigQuery to carry out one-off predictions is not very efficient. However, what if we want to apply the machine learning model to every complaint in the CFPB database?[1] We can simply adapt the query above, making sure to alias the `consumer_complaint_narrative` column in the inner SELECT as the `reviews` to be assessed:

```
SELECT * FROM ML.PREDICT(MODEL mlpatterns.imdb_sentiment,
  (SELECT consumer_complaint_narrative AS reviews
   FROM `bigquery-public-data`.cfpb_complaints.complaint_database
   WHERE consumer_complaint_narrative IS NOT NULL
   )
)
```

The database has more than 1.5 million complaints, but they get processed in about 30 seconds, proving the benefits of using a distributed data processing framework.

Why It Works

The Stateless Serving Function design pattern is set up for low-latency serving to support thousands of simultaneous queries. Using such a framework for occasional or periodic processing of millions of items can get quite expensive. If these requests are not latency-sensitive, it is more cost effective to use a distributed data processing architecture to invoke machine learning models on millions of items. The reason is that invoking an ML model on millions of items is an embarrassingly parallel problem—it is possible to take the million items, break them down into 1,000 groups of 1,000 items each, send each group of items to a machine, then combine the results. The result of the machine learning model on item number 2,000 is completely independent of the result of the machine learning model on item number 3,000, and so it is possible to divide up the work and conquer it.

Take, for example, the query to find the five most positive complaints:

```
WITH all_complaints AS (
SELECT * FROM ML.PREDICT(MODEL mlpatterns.imdb_sentiment,
  (SELECT consumer_complaint_narrative AS reviews
```

---

1 Curious what a "positive" complaint looks like? Here you go:
"I get phone calls morning XXXX and night. I have told them to stop so many calls but they still call even on Sunday in the morning. I had two calls in a row on a Sunday morning from XXXX XXXX. I received nine calls on Saturday. I receive about nine during the week day every day as well.
The only hint that the complainer is unhappy is that they have asked the callers to stop. Otherwise, the rest of the statements might well be about someone bragging about how popular they are!"

```
    FROM `bigquery-public-data`.cfpb_complaints.complaint_database
    WHERE consumer_complaint_narrative IS NOT NULL
    )
  )
  )
  SELECT * FROM all_complaints
  ORDER BY positive_review_probability DESC LIMIT 5
```

Looking at the execution details in the BigQuery web console, we see that the entire query took 35 seconds (see the box marked #1 in Figure 5-1).

Figure 5-1. The first two steps of a query to find the five most "positive" complaints in the Consumer Financial Protection Bureau dataset of consumer complaints.

The first step (see box #2 in Figure 5-1) reads the `consumer_complaint_narrative` column from the BigQuery public dataset where the complaint narrative is not NULL. From the number of rows highlighted in box #3, we learn that this involves reading 1,582,045 values. The output of this step is written into 10 shards (see box #4 of Figure 5-1).

The second step reads the data from this shard (note the `$12:shard` in the query), but also obtains the `file_path` and `file_contents` of the machine learning model `imdb_sentiment` and applies the model to the data in each shard. The way Map-Reduce works is that each shard is processed by a worker, so the fact that there are 10 shards indicates that the second step is being done by 10 workers. The original 1.5 million rows would have been stored over many files, and so the first step was likely to have been processed by as many workers as the number of files that comprised that dataset.

The remaining steps are shown in Figure 5-2.

```
SORT                    $20 DESC
                        LIMIT 5

COMPUTE                 $20 := STRUCT_FIELD_OP(1, $30)
                        $21 := STRUCT_FIELD_OP(0, $30)

BUFFERING_COMPUTE       $30 := TENSORFLOW_PREDICT_SIGNATURE_ID(MAKE_STRUCT(STRUCT_FIELD_OP(0, $61), STRUCT_FIELD_OP(1, $61),
                        STRUCT_FIELD_OP(2, $61), STRUCT_FIELD_OP(-1, $61)), $60, NULL, ...)

JOIN                    CROSS EACH WITH EACH

COMPUTE                 $40 := TENSORFLOW_LOAD_TYPE_MODEL_ID(MAKE_STRUCT($52, $51, $50), NULL, '', '', 'mlpatterns.imdb_sentiment',
                        ARRAY<...>)

AGGREGATE               $50 := ARRAY_AGG($12)
                        $51 := ARRAY_AGG($11)
                        $52 := ARRAY_AGG($10)

WRITE                   $80, $81, $82
                        TO __stage00_output
```

Figure 5-2. Third and subsequent steps of the query to find the five most "positive" complaints.

The third step sorts the dataset in descending order and takes five. This is done on each worker, so each of the 10 workers finds the 5 most positive complaints in "their" shard. The remaining steps retrieve and format the remaining bits of data and write them to the output.

The final step (not shown) takes the 50 complaints, sorts them, and selects the 5 that form the actual result. The ability to separate work in this way across many workers is what enables BigQuery to carry out the entire operation on 1.5 million complaint documents in 35 seconds.

Trade-Offs and Alternatives

The Batch Serving design pattern depends on the ability to split a task across multiple workers. So, it is not restricted to data warehouses or even to SQL. Any MapReduce framework will work. However, SQL data warehouses tend to be the easiest and are often the default choice, especially when the data is structured in nature.

Even though batch serving is used when latency is not a concern, it is possible to incorporate precomputed results and periodic refreshing to use this in scenarios where the space of possible prediction inputs is limited.

Batch and stream pipelines

Frameworks like Apache Spark or Apache Beam are useful when the input needs pre-processing before it can be supplied to the model, if the machine learning model outputs require postprocessing, or if either the preprocessing or postprocessing are hard to express in SQL. If the inputs to the model are images, audio, or video, then SQL is not an option and it is necessary to use a data processing framework that can handle unstructured data. These frameworks can also take advantage of accelerated hardware like TPUs and GPUs to carry out preprocessing of the images.

Another reason to use a framework like Apache Beam is if the client code needs to maintain state. A common reason that the client needs to maintain state is if one of the inputs to the ML model is a time-windowed average. In that case, the client code has to carry out moving averages of the incoming stream of data and supply the moving average to the ML model.

Imagine that we are building a comment moderation system and we wish to reject people who comment more than two times a day about a specific person. For example, the first two times that a commenter writes something about President Obama, we will let it go but block all attempts by that commenter to mention President Obama for the rest of the day. This is an example of postprocessing that needs to maintain state because we need a counter of the number of times that each commenter has mentioned a particular celebrity. Moreover, this counter needs to be over a rotating time period of 24 hours.

We can do this using a distributed data processing framework that can maintain state. Enter Apache Beam. Invoking an ML model to identify mentions of a celebrity and tying them to a canonical knowledge graph (so that a mention of Obama and a mention of President Obama both tie to *en.wikipedia.org/wiki/Barack_Obama*) from Apache Beam can be accomplished using the following (see this notebook in GitHub (*https://github.com/GoogleCloudPlatform/ml-design-patterns/blob/master/05_resil ience/nlp_api.ipynb*) for complete code):

```
| beam.Map(lambda x : nlp.Document(x, type='PLAIN_TEXT'))
| nlp.AnnotateText(features)
| beam.Map(parse_nlp_result)
```

where `parse_nlp_result` parses the JSON request that goes through the `Annotate Text` transform which, beneath the covers, invokes an NLP API.

Cached results of batch serving

We discussed batch serving as a way to invoke a model over millions of items when the model is normally served online using the Stateless Serving Function design pattern. Of course, it is possible for batch serving to work even if the model does not support online serving. What matters is that the machine learning framework doing inference is capable of taking advantage of embarrassingly parallel processing.

Recommendation engines, for example, need to fill out a sparse matrix consisting of every user–item pair. A typical business might have 10 million all-time users and 10,000 items in the product catalog. In order to make a recommendation for a user, recommendation scores have to be computed for each of the 10,000 items, ranked, and the top 5 presented to the user. This is not feasible to do in near real time off a serving function. Yet, the near real-time requirement means that simply using batch serving will not work either.

In such cases, use batch serving to precompute recommendations for all 10 million users:

```
SELECT
  *
FROM
  ML.RECOMMEND(MODEL mlpatterns.recommendation_model)
```

Store it in a relational database such as MySQL, Datastore, or Cloud Spanner (there are pre-built transfer services and Dataflow templates (*https://github.com/Google CloudPlatform/DataflowTemplates/blob/master/src/main/java/com/google/cloud/tele port/templates/BigQueryToDatastore.java*) that can do this). When any user visits, the recommendations for that user are pulled from the database and served immediately and at very low latency.

In the background, the recommendations are refreshed periodically. For example, we might retrain the recommendation model hourly based on the latest actions on the website. We can then carry out inference for just those users who visited in the last hour:

```
SELECT
  *
FROM
  ML.RECOMMEND(MODEL mlpatterns.recommendation_model,
    (
    SELECT DISTINCT
      visitorId
    FROM
      mlpatterns.analytics_session_data
    WHERE
      visitTime > TIME_DIFF(CURRENT_TIME(), 1 HOUR)
    ))
```

We can then update the corresponding rows in the relational database used for serving.

Lambda architecture

A production ML system that supports both online serving and batch serving is called a *Lambda architecture* (*https://oreil.ly/jLZ46*)—such a production ML system allows ML practitioners to trade-off between latency (via the Stateless Serving Function pattern) and throughput (via the Batch Serving pattern).

AWS Lambda (*https://oreil.ly/RqPan*), in spite of its name, is not a Lambda architecture. It is a serverless framework for scaling state-less functions, similar to Google Cloud Functions or Azure Functions.

Typically, a Lambda architecture is supported by having separate systems for online serving and batch serving. In Google Cloud, for example, the online serving infrastructure is provided by Cloud AI Platform Predictions and the batch serving infrastructure is provided by BigQuery and Cloud Dataflow (Cloud AI Platform Predictions provides a convenient interface so that users don't have to explicitly use Dataflow). It is possible to take a TensorFlow model and import it into BigQuery for batch serving. It is also possible to take a trained BigQuery ML model and export it as a TensorFlow SavedModel for online serving. This two-way compatibility enables users of Google Cloud to hit any point in the spectrum of latency–hroughput trade-off.

Design Pattern 18: Continued Model Evaluation

The Continued Model Evaluation design pattern handles the common problem of needing to detect and take action when a deployed model is no longer fit-for-purpose.

Problem

So, you've trained your model. You collected the raw data, cleaned it up, engineered features, created embedding layers, tuned hyperparameters, the whole shebang. You're able to achieve 96% accuracy on your hold-out test set. Amazing! You've even gone through the painstaking process of deploying your model, taking it from a Jupyter notebook to a machine learning model in production, and are serving predictions via a REST API. Congratulations, you've done it. You're finished!

Well, not quite. Deployment is not the end of a machine learning model's life cycle. How do you know that your model is working as expected in the wild? What if there are unexpected changes in the incoming data? Or the model no longer produces accurate or useful predictions? How will these changes be detected?

The world is dynamic, but developing a machine learning model usually creates a static model from historical data. This means that once the model goes into production, it can start to degrade and its predictions can grow increasingly unreliable. Two of the main reasons models degrade over time are concept drift and data drift.

Concept drift occurs whenever the relationship between the model inputs and target have changed. This often happens because the underlying assumptions of your model have changed, such as models trained to learn adversarial or competitive behavior like fraud detection, spam filters, stock market trading, online ad bidding, or cyberse-curity. In these scenarios, a predictive model aims to identify patterns that are characteristic of desired (or undesired) activity, while the adversary learns to adapt and may modify their behavior as circumstances change. Think for example of a model developed to detect credit card fraud. The way people use credit cards has changed over time and thus the common characteristics of credit card fraud have also changed. For instance, when "Chip and Pin" technology was introduced, fraudulent transactions began to move more online. As fraudulent behavior adapted, the performance of a model that had been developed before this technology would suddenly begin to suffer and model predictions would be less accurate.

Another reason for a model's performance to degrade over time is data drift. We introduced the problem of data drift in "Common Challenges in Machine Learning" on page 11 in Chapter 1. Data drift refers to any change that has occurred to the data being fed to your model for prediction as compared to the data that was used for training. Data drift can occur for a number of reasons: the input data schema changes at the source (for example, fields are added or deleted upstream), feature distributions change over time (for example, a hospital might start to see more younger adults because a ski resort opened nearby), or the meaning of the data changes even if the structure/schema hasn't (for example, whether a patient is considered "over-weight" may change over time). Software updates could introduce new bugs or the business scenario changes and creates a new product label previously not available in the training data. ETL pipelines for building, training, and predicting with ML models can be brittle and opaque, and any of these changes would have drastic effects on the performance of your model.

Model deployment is a continuous process, and to solve for concept drift or data drift, it is necessary to update your training dataset and retrain your model with fresh data to improve predictions. But how do you know when retraining is necessary? And how often should you retrain? Data preprocessing and model training can be costly both in time and money and each step of the model development cycle adds additional overhead of development, monitoring, and maintenance.

Solution

The most direct way to identify model deterioration is to continuously monitor your model's predictive performance over time, and assess that performance with the same evaluation metrics you used during development. This kind of continuous model evaluation and monitoring is how we determine whether the model, or any changes we've made to the model, are working as they should.

Concept

Continuous evaluation of this kind requires access to the raw prediction request data and the predictions the model generated as well as the ground truth, all in the same place. Google Cloud AI Platform provides the ability to configure the deployed model version so that the online prediction input and output are regularly sampled and saved to a table in BigQuery. In order to keep the service performant to a large number of requests per second, we can customize how much data is sampled by specifying a percentage of the number of input requests. In order to measure performance metrics, it is necessary to combine this saved sample of predictions against the ground truth.

In most situations, it may take time before the ground truth labels become available. For example, for a churn model, it may not be known until the next subscription cycle which customers have discontinued their service. Or, for a financial forecasting model, the true revenue isn't known until after that quarter's close and earnings report. In either of these cases, evaluation cannot take place until ground truth data is available.

To see how continuous evaluation works, we'll deploy a text classification model trained on the HackerNews dataset to Google Cloud AI Platform. The full code for this example can be found in the continuous evaluation notebook (*https:// github.com/GoogleCloudPlatform/ml-design-patterns/blob/master/05_resilience/contin uous_eval.ipynb*) in the repository accompanying this book.

Deploying the model

The input for our training dataset is an article title and its associated label is the news source where the article originated, either `nytimes`, `techcrunch`, or `github`. As news trends evolve over time, the words associated with a *New York Times* headline will change. Similarly, releases of new technology products will affect the words to be found in TechCrunch. Continuous evaluation allows us to monitor model predictions to track how those trends affect our model performance and kick off retraining if necessary.

Suppose that the model is exported with a custom serving input function as described in "Design Pattern 16: Stateless Serving Function" on page 201:

```
@tf.function(input_signature=[tf.TensorSpec([None], dtype=tf.string)])
def source_name(text):
    labels = tf.constant(['github', 'nytimes', 'techcrunch'],dtype=tf.string)
    probs = txtcls_model(text, training=False)
    indices = tf.argmax(probs, axis=1)
    pred_source = tf.gather(params=labels, indices=indices)
    pred_confidence = tf.reduce_max(probs, axis=1)
    return {'source': pred_source,
            'confidence': pred_confidence}
```

After deploying this model, when we make an online prediction, the model will return the predicted news source as a string value and a numeric score of that prediction label related to how confident the model is. For example, we can create an online prediction by writing an input JSON example to a file called *input.json* to send for prediction:

```
%%writefile input.json
{"text":
"YouTube introduces Video Chapters to make it easier to navigate longer videos"}
```

This returns the following prediction output:

```
CONFIDENCE  SOURCE
0.918685    techcrunch
```

Saving predictions

Once the model is deployed, we can set up a job to save a sample of the prediction requests—the reason to save a sample, rather than all requests, is to avoid unnecessarily slowing down the serving system. We can do this in the Continuous Evaluation section of the Google Cloud AI Platform (CAIP) console by specifying the LabelKey (the column that is the output of the model, which in our case will be source since we are predicting the source of the article), a ScoreKey in the prediction outputs (a numeric value, which in our case is confidence), and a table in BigQuery where a portion of the online prediction requests are stored. In our example code, the table is called txtcls_eval.swivel. Once this has been configured, whenever online predictions are made, CAIP streams the model name, the model version, the timestamp of the prediction request, the raw prediction input, and the model's output to the specified BigQuery table, as shown in Table 5-1.

Table 5-1. A proportion of the online prediction requests and the raw prediction output is saved to a table in BigQuery

Row	model	model_version	time	raw_data	raw_prediction	groundtruth
1	txtcls	swivel	2020-06-10 01:40:32 UTC	{"instances": [{"text": "Astronauts Dock With Space Station After Historic SpaceX Launch"}]}	{"predictions": [{"source": "github", "confidence": 0.9994275569915771}]}	*null*
2	txtcls	swivel	2020-06-10 01:37:46 UTC	{"instances": [{"text": "Senate Confirms First Black Air Force Chief"}]}	{"predictions": [{"source": "nytimes", "confidence": 0.9989787340164185}]}	*null*
3	txtcls	swivel	2020-06-09 21:21:47 UTC	{"instances": [{"text": "A native Mac app wrapper for WhatsApp Web"}]}	{"predictions": [{"source": "github", "confidence": 0.745254397392273}]}	*null*

Capturing ground truth

It is also necessary to capture the ground truth for each of the instances sent to the model for prediction. This can be done in a number of ways depending on the use case and data availability. One approach would be to use a human labeling service— all instances sent to the model for prediction, or maybe just the ones for which the model has marginal confidence, are sent out for human annotation. Most cloud providers offer some form of a human labeling service to enable labeling instances at scale in this way.

Ground truth labels can also be derived from how users interact with the model and its predictions. By having users take a specific action, it is possible to obtain implicit feedback for a model's prediction or to produce a ground truth label. For example, when a user chooses one of the proposed alternate routes in Google Maps, the chosen route serves as an implicit ground truth. More explicitly, when a user rates a recommended movie, this is a clear indication of the ground truth for a model that is built to predict user ratings in order to surface recommendations. Similarly, if the model allows the user to change the prediction, for example, as in medical settings when a doctor is able to change a model's suggested diagnosis, this provides a clear signal for the ground truth.

 It is important to keep in mind how the feedback loop of model predictions and capturing ground truth might affect training data down the road. For example, suppose you've built a model to predict when a shopping cart will be abandoned. You can even check the status of the cart at routine intervals to create ground truth labels for model evaluation. However, if your model suggests a user will abandon their shopping cart and you offer them free shipping or some discount to influence their behavior, then you'll never know if the original model prediction was correct. In short, you've violated the assumptions of the model evaluation design and will need to determine ground truth labels some other way. This task of estimating a particular outcome under a different scenario is referred to as counterfactual reasoning and often arises in use cases like fraud detection, medicine, and advertising where a model's predictions likely lead to some intervention that can obscure learning the actual ground truth for that example.

Evaluating model performance

Initially, the `groundtruth` column of the `txtcls_eval.swivel` table in BigQuery is left empty. We can provide the ground truth labels once they are available by updating the value directly with a SQL command. Of course, we should make sure the ground truth is available before we run an evaluation job. Note that the ground truth adheres to the same JSON structure as the prediction output from the model:

```
UPDATE
  txtcls_eval.swivel
SET
  groundtruth = '{"predictions": [{"source": "techcrunch"}]}'
WHERE
  raw_data = '{"instances":
[{"text": "YouTube introduces Video Chapters to help navigate longer
videos"}]}'
```

To update more rows, we'd use a MERGE statement instead of an UPDATE. Once the ground truth has been added to the table, it's possible to easily examine the text input and your model's prediction and compare with the ground truth as in Table 5-2:

```
SELECT
  model,
  model_version,
  time,
  REGEXP_EXTRACT(raw_data, r'.*"text": "(.*)"') AS text,
  REGEXP_EXTRACT(raw_prediction, r'.*"source": "(.*?)"') AS prediction,
  REGEXP_EXTRACT(raw_prediction, r'.*"confidence": (0.\d{2}).*') AS confidence,
  REGEXP_EXTRACT(groundtruth, r'.*"source": "(.*?)"') AS groundtruth,
FROM
  txtcls_eval.swivel
```

Table 5-2. Once ground truth is available, it can be added to the original BigQuery table and the performance of the model can be evaluated

Row	model	model_version	time	text	prediction	confidence	groundtruth
1	txtcls	swivel	2020-06-10 01:38:13 UTC	A native Mac app wrapper for WhatsApp Web	github	0.77	github
2	txtcls	swivel	2020-06-10 01:37:46 UTC	Senate Confirms First Black Air Force Chief	nytimes	0.99	nytimes
3	txtcls	swivel	2020-06-10 01:40:32 UTC	Astronauts Dock With Space Station After Historic SpaceX Launch	github	0.99	nytimes
4	txtcls	swivel	2020-06-09 21:21:44 UTC	YouTube introduces Video Chapters to make it easier to navigate longer videos	techcrunch	0.77	techcrunch

With this information accessible in BigQuery, we can load the evaluation table into a dataframe, df_evals, and directly compute evaluation metrics for this model version. Since this is a multiclass classification, we can compute the precision, recall, and F1-score for each class. We can also create a confusion matrix, which helps to analyze where model predictions within certain categorical labels may suffer. Figure 5-3 shows the confusion matrix comparing this model's predictions with the ground truth.

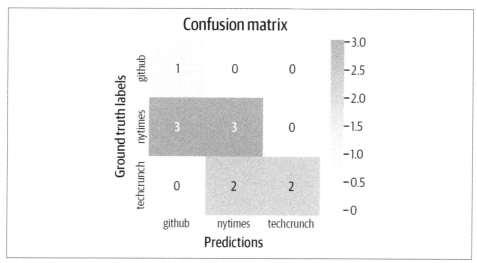

Figure 5-3. A confusion matrix shows all pairs of ground truth labels and predictions so you can explore your model performance within different classes.

Continuous evaluation

We should make sure the output table also captures the model version and the time-stamp of prediction requests so that we can use the same table for continuous evaluation of two different model versions for comparing metrics between the models. For example, if we deploy a newer version of our model, called swivel_v2, that is trained on more recent data or has different hyperparameters, we can compare their performance by slicing the evaluation dataframe according to the model version:

```
df_v1 = df_evals[df_evals.version == "swivel"]
df_v2 = df_evals[df_evals.version == "swivel_v2"]
```

Similarly, we can create evaluation slices in time, focusing only on model predictions within the last month or the last week:

```
today = pd.Timestamp.now(tz='UTC')
one_month_ago = today - pd.DateOffset(months=1)
one_week_ago = today - pd.DateOffset(weeks=1)

df_prev_month = df_evals[df_evals.time >= one_month_ago]
df_prev_week = df_evals[df_evals.time >= one_week_ago]
```

To carry out the above evaluations continuously, the notebook (or a containerized form) can be scheduled. We can set it up to trigger a model retraining if the evaluation metric falls below some threshold.

Why It Works

When developing machine learning models, there is an implicit assumption that the train, validation, and test data come from the same distribution, as shown in Figure 5-4. When we deploy models to production, this assumption implies that future data will be similar to past data. However, once the model is in production "in the wild," this static assumption on the data may no longer be valid. In fact, many production ML systems encounter rapidly changing, nonstationary data, and models become stale over time, which negatively impacts the quality of predictions.

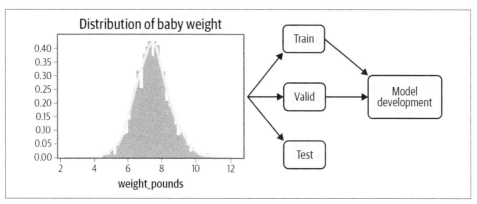

Figure 5-4. When developing a machine learning model, the train, validation, and test data come from the same data distribution. However, once the model is deployed, that distribution can change, severely affecting model performance.

Continuous model evaluation provides a framework to evaluate a deployed model's performance *exclusively* on new data. This allows us to detect model staleness as early as possible. This information helps determine how frequently to retrain a model or when to replace it with a new version entirely.

By capturing prediction inputs and outputs and comparing with ground truth, it's possible to quantifiably track model performance or measure how different model versions perform with A/B testing in the current environment, without regard to how the versions performed in the past.

Trade-Offs and Alternatives

The goal of continuous evaluation is to provide a means to monitor model performance and keep models in production fresh. In this way, continuous evaluation provides a trigger for when to retrain the model. In this case, it is important to consider tolerance thresholds for model performance, the trade-offs they pose, and the role of scheduled retraining. There are also techniques and tools, like TFX, to help detect data and concept drift preemptively by monitoring input data distributions directly.

Triggers for retraining

Model performance will usually degrade over time. Continuous evaluation allows you to measure precisely how much in a structured way and provides a trigger to retrain the model. So, does that mean you should retrain your model as soon as performance starts to dip? It depends. The answer to this question is heavily tied to the business use case and should be discussed alongside evaluation metrics and model assessment. Depending on the complexity of the model and ETL pipelines, the cost of retraining could be expensive. The trade-off to consider is what amount of deterioration of performance is acceptable in relation to this cost.

Serverless Triggers

Cloud Functions, AWS Lambda, and Azure Functions provide serverless ways to automate retraining via triggers. The trigger type determines how and when your function executes. These triggers could be messages published to a message queue, a change notification from a cloud storage bucket indicating a new file has been added, changes to data in a database, or even an HTTPS request. Once the event has fired, the function code is executed.

In the context of retraining, the cloud event trigger would be a significant change or dip in model accuracy. The function, or action taken, would be to invoke the training pipeline to retrain the model and deploy the new version. "Design Pattern 25: Workflow Pipeline" on page 282 describes how this can be accomplished. Workflow pipelines containerize and orchestrate the end-to-end machine learning workflow from data collection and validation to model building, training, and deployment. Once the new model version has been deployed, it can then be compared against the current version to determine if it should be replaced.

The threshold itself could be set as an absolute value; for example, model retraining occurs once model accuracy falls below 95%. Or the threshold could be set as a rate of change of performance, for example, once performance begins to experience a downward trajectory. Whichever approach, the philosophy for choosing the threshold is similar to that for checkpointing models during training. With a higher, more sensitive threshold, models in production remain fresh, but there is a higher cost for frequent retraining as well as technical overhead of maintaining and switching between different model versions. With a lower threshold, training costs decrease but models in production are more stale. Figure 5-5 shows this trade-off between the performance threshold and how it affects the number of model retraining jobs.

If the model retraining pipeline is automatically triggered by such a threshold, it is important to track and validate the triggers as well. Not knowing when your model has been retrained inevitably leads to issues. Even if the process is automated, you

---

should always have control of the retraining of your model to better understand and debug the model in the production.

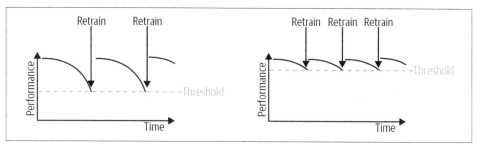

Figure 5-5. Setting a higher threshold for model performance ensures a higher-quality model in production but will require more frequent retraining jobs, which can be costly.

Scheduled retraining

Continuous evaluation provides a crucial signal for knowing when it's necessary to retrain your model. This process of retraining is often carried out by fine-tuning the previous model using any newly collected training data. Where continued evaluation may happen every day, scheduled retraining jobs may occur only every week or every month (Figure 5-6).

Once a new version of the model is trained, its performance is compared against the current model version. The updated model is deployed as a replacement only if it outperforms the previous model with respect to a test set of current data.

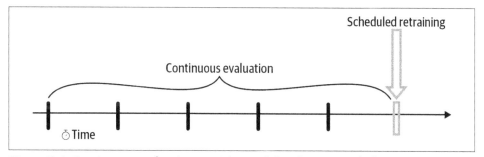

Figure 5-6. Continuous evaluation provides model evaluation each day as new data is collected. Periodic retraining and model comparison provides evaluation at discrete time points.

So how often should you schedule retraining? The timeline for retraining will depend on the business use case, prevalence of new data, and the cost (in time and money) of executing the retraining pipeline. Sometimes, the time horizon of the model naturally determines when to schedule retraining jobs. For example, if the goal of the model is to predict next quarter's earnings, since you will get new ground truth labels only once each quarter, it doesn't make sense to train more frequently than that. However, if the volume and occurrence of new data is high, then it would be beneficial to retrain more frequently. The most extreme version of this is online machine learning (*https://oreil.ly/Mj-DA*). Some machine learning applications, such as ad placement or newsfeed recommendation, require online, real-time decision, and can continuously improve performance by retraining and updating parameter weights with each new training example.

In general, the optimal time frame is something you as a practitioner will determine through experience and experimentation. If you are trying to model a rapidly moving task, such as adversary or competitive behavior, then it makes sense to set a more frequent retraining schedule. If the problem is fairly static, like predicting a baby's birth weight, then less frequent retrainings should suffice.

In either case, it is helpful to have an automated pipeline set up that can execute the full retraining process with a single API call. Tools like Cloud Composer/Apache Airflow and AI Platform Pipelines are useful to create, schedule, and monitor ML workflows from preprocessing raw data and training to hyperparameter tuning and deployment. We discuss this further in the next chapter in "Design Pattern 25: Workflow Pipeline".

Data validation with TFX

Data distributions can change over time, as shown in Figure 5-7. For example, consider the natality birth weight dataset. As medicine and societal standards change over time, the relationship between model features, such as the mother's age or the number of gestation weeks, change with respect to the model label, the weight of the baby. This data drift negatively impacts the model's ability to generalize to new data. In short, your model has gone *stale,* and it needs to be retrained on fresh data.

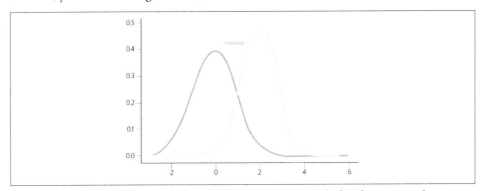

Figure 5-7. Data distributions can change over time. Data drift refers to any change that has occurred to the data being fed to your model for prediction as compared to the data used for training.

While continuous evaluation provides a post hoc way of monitoring a deployed model, it is also valuable to monitor the new data that is received during serving and preemptively identify changes in data distributions.

TFX's Data Validation is a useful tool to accomplish this. TFX (*https://oreil.ly/RP2e9*) is an end-to-end platform for deploying machine learning models open sourced by Google. The Data Validation library can be used to compare the data examples used in training with those collected during serving. Validity checks detect anomalies in the data, training-serving skew, or data drift. TensorFlow Data Validation creates data visualizations using Facets (*https://oreil.ly/NE-SQ*), an open source visualization tool for machine learning. The Facets Overview gives a high-level look at the distributions of values across various features and can uncover several common and uncommon issues like unexpected feature values, missing feature values, and training-serving skew.

Estimating retraining interval

A useful and relatively cheap tactic to understand how data and concept drift affect your model is to train a model using only stale data and assess the performance of that model on more current data (Figure 5-8). This mimics the continued model evaluation process in an offline environment. That is, collect data from six months or a year ago and go through the usual model development workflow, generating features, optimizing hyperparameters, and capturing relevant evaluation metrics. Then, compare those evaluation metrics against the model predictions for more recent data collected from only a month prior. How much worse does your stale model perform on the current data? This gives a good estimate of the rate at which a model's performance falls off over time and how often it might be necessary to retrain.

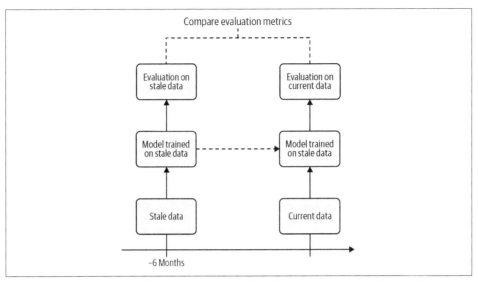

Figure 5-8. Training a model on stale data and evaluating on current data mimics the continued model evaluation process in an offline environment.

Design Pattern 19: Two-Phase Predictions

The Two-Phase Predictions design pattern provides a way to address the problem of keeping large, complex models performant when they have to be deployed on distributed devices by splitting the use cases into two phases, with only the simpler phase being carried out on the edge.

Problem

When deploying machine learning models, we cannot always rely on end users having reliable internet connections. In such situations, models are deployed at the *edge* —meaning they are loaded on a user's device and don't require an internet connection to generate predictions. Given device constraints, models deployed on the edge typically need to be smaller than models deployed in the cloud, and consequently require balancing trade-offs between model complexity and size, update frequency, accuracy, and low latency.

There are various scenarios where we'd want our model deployed on an edge device. One example is a fitness tracking device, where a model makes recommendations for users based on their activity, tracked through accelerometer and gyroscope movement. It's likely that a user could be exercising in a remote outdoor area without connectivity. In these cases, we'd still want our application to work. Another example is an environmental application that uses temperature and other environmental data to make predictions on future trends. In both of these examples, even if we have internet connectivity, it may be slow and expensive to continuously generate predictions from a model deployed in the cloud.

To convert a trained model into a format that works on edge devices, models often go through a process known as *quantization*, where learned model weights are represented with fewer bytes. TensorFlow, for example, uses a format called TensorFlow Lite to convert (*https://oreil.ly/UaMq7*) saved models into a smaller format optimized for serving at the edge. In addition to quantization, models intended for edge devices may also start out smaller to fit into stringent memory and processor constraints.

Quantization and other techniques employed by TF Lite significantly reduce the size and prediction latency of resulting ML models, but with that may come reduced model accuracy. Additionally, since we can't consistently rely on edge devices having connectivity, deploying new model versions to these devices in a timely manner also presents a challenge.

We can see how these trade-offs play out in practice by looking at the options for training edge models in Cloud AutoML Vision (*https://oreil.ly/MWsQH*) in Figure 5-9.

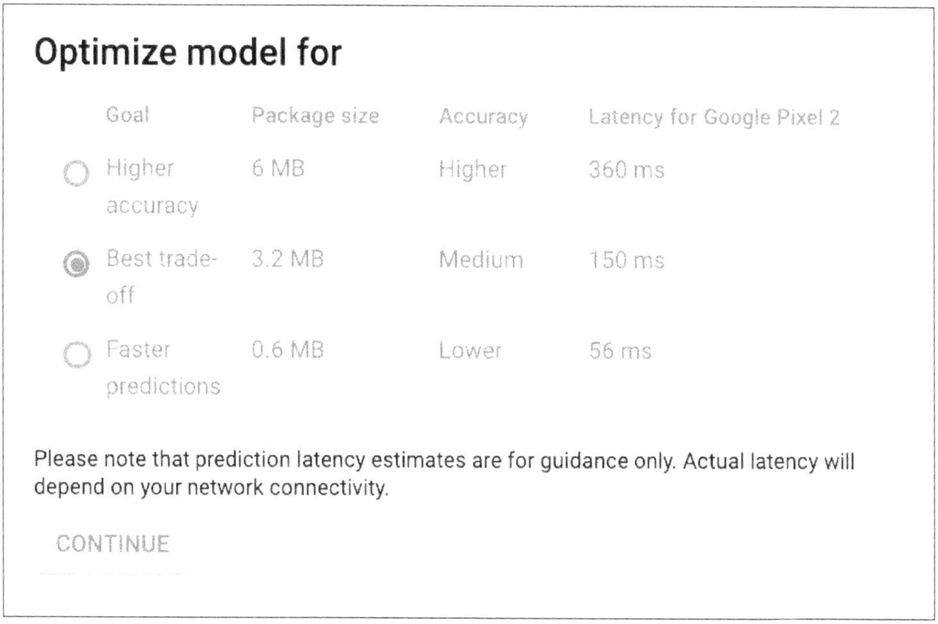

Optimize model for

	Goal	Package size	Accuracy	Latency for Google Pixel 2
○	Higher accuracy	6 MB	Higher	360 ms
◉	Best trade-off	3.2 MB	Medium	150 ms
○	Faster predictions	0.6 MB	Lower	56 ms

Please note that prediction latency estimates are for guidance only. Actual latency will depend on your network connectivity.

CONTINUE

Figure 5-9. Making trade-offs between accuracy, model size, and latency for models deployed at the edge in Cloud AutoML Vision.

To account for these trade-offs, we need a solution that balances the reduced size and latency of edge models against the added sophistication and accuracy of cloud models.

Solution

With the Two-Phase Predictions design pattern, we split our problem into two parts. We start with a smaller, cheaper model that can be deployed on-device. Because this model typically has a simpler task, it can accomplish this task on-device with relatively high accuracy. This is followed by a second, more complex model deployed in the cloud and triggered only when needed. Of course, this design pattern requires you to have a problem that can be split into two parts with varying levels of complexity. One example of such a problem is smart devices like Google Home (*https://oreil.ly/3ROKg*), which are activated by a wake word and can then answer questions and respond to commands related to setting alarms, reading the news, and interacting with integrated devices like lights and thermostats. Google Home, for example, is activated by saying "OK Google" or "Hey Google." Once the device recognizes a wake word, users can ask more complex questions like, "Can you schedule a meeting with Sara at 10 a.m.?"

This problem can be broken into two distinct parts: an initial model that listens for a wake word, and a more complex model that can understand and respond to any other user query. Both models will perform audio recognition. The first model, however, will only need to perform binary classification: does the sound it just heard match the wake word or not? Although this model is simpler in complexity, it needs to be constantly running, which will be expensive if it's deployed to the cloud. The second model will require audio recognition *and* natural language understanding in order to parse the user's query. This model only needs to run when a user asks a question, but places more emphasis on high accuracy. The Two-Phase Predictions pattern can solve this by deploying the wake word model on-device and the more complex model in the cloud.

In addition to this smart device use case, there are many other situations where the Two-Phase Predictions pattern can be employed. Let's say you work on a factory floor where many different machines are running at a given time. When a machine stops working correctly, it typically makes a noise that can be associated with a malfunction. There are different noises corresponding with each distinct machine and the different ways a machine could be broken. Ideally, you can build a model to flag problematic noises and identify what they mean. With Two-Phase Predictions, you could build one offline model to detect anomalous sounds. A second cloud model could then be used to identify whether the usual sound is indicative of some malfunctioning condition.

You could also use the Two-Phase Predictions pattern for an image-based scenario. Let's say you have cameras deployed in the wild to identify and track endangered species. You can have one model on the device that detects whether the latest image captured contains an endangered animal. If it does, this image can then be sent to a cloud model that determines the specific type of animal in the image.

To illustrate the Two-Phase Predictions pattern, let's employ a general-purpose audio recognition dataset from Kaggle (*https://oreil.ly/I89Pr*). The dataset contains around 9,000 audio samples of familiar sounds with a total of 41 label categories, including "cello," "knock," "telephone," "trumpet," and more. The first phase of our solution will be a model that predicts whether or not the given sound is a musical instrument. Then, for sounds that the first model predicts are an instrument, we'll get a prediction from a model deployed in the cloud to predict the specific instrument from a total of 18 possible options. Figure 5-10 shows the two-phased flow for this example.

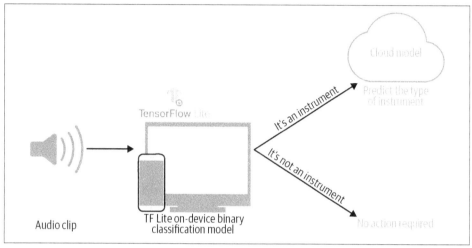

Figure 5-10. Using the Two-Phase Predictions pattern to identify instrument sounds.

To build each of these models, we'll convert the audio data to spectrograms, which are visual representations of sound. This will allow us to use common image model architectures along with the Transfer Learning design pattern to solve this problem. See Figure 5-11 for a spectrogram of a saxophone audio clip from our dataset.

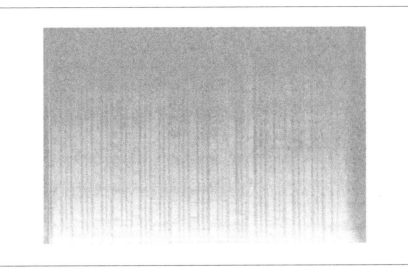

Figure 5-11. The image representation (spectrogram) of a saxophone audio clip from our training dataset. Code for converting .wav files to spectrograms can be found in the GitHub repository (https://github.com/GoogleCloudPlatform/ml-design-patterns/blob/master/05_resilience/audio_to_spectro.ipynb).

Phase 1: Building the offline model

The first model in our Two-Phase Predictions solution should be small enough that it can be loaded on a mobile device for quick inference without relying on internet connectivity. Building on the instrument example introduced above, we'll provide an example of the first prediction phase by building a binary classification model optimized for on-device inference.

The original sound dataset has 41 labels for different types of audio clips. Our first model will only have two labels: "instrument" or "not instrument." We'll build our model using the MobileNetV2 (*https://oreil.ly/zvbzR*) model architecture trained on the ImageNet dataset. MobileNetV2 is available directly in Keras and is an architecture optimized for models that will be served on-device. For our model, we'll freeze the MobileNetV2 weights and load it *without* the top so that we can add our own binary classification output layer:

```
mobilenet = tf.keras.applications.MobileNetV2(
    input_shape=((128,128,3)),
    include_top=False,
    weights='imagenet'
)
mobilenet.trainable = False
```

If we organize our spectrogram images into directories with the corresponding label name, we can use Keras's `ImageDataGenerator` class to create our training and validation datasets:

```
train_data_gen = image_generator.flow_from_directory(
        directory=data_dir,
        batch_size=32,
        shuffle=True,
        target_size=(128,128),
        classes = ['not_instrument','instrument'],
        class_mode='binary')
```

With our training and validation datasets ready, we can train the model as we normally would. The typical approach for exporting trained models for serving is to use TensorFlow's `model.save()` method. However, remember that this model will be served on-device, and as a result we want to keep it as small as possible. To build a model that fits these requirements, we'll use TensorFlow Lite (*https://oreil.ly/dyx93*), a library optimized for building and serving models directly on mobile and embedded devices that may not have reliable internet connectivity. TF Lite has some built-in utilities for quantizing models both during and after training.

To prepare the trained model for edge serving, we use TF Lite to export it in an optimized format:

```
converter = tf.lite.TFLiteConverter.from_keras_model(model)
converter.optimizations = [tf.lite.Optimize.DEFAULT]
tflite_model = converter.convert()
open('converted_model.tflite', 'wb').write(tflite_model)
```

This is the fastest way to quantize a model *after* training. Using the TF Lite optimization defaults, it will reduce our model's weights to their 8-bit representation. It will also quantize inputs at inference time when we make predictions on our model. By running the code above, the resulting exported TF Lite model is one-fourth the size it would have been if we had exported it without quantization.

> To further optimize your model for offline inference, you can also quantize your model's weights *during* training or quantize all of your model's math operations in addition to weights. At the time of writing, quantization-optimized training for TensorFlow 2 models is on the roadmap (*https://oreil.ly/RuONn*).

To generate a prediction on a TF Lite model, you use the TF Lite interpreter, which is optimized for low latency. You'll likely want to load your model on an Android or iOS device and generate predictions directly from your application code. There are APIs for both platforms, but we'll show the Python code for generating predictions here so that you can run it from the same notebook where you created your model.

First, we create an instance of TF Lite's interpreter and get details on the input and output format it's expecting:

```
interpreter = tf.lite.Interpreter(model_path="converted_model.tflite")
interpreter.allocate_tensors()

input_details = interpreter.get_input_details()
output_details = interpreter.get_output_details()
```

For the MobileNetV2 binary classification model we trained above, input_details looks like the following:

```
[{'dtype': numpy.float32,
  'index': 0,
  'name': 'mobilenetv2_1.00_128_input',
  'quantization': (0.0, 0),
  'quantization_parameters': {'quantized_dimension': 0,
  'scales': array([], dtype=float32),
  'zero_points': array([], dtype=int32)},
  'shape': array([  1, 128, 128,   3], dtype=int32),
  'shape_signature': array([  1, 128, 128,   3], dtype=int32),
  'sparsity_parameters': {}}]
```

We'll then pass the first image from our validation batch to the loaded TF Lite model for prediction, invoke the interpreter, and get the output:

```
input_data = np.array([image_batch[21]], dtype=np.float32)
interpreter.set_tensor(input_details[0]['index'], input_data)

interpreter.invoke()
output_data = interpreter.get_tensor(output_details[0]['index'])
print(output_data)
```

The resulting output is a sigmoid array with a single value in the [0,1] range indicating whether or not the given input sound is an instrument.

> Depending on how costly it is to call your cloud model, you can change what metric you're optimizing for when you train the on-device model. For example, you might choose to optimize for precision over recall if you care more about avoiding false positives.

With our model now working on-device, we can get fast predictions without having to rely on internet connectivity. If the model is confident that a given sound is not an instrument, we can stop here. If the model predicts "instrument," it's time to proceed by sending the audio clip to a more complex cloud-hosted model.

What Models Are Suitable on the Edge?

How should you determine whether a model is a good fit for the edge? There are a few considerations related to model size, complexity, and available hardware. As a general rule of thumb, smaller, less complex models are better optimized for running on-device. This is because edge models are constrained by the available device storage. Often, when models are scaled down—through quantization or other techniques—this is done at the expense of accuracy. As such, models with a simpler prediction task and model architecture are the best fit for edge devices. By "simpler," we mean trade-offs like favoring binary classification over multiclass or choosing a less complex model architecture (like a decision tree or linear regression model) when possible.

When you need to deploy models to the edge while still adhering to certain model size and complexity constraints, it's worth looking at edge hardware designed specifically with ML inference in mind. For example, the Coral Edge TPU (*https://oreil.ly/N2NOs*) board provides a custom ASIC optimized for high-performance, offline ML inference on TensorFlow Lite models. Similarly, NVIDIA offers the Jetson Nano (*https://oreil.ly/GUOQc*) for edge-optimized, low-power ML inference. The hardware support for ML inference is rapidly evolving as embedded, on-device ML becomes more common.

Phase 2: Building the cloud model

Since our cloud-hosted model doesn't need to be optimized for inference without a network connection, we can follow a more traditional approach for training, exporting, and deploying this model. Depending on your Two-Phase Prediction use case, this second model could take many different forms. In the Google Home example, phase 2 might include multiple models: one that converts a speaker's audio input to text, and a second one that performs NLP to understand the text and route the user's query. If the user asks for something more complex, there could even be a third model to provide a recommendation based on user preferences or past activity.

In our instrument example, the second phase of our solution will be a multiclass model that classifies sounds into one of 18 possible instrument categories. Since this model doesn't need to be deployed on-device, we can use a larger model architecture like VGG as a starting point and then follow the Transfer Learning design pattern outlined in Chapter 4.

We'll load VGG trained on the ImageNet dataset, specify the size of our spectrogram images in the `input_shape` parameter, and freeze the model's weights before adding our own softmax classification output layer:

```
vgg_model = tf.keras.applications.VGG19(
    include_top=False,
```

```
        weights='imagenet',
        input_shape=((128,128,3))
)

vgg_model.trainable = False
```

Our output will be an 18-element array of softmax probabilities:

```
prediction_layer = tf.keras.layers.Dense(18, activation='softmax')
```

We'll limit our dataset to only the audio clips of instruments, then transform the instrument labels to 18-element one-hot vectors. We can use the same `image_generator` approach above to feed our images to our model for training. Instead of exporting this as a TF Lite model, we can use `model.save()` to export our model for serving.

To demonstrate deploying the phase 2 model to the cloud, we'll use Cloud AI Platform Prediction (*https://oreil.ly/P5Cn9*). We'll need to upload our saved model assets to a Cloud Storage bucket, then deploy the model by specifying the framework and pointing AI Platform Prediction to our storage bucket.

 You can use any cloud-based custom model deployment tool for the second phase of the Two-Phase Predictions design pattern. In addition to Google Cloud's AI Platform Prediction, AWS Sage-Maker (*https://oreil.ly/zIHey*) and Azure Machine Learning (*https://oreil.ly/dCxHE*) both offer services for deploying custom models.

When we export our model as a TensorFlow SavedModel, we can pass a Cloud Storage bucket URL directly to the save model method:

```
model.save('gs://your_storage_bucket/path')
```

This will export our model in the TF SavedModel format and upload it to our Cloud Storage bucket.

In AI Platform, a model resource contains different versions of your model. Each model can have hundreds of versions. We'll first create the model resource using gcloud, the Google Cloud CLI:

```
gcloud ai-platform models create instrument_classification
```

There are a few ways to deploy your model. We'll use gcloud and point AI Platform at the storage subdirectory that contains our saved model assets:

```
gcloud ai-platform versions create v1 \
  --model instrument_classification \
  --origin 'gs://your_storage_bucket/path/model_timestamp' \
  --runtime-version=2.1 \
```

```
--framework='tensorflow' \
--python-version=3.7
```

We can now make prediction requests to our model via the AI Platform Prediction API, which supports online and batch prediction. Online prediction lets us get predictions in near real time on a few examples at once. If we have hundreds or thousands of examples we want to send for prediction, we can create a batch prediction job that will run asynchronously in the background and output the prediction results to a file when complete.

To handle cases where the device calling our model may not always be connected to the internet, we could store audio clips for instrument prediction on the device while it is offline. When it regains connectivity, we could then send these clips to the cloud-hosted model for prediction.

Trade-Offs and Alternatives

While the Two-Phase Predictions pattern works for many cases, there are situations where your end users may have very little internet connectivity and you therefore cannot rely on being able to call a cloud-hosted model. In this section, we'll discuss two offline-only alternatives, a scenario where a client needs to make many prediction requests at a time, and suggestions on how to run continuous evaluation for offline models.

Standalone single-phase model

Sometimes, the end users of your model may have little to no internet connectivity. Even though these users' devices won't be able to reliably access a cloud model, it's still important to give them a way to access your application. For this case, rather than relying on a two-phase prediction flow, you can make your first model robust enough that it can be self-sufficient.

To do this, we can create a smaller version of our complex model, and give users the option to download this simpler, smaller model for use when they are offline. These offline models may not be quite as accurate as their larger online counterparts, but this solution is infinitely better than having no offline support at all. To build more complex models designed for offline inference, it's best to use a tool that allows you to quantize your model's weights and other math operations both during and after training. This is known as *quantization aware training* (*https://oreil.ly/ABd8r*).

One example of an application that provides a simpler offline model is Google Translate (*https://oreil.ly/uEWAM*). Google Translate is a robust, online translation service available in hundreds of languages. However, there are many scenarios where you'd need to use a translation service without internet access. To handle this, Google translate lets you download offline translations in over 50 different languages. These offline models are small, around 40 to 50 megabytes, and come close in accuracy to the more complex online versions. Figure 5-12 shows a quality comparison of on-device and online translation models.

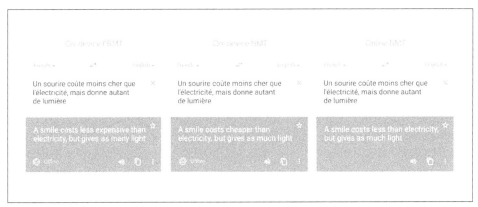

Figure 5-12. A comparison of on-device phrase-based and (newer) neural-machine translation models and online neural machine translation (source: The Keyword (https://oreil.ly/S_woM)).

Another example of a standalone single-phase model is Google Bolo (*https://oreil.ly/ zTy79*), a speech-based language learning app for children. The app works entirely offline and was developed with the intention of helping populations where reliable internet access is not always available.

Offline support for specific use cases

Another solution for making your application work for users with minimal internet connectivity is to make only certain parts of your app available offline. This could involve enabling a few common features offline or caching the results of an ML model's prediction for later use offline. With this alternative, we're still employing two prediction phases, but we're limiting the use cases covered by our offline model. In this approach, the app works sufficiently offline, but provides full functionality when it regains connectivity.

Google Maps, for example, lets you download maps and directions in advance. To avoid having directions take up too much space on a mobile device, only driving directions might be made available offline (not walking or biking). Another example could be a fitness application that tracks your steps and makes recommendations for future activity. Let's say the most common use of this app is checking how many steps you have walked on the current day. To support this use case offline, we could sync the fitness tracker's data to a user's device over Bluetooth to enable checking the current day's fitness status offline. To optimize our app's performance, we might decide to make fitness history and recommendations only available online.

We could further build upon this by storing the user's queries while their device is offline and sending them to a cloud model when they regain connectivity to provide more detailed results. Additionally, we could even provide a basic recommendation model available offline, with the intention of complementing this with improved results when the app is able to send the user's queries to a cloud-hosted model. With this solution, the user still gets some functionality when they aren't connected. When they come back online, they can then benefit from a full-featured app and robust ML model.

Handling many predictions in near real time

In other cases, end users of your ML model may have reliable connectivity but might need to make hundreds or even thousands of predictions to your model at once. If you only have a cloud-hosted model and each prediction requires an API call to a hosted service, getting prediction responses on thousands of examples at once will take too much time.

To understand this, let's say we have embedded devices deployed in various areas throughout a user's house. These devices are capturing data on temperature, air pressure, and air quality. We have a model deployed in the cloud for detecting anomalies from this sensor data. Because the sensors are continuously collecting new data, it would be inefficient and expensive to send every incoming data point to our cloud model. Instead, we can have a model deployed directly on the sensors to identify possible anomaly candidates from incoming data. We can then send only the potential anomalies to our cloud model for consolidated verification, taking sensor readings from all the locations into account. This is a variation of the Two-Phase Predictions pattern described earlier, the main difference being that both the offline and cloud models perform the same prediction task but with different inputs. In this case, models also end up throttling the number of prediction requests sent to the cloud model at one time.

Continuous evaluation for offline models

How can we ensure our on-device models stay up to date and don't suffer from data drift? There are a few options for performing continuous evaluation on models that do not have network connectivity. First, we could save a subset of predictions that are received on-device. We could then periodically evaluate our model's performance on these examples and determine if the model needs retraining. In the case of our two-phase model, it's important we do this evaluation regularly since it's likely that many calls to our on-device model will not go onto the second-phase cloud model. Another option is to create a replica of our on-device model to run *online*, only for continuous evaluation purposes. This solution is preferred if our offline and cloud models are running similar prediction tasks, like in the translation case mentioned previously.

Design Pattern 20: Keyed Predictions

Normally, you train your model on the same set of input features that the model will be supplied in real time when it is deployed. In many situations, however, it can be advantageous for your model to also pass through a client-supplied key. This is called the Keyed Predictions design pattern, and it is a necessity to scalably implement several of the design patterns discussed in this chapter.

Problem

If your model is deployed as a web service and accepts a single input, then it is quite clear which output corresponds to which input. But what if your model accepts a file with a million inputs and sends back a file with a million output predictions?

You might think that it should be obvious that the first output instance corresponds to the first input instance, the second output instance to the second input instance, etc. However, with a 1:1 relationship, it is necessary for each server node to process the full set of inputs serially. It would be much more advantageous if you use a distributed data processing system and farm out instances to multiple machines, collect all the resulting outputs, and send them back. The problem with this approach is that the outputs are going to be jumbled. Requiring that the outputs be ordered the same way poses scalability challenges, and providing the outputs in an unordered manner requires the clients to somehow know which output corresponds to which input.

This same problem occurs if your online serving system accepts an array of instances as discussed in the Stateless Serving Function pattern. The problem is that processing a large number of instances locally will lead to hot spots. Server nodes that receive only a few requests will be able to keep up, but any server node that receives a particularly large array will start to fall behind. These hot spots will force you to make your server machines more powerful than they need to be. Therefore, many online serving systems will impose a limit on the number of instances that can be sent in one request. If there is no such limit, or if the model is so computationally expensive that requests with fewer instances than this limit can overload the server, you will run into the problem of hot spots. Therefore, any solution to the batch serving problem will also address the problem of hot spots in online serving.

Solution

The solution is to use pass-through keys. Have the client supply a key associated with each input. For example (see Figure 5-13), suppose your model is trained with three inputs (a, b, c), shown on the left, to produce the output d, shown on the right. Make your clients supply (k, a, b, c) to your model where k is a key with a unique identifier. The key could be as simple as numbering the input instances 1, 2, 3, …, etc. Your model will then return (k, d), and so the client will be able to figure out which output instance corresponds to which input instance.

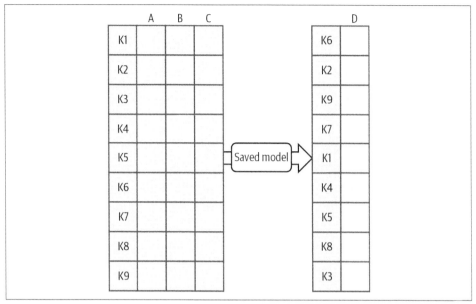

Figure 5-13. The client supplies a unique key with each input instance. The serving system attaches those keys to the corresponding prediction. This allows the client to retrieve the correct prediction for each input even if outputs are out of order.

How to pass through keys in Keras

In order to get your Keras model to pass through keys, supply a serving signature when exporting the model.

For example, this is the code to take a model that would otherwise take four inputs (is_male, mother_age, plurality, and gestation_weeks) and have it also take a key that it will pass through to the output along with the original output of the model (the babyweight):

```
# Serving function that passes through keys
@tf.function(input_signature=[{
    'is_male': tf.TensorSpec([None,], dtype=tf.string, name='is_male'),
    'mother_age': tf.TensorSpec([None,], dtype=tf.float32,
name='mother_age'),
    'plurality': tf.TensorSpec([None,], dtype=tf.string, name='plurality'),
    'gestation_weeks': tf.TensorSpec([None,], dtype=tf.float32,

name='gestation_weeks'),
    'key': tf.TensorSpec([None,], dtype=tf.string, name='key')
}])
def keyed_prediction(inputs):
```

```
feats = inputs.copy()
key = feats.pop('key') # get the key from input
output = model(feats) # invoke model
return {'key': key, 'babyweight': output}
```

This model is then saved as discussed in the Stateless Serving Function design pattern:

```
model.save(EXPORT_PATH,
           signatures={'serving_default': keyed_prediction})
```

Adding keyed prediction capability to an existing model

Note that the code above works even if the original model was not saved with a serving function. Simply load the model using `tf.saved_model.load()`, attach a serving function, and use the code snippet above, as shown in Figure 5-14.

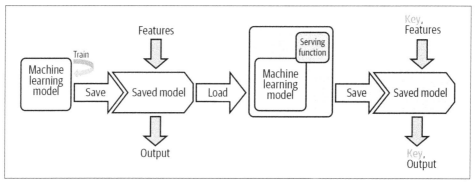

Figure 5-14. Load a SavedModel, attach a nondefault serving function, and save it.

When doing so, it is preferable to provide a serving function that replicates the older, no-key behavior:

```
# Serving function that does not require a key
@tf.function(input_signature=[{
    'is_male': tf.TensorSpec([None,], dtype=tf.string, name='is_male'),
    'mother_age': tf.TensorSpec([None,],  dtype=tf.float32,
name='mother_age'),
    'plurality': tf.TensorSpec([None,], dtype=tf.string, name='plurality'),
    'gestation_weeks': tf.TensorSpec([None,], dtype=tf.float32,
name='gestation_weeks')
}])
def nokey_prediction(inputs):
    output = model(inputs) # invoke model
    return {'babyweight': output}
```

Use the previous behavior as the default and add the `keyed_prediction` as a new serving function:

```
model.save(EXPORT_PATH,
          signatures={'serving_default': nokey_prediction,
                      'keyed_prediction': keyed_prediction
})
```

Trade-Offs and Alternatives

Why can't the server just assign keys to the inputs it receives? For online prediction, it is possible for servers to assign unique request IDs that lack any semantic information. For batch prediction, the problem is that the inputs need to be associated with the outputs, so the server assigning a unique ID is not enough since it can't be joined back to the inputs. What the server has to do is to assign keys to the inputs it receives before it invokes the model, use the keys to order the outputs, and then remove the keys before sending along the outputs. The problem is that ordering is computationally very expensive in distributed data processing.

In addition, there are a couple of other situations where client-supplied keys are useful—asynchronous serving and evaluation. Given these two situations, it is preferable that what constitutes a key becomes use case specific and needs to be identifiable. Therefore, asking clients to supply a key makes the solution simpler.

Asynchronous serving

Many production machine learning models these days are neural networks, and neural networks involve matrix multiplications. Matrix multiplication on hardware like GPUs and TPUs is more efficient if you can ensure that the matrices are within certain size ranges and/or multiples of a certain number. It can, therefore, be helpful to accumulate requests (up to a maximum latency of course) and handle the incoming requests in chunks. Since the chunks will consist of interleaved requests from multiple clients, the key, in this case, needs to have some sort of client identifier as well.

Continuous evaluation

If you are doing continuous evaluation, it can be helpful to log metadata about the prediction requests so that you can monitor whether performance drops across the board, or only in specific situations. Such slicing is made much easier if the key identifies the situation in question. For example, suppose that we need to apply a Fairness Lens (see Chapter 7) to ensure that our model's performance is fair across different customer segments (age of customer and/or race of customer, for example). The model will not use the customer segment as an input, but we need to evaluate the performance of the model sliced by the customer segment. In such cases, having the customer segment(s) be embedded in the key (an example key might be 35-Black-Male-34324323) makes slicing easier.

An alternate solution is to have the model ignore unrecognized inputs and send back not just the prediction outputs but also all inputs, including the unrecognized ones. This allows the client to match inputs to outputs, but is more expensive in terms of bandwidth and client-side computation.

Because high-performance servers will support multiple clients, be backed by a cluster, and batch up requests to gain performance benefits, it's better to plan ahead for this—ask that clients supply keys with every prediction and for clients to specify keys that will not cause a collision with other clients.

Summary

In this chapter, we looked at techniques for operationalizing machine learning models to ensure they are resilient and can scale to handle production load. Each resilience pattern we discussed relates to the deployment and serving steps in a typical ML workflow.

We started this chapter by looking at how to encapsulate your trained machine learning model as a stateless function using the *Stateless Serving Function* design pattern. A serving function decouples your model's training and deployment environments by defining a function that performs inference on an exported version of your model, and is deployed to a REST endpoint. Not all production models require immediate prediction results, as there are situations where you need to send a large batch of data to your model for prediction but don't need results right away. We saw how the *Batch Serving* design pattern solves this by utilizing distributed data processing infrastructure designed to run many model prediction requests asynchronously as a background job, with output written to a specified location.

Next, with the *Continued Model Evaluation* design pattern, we looked at an approach to verifying that your deployed model is still performing well on new data. This pattern addresses the problems of data and concept drift by regularly evaluating your model and using these results to determine if retraining is necessary. In the *Two-Phase Predictions* design pattern, we solved for specific use cases where models need to be deployed at the edge. When you can break a problem into two logical parts, this pattern first creates a simpler model that can be deployed on-device. This edge model is connected to a more complex model hosted in the cloud. Finally, in the *Keyed Prediction* design pattern, we discussed why it can be beneficial to supply a unique key with each example when making prediction requests. This ensures that your client associates each prediction output with the correct input example.

In the next chapter, we'll look at *reproducibility* patterns. These patterns address challenges associated with the inherent randomness present in many aspects of machine learning and focus on enabling reliable, consistent results each time a machine learning process runs.

## CHAPTER 6

Reproducibility Design Patterns

Software best practices such as unit testing assume that if we run a piece of code, it produces deterministic output:

```
def sigmoid(x):
    return 1.0 / (1 + np.exp(-x))

class TestSigmoid(unittest.TestCase):
    def test_zero(self):
        self.assertAlmostEqual(sigmoid(0), 0.5)

    def test_neginf(self):
        self.assertAlmostEqual(sigmoid(float("-inf")), 0)

    def test_inf(self):
        self.assertAlmostEqual(sigmoid(float("inf")), 1)
```

This sort of reproducibility is difficult in machine learning. During training, machine learning models are initialized with random values and then adjusted based on training data. A simple k-means algorithm implemented by scikit-learn requires setting the random_state in order to ensure the algorithm returns the same results each time:

```
def cluster_kmeans(X):
    from sklearn import cluster
    k_means = cluster.KMeans(n_clusters=10, random_state=10)
    labels = k_means.fit(X).labels_[::]
    return labels
```

Beyond the random seed, there are many other artifacts that need to be fixed in order to ensure reproducibility during training. In addition, machine learning consists of different stages, such as training, deployment, and retraining. It is often important that some things are reproducible across these stages as well.

In this chapter, we'll look at design patterns that address different aspects of reproducibility. The *Transform* design pattern captures data preparation dependencies from the model training pipeline to reproduce them during serving. *Repeatable Splitting* captures the way data is split among training, validation, and test datasets to ensure that a training example that is used in training is never used for evaluation or testing even as the dataset grows. The *Bridged Schema* design pattern looks at how to ensure reproducibility when the training dataset is a hybrid of data conforming to different schema. The *Workflow Pipeline* design pattern captures all the steps in the machine learning process to ensure that as the model is retrained, parts of the pipeline can be reused. The *Feature Store* design pattern addresses reproducibility and reusability of features across different machine learning jobs. The *Windowed Inference* design pattern ensures that features that are calculated in a dynamic, time-dependent way can be correctly repeated between training and serving. *Versioning* of data and models is a prerequisite to handle many of the design patterns in this chapter.

Design Pattern 21: Transform

The Transform design pattern makes moving an ML model to production much easier by keeping inputs, features, and transforms carefully separate.

Problem

The problem is that the *inputs* to a machine learning model are not the *features* that the machine learning model uses in its computations. In a text classification model, for example, the inputs are the raw text documents and the features are the numerical embedding representations of this text. When we train a machine learning model, we train it with features that are extracted from the raw inputs. Take this model that is trained to predict the duration of bicycle rides in London using BigQuery ML:

```
CREATE OR REPLACE MODEL ch09eu.bicycle_model
OPTIONS(input_label_cols=['duration'],
        model_type='linear_reg')
AS
SELECT
 duration
 , start_station_name
 , CAST(EXTRACT(dayofweek from start_date) AS STRING)
 as dayofweek
 , CAST(EXTRACT(hour from start_date) AS STRING)
 as hourofday
FROM
 `bigquery-public-data.london_bicycles.cycle_hire`
```

This model has three features (start_station_name, dayofweek, and hourofday) computed from two inputs, start_station_name and start_date, as shown in Figure 6-1.

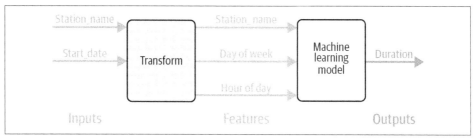

Figure 6-1. The model has three features computed from two inputs.

But the SQL code above mixes up the inputs and features and doesn't keep track of the transformations that were carried out. This comes back to bite when we try to predict with this model. Because the model was trained on three features, this is what the prediction signature has to look like:

```
SELECT * FROM ML.PREDICT(MODEL ch09eu.bicycle_model,(
    'Kings Cross' AS start_station_name
  , '3' as dayofweek
  , '18' as hourofday
))
```

Note that, at inference time, we have to know what features the model was trained on, how they should be interpreted, and the details of the transformations that were applied. We have to know that we need to send in '3' for dayofweek. That '3' …is that Tuesday or Wednesday? Depends on which library was used by the model, or what we consider the start of a week!

Training-serving skew, caused by differences in any of these factors between the training and serving environments, is one of the key reasons why productionization of ML models is so hard.

Solution

The solution is to explicitly capture the transformations applied to convert the model inputs into features. In BigQuery ML, this is done using the TRANSFORM clause. Using TRANSFORM ensures that these transformations are automatically applied during ML.PREDICT.

Given the support for TRANSFORM, the model above should be rewritten as:

```
CREATE OR REPLACE MODEL ch09eu.bicycle_model
OPTIONS(input_label_cols=['duration'],
        model_type='linear_reg')
TRANSFORM(
 SELECT * EXCEPT(start_date)
 , CAST(EXTRACT(dayofweek from start_date) AS STRING)
 as dayofweek -- feature1
 , CAST(EXTRACT(hour from start_date) AS STRING)
```

```
  as hourofday -- feature2
)
AS
SELECT
 duration, start_station_name, start_date -- inputs
FROM
  `bigquery-public-data.london_bicycles.cycle_hire`
```

Notice how we have clearly separated out the inputs (in the SELECT clause) from the features (in the TRANSFORM clause). Now, prediction is much easier. We can simply send to the model the station name and a timestamp (the inputs):

```
SELECT * FROM ML.PREDICT(MODEL ch09eu.bicycle_model,(
    'Kings Cross' AS start_station_name
, CURRENT_TIMESTAMP() as start_date
))
```

The model will then take care of carrying out the appropriate transformations to create the necessary features. It does so by capturing both the transformation logic and artifacts (such as scaling constants, embedding coefficients, lookup tables, and so on) to carry out the transformation.

As long as we carefully use only the raw inputs in the SELECT statement and put all subsequent processing of the input in the TRANSFORM clause, BigQuery ML will automatically apply these transformations during prediction.

Trade-Offs and Alternatives

The solution described above works because BigQuery ML keeps track of the transformation logic and artifacts for us, saves them in the model graph, and automatically applies the transformations during prediction.

If we are using a framework where support for the Transform design pattern is not built in, we should design our model architecture in such a way that the transformations carried out during training are easy to reproduce during serving. We can do this by making sure to save the transformations in the model graph or by creating a repository of transformed features ("Design Pattern 26: Feature Store" on page 295).

Transformations in TensorFlow and Keras

Assume that we are training an ML model to estimate taxi fare in New York and have six inputs (pickup latitude, pickup longitude, dropoff latitude, dropoff longitude, passenger count, and pickup time). TensorFlow supports the concept of feature columns, which are saved in the model graph. However, the API is designed assuming that the raw inputs are the same as the features.

Let's say that we want to scale the latitudes and longitudes (see "Simple Data Representations" on page 22 in Chapter 2 for details), create a transformed feature that is the Euclidean distance, and extract the hour of day from the timestamp. We have to carefully design the model graph (see Figure 6-2), keeping the Transform concept firmly in mind. As we walk through the code below, notice how we set things up so that we clearly design three separate layers in our Keras model—the Inputs layer, the Transform layer, and a `DenseFeatures` layer.

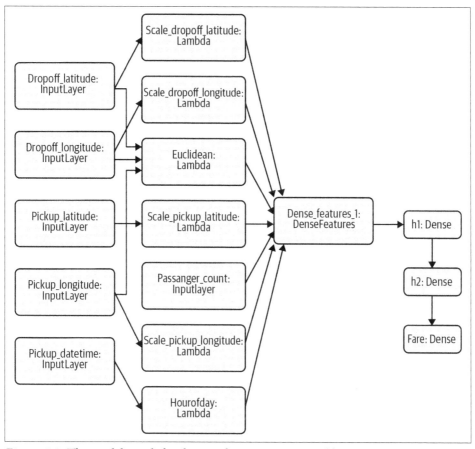

Figure 6-2. The model graph for the taxi fare estimation problem in Keras.

First, make every input to the Keras model an `Input` layer (the full code is in a note-book on GitHub (*https://github.com/GoogleCloudPlatform/training-data-analyst/blob/master/quests/serverlessml/06_feateng_keras/solution/taxifare_fc.ipynb*)):

```
inputs = {
        colname : tf.keras.layers.Input(
                        name=colname, shape=(), dtype='float32')
            for colname in ['pickup_longitude', 'pickup_latitude',
                            'dropoff_longitude', 'dropoff_latitude']
}
```

In Figure 6-2, these are the boxes marked dropoff_latitude, dropoff_longitude, and so on.

Second, maintain a dictionary of transformed features, and make every transformation either a Keras Preprocessing layer or a Lambda layer. Here, we scale the inputs using Lambda layers:

```
transformed = {}
for lon_col in ['pickup_longitude', 'dropoff_longitude']:
        transformed[lon_col] = tf.keras.layers.Lambda(
            lambda x: (x+78)/8.0,
            name='scale_{}'.format(lon_col)
        )(inputs[lon_col])
for lat_col in ['pickup_latitude', 'dropoff_latitude']:
        transformed[lat_col] = tf.keras.layers.Lambda(
            lambda x: (x-37)/8.0,
            name='scale_{}'.format(lat_col)
        )(inputs[lat_col])
```

In Figure 6-2, these are the boxes marked scale_dropoff_latitude, scale_drop off_longitude, and so on.

We will also have one Lambda layer for the Euclidean distance, which is computed from four of the Input layers (see Figure 6-2):

```
def euclidean(params):
    lon1, lat1, lon2, lat2 = params
    londiff = lon2 - lon1
    latdiff = lat2 - lat1
    return tf.sqrt(londiff*londiff + latdiff*latdiff)
transformed['euclidean'] = tf.keras.layers.Lambda(euclidean, name='euclidean')([
        inputs['pickup_longitude'],
        inputs['pickup_latitude'],
        inputs['dropoff_longitude'],
        inputs['dropoff_latitude']
    ])
```

Similarly, the column to create the hour of day from the timestamp is a Lambda layer:

```
transformed['hourofday'] = tf.keras.layers.Lambda(
        lambda x: tf.strings.to_number(tf.strings.substr(x, 11, 2),
                                       out_type=tf.dtypes.int32),
        name='hourofday'
    )(inputs['pickup_datetime'])
```

Third, all these transformed layers will be concatenated into a DenseFeatures layer:

```
dnn_inputs = tf.keras.layers.DenseFeatures(feature_columns.values())(transformed)
```

Because the constructor for DenseFeatures requires a set of feature columns, we will have to specify how to take each of the transformed values and convert them into an input to the neural network. We might use them as is, one-hot encode them, or choose to bucketize the numbers. For simplicity, let's just use them all as is:

```
feature_columns = {
        colname: tf.feature_column.numeric_column(colname)
            for colname in ['pickup_longitude', 'pickup_latitude',
                            'dropoff_longitude', 'dropoff_latitude']
}
feature_columns['euclidean'] = \
            tf.feature_column.numeric_column('euclidean')
```

Once we have a `DenseFeatures` input layer, we can build the rest of our Keras model as usual:

```
h1 = tf.keras.layers.Dense(32, activation='relu', name='h1')(dnn_inputs)
h2 = tf.keras.layers.Dense(8, activation='relu', name='h2')(h1)
output = tf.keras.layers.Dense(1, name='fare')(h2)
model = tf.keras.models.Model(inputs, output)
model.compile(optimizer='adam', loss='mse', metrics=['mse'])
```

The complete example (*https://github.com/GoogleCloudPlatform/training-data-analyst/blob/master/quests/serverlessml/06_feateng_keras/solution/taxifare_fc.ipynb*) is on GitHub.

Notice how we set things up so that the first layer of the Keras model was `Inputs`. The second layer was the `Transform` layer. The third layer was the `DenseFeatures` layer that combined them. After this sequence of layers, the usual model architecture starts. Because the `Transform` layer is part of the model graph, the usual Serving Function and Batch Serving solutions (see Chapter 5) will work as is.

Efficient transformations with tf.transform

One drawback to the above approach is that the transformations will be carried out during each iteration of training. This is not such a big deal if all we are doing is scaling by known constants. But what if our transformations are more computationally expensive? What if we want to scale using the mean and variance, in which case, we need to pass through all the data first to compute these variables?

It is helpful to differentiate between *instance-level* transformations that can be part of the model directly (where the only drawback is applying them on each training iteration) and *dataset-level* transformations, where we need a full pass to compute overall statistics or the vocabulary of a categorical variable. Such dataset-level transformations cannot be part of the model and have to be applied as a scalable preprocessing step, which produces the Transform, capturing the logic and the artifacts (mean, variance, vocabulary, and so on) to be attached to the model. For dataset-level transformations, use `tf.transform`.

The tf.transform library (which is part of TensorFlow Extended (*https://oreil.ly/OznI3*)) provides an efficient way of carrying out transformations over a preprocessing pass through the data and saving the resulting features and transformation artifacts so that the transformations can be applied by TensorFlow Serving during prediction time.

The first step is to define the transformation function. For example, to scale all the inputs to be zero mean and unit variance and bucketize them, we would create this preprocessing function (see the full code (*https://github.com/tensorflow/tfx/blob/master/tfx/examples/chicago_taxi_pipeline/taxi_utils_native_keras.py*) on GitHub):

```
def preprocessing_fn(inputs):
    outputs = {}
    for key in ...:
        outputs[key + '_z'] = tft.scale_to_z_score(inputs[key])
        outputs[key + '_bkt'] = tft.bucketize(inputs[key], 5)
    return outputs
```

Before training, the raw data is read and transformed using the prior function in Apache Beam:

```
transformed_dataset, transform_fn = (raw_dataset |
    beam_impl.AnalyzeAndTransformDataset(preprocessing_fn))
transformed_data, transformed_metadata = transformed_dataset
```

The transformed data is then written out in a format suitable for reading by the training pipeline:

```
transformed_data | tfrecordio.WriteToTFRecord(
    PATH_TO_TFT_ARTIFACTS,
    coder=example_proto_coder.ExampleProtoCoder(
        transformed_metadata.schema))
```

The Beam pipeline also stores the preprocessing function that needs to be run, along with any artifacts the function needs, into an artifact in TensorFlow graph format. In the case above, for example, this artifact would include the mean and variance for scaling the numbers, and the bucket boundaries for bucketizing numbers. The training function reads transformed data and, therefore, the transformations do not have to be repeated within the training loop.

The serving function needs to load in these artifacts and create a Transform layer:

```
tf_transform_output = tft.TFTransformOutput(PATH_TO_TFT_ARTIFACTS)
tf_transform_layer = tf_transform_output.transform_features_layer()
```

Then, the serving function can apply the Transform layer to the parsed input features and invoke the model with the transformed data to calculate the model output:

```
@tf.function
def serve_tf_examples_fn(serialized_tf_examples):
    feature_spec = tf_transform_output.raw_feature_spec()
    feature_spec.pop(_LABEL_KEY)
```

```
parsed_features = tf.io.parse_example(serialized_tf_examples, feature_spec)

transformed_features = tf_transform_layer(parsed_features)
return model(transformed_features)
```

In this way, we are making sure to insert the transformations into the model graph for serving. At the same time, because the model training happens on the transformed data, our training loop does not have to carry out these transformations during each epoch.

Text and image transformations

In text models, it is common to preprocess input text (such as to remove punctuation, stop words, capitalization, stemming, and so on) before providing the cleaned text as a feature to the model. Other common feature engineering on text inputs includes tokenization and regular expression matching. It is essential that the same cleanup or extraction steps be carried out at inference time.

The need to capture transformations is important even if there is no explicit feature engineering as when using deep learning with images. Image models usually have an Input layer that accepts images of a specific size. Image inputs of a different size will have to be cropped, padded, or resampled to this fixed size before being fed into the model. Other common transformations in image models include color manipulations (gamma correction, grayscale conversion, and so on) and orientation correction. It is essential that such transformations be identical between what was carried out on the training dataset and what will be carried out during inference. The Transform pattern helps ensure this reproducibility.

With image models, there are some transformations (such as data augmentation by random cropping and zooming) that are applied only during training. These transformations do not need to be captured during inference. Such transformations will not be part of the Transform pattern.

Alternate pattern approaches

An alternative approach to solving the training-serving skew problem is to employ the Feature Store pattern. The feature store comprises a coordinated computation engine and repository of transformed feature data. The computation engine supports low-latency access for inference and batch creation of transformed features while the data repository provides quick access to transformed features for model training. The advantage of a feature store is there is no requirement for the transformation operations to fit into the model graph. For example, as long as the feature store supports Java, the preprocessing operations could be carried out in Java while the model itself could be written in PyTorch. The disadvantage of a feature store is that it makes the model dependent on the feature store and makes the serving infrastructure much more complex.

Another way to separate out the programming language and framework used for transformation of the features from the language used to write the model is to carry out the preprocessing in containers and use these custom containers as part of both the training and serving. This is discussed in "Design Pattern 25: Workflow Pipeline" on page 282 and is adopted in practice by Kubeflow Serving.

Design Pattern 22: Repeatable Splitting

To ensure that sampling is repeatable and reproducible, it is necessary to use a well-distributed column and a deterministic hash function to split the available data into training, validation, and test datasets.

Problem

Many machine learning tutorials will suggest splitting data randomly into training, validation, and test datasets using code similar to the following:

```
df = pd.DataFrame(...)
rnd = np.random.rand(len(df))
train = df[ rnd < 0.8  ]
valid = df[ rnd >= 0.8 & rnd < 0.9 ]
test  = df[ rnd >= 0.9 ]
```

Unfortunately, this approach fails in many real-world situations. The reason is that it is rare that the rows are independent. For example, if we are training a model to predict flight delays, the arrival delays of flights on the same day will be highly correlated with one another. This leads to leakage of information between the training and testing dataset when some of the flights on any particular day are in the training dataset and some other flights on the same day are in the testing dataset. This leakage due to correlated rows is a frequently occurring problem, and one that we have to avoid when doing machine learning.

In addition, the rand function orders data differently each time it is run, so if we run the program again, we will get a different 80% of rows. This can play havoc if we are experimenting with different machine learning models with the goal of choosing the best one—we need to compare the model performance on the same test dataset. In order to address this, we need to set the random seed in advance or store the data after it is split. Hardcoding how the data is to be split is not a good idea because, when carrying out techniques like jackknifing, bootstrapping, cross-validation, and hyperparameter tuning, we will need to change this data split and do so in a way that allows us to do individual trials.

For machine learning, we want lightweight, repeatable splitting of the data that works regardless of programming language or random seeds. We also want to ensure that correlated rows fall into the same split. For example, we do not want flights on January 2, 2019 in the test dataset if flights on that day are in the training dataset.

Solution

First, we identify a column that captures the correlation relationship between rows. In our airline delay dataset, this is the date column. Then, we use the last few digits of a hash function on that column to split the data. For the airline delay problem, we can use the Farm Fingerprint hashing algorithm on the date column to split the available data into training, validation, and testing datasets.

 For more on the Farm Fingerprint algorithm, support for other frameworks and languages, and the relationship between hashing and cryptography, please see "Design Pattern 1: Hashed Feature" on page 32 in Chapter 2. In particular, open source wrappers of the Farm Hash (*https://github.com/google/farmhash*) algorithm are available in a number of languages (including Python (*https:// oreil.ly/526Dc*)), and so this pattern can be applied even if data is not in a data warehouse that supports a repeatable hash out of the box.

Here is how to split the dataset based on the hash of the date column:

```
SELECT
  airline,
  departure_airport,
  departure_schedule,
  arrival_airport,
  arrival_delay
FROM
  `bigquery-samples`.airline_ontime_data.flights
WHERE
  ABS(MOD(FARM_FINGERPRINT(date), 10)) < 8 -- 80% for TRAIN
```

To split on the date column, we compute its hash using the FARM_FINGERPRINT function and then use the modulo function to find an arbitrary 80% subset of the rows. This is now repeatable—because the FARM_FINGERPRINT function returns the same value any time it is invoked on a specific date, we can be sure we will get the same 80% of data each time. As a result, all the flights on any given date will belong to the same split—train, validation, or test. This is repeatable regardless of the random seed.

If we want to split our data by arrival_airport (so that 80% of airports are in the training dataset, perhaps because we are trying to predict something about airport amenities), we would compute the hash on arrival_airport instead of date.

It is also straightforward to get the validation data: change the < 8 in the query above to =8, and for testing data, change it to =9. This way, we get 10% of samples in validation and 10% in testing.

What are the considerations for choosing the column to split on? The `date` column has to have several characteristics for us to be able to use it as the splitting column:

- Rows at the same date tend to be correlated—again, this is the key reason why we want to ensure that all rows on the same date are in the same split.

- `date` is not an input to the model even though it is used as a criteria for splitting. Features extracted from `date` such as day of week or hour of day can be inputs, but we can't use an actual input as the field with which to split because the trained model will not have seen 20% of the possible input values for the `date` column if we use 80% of the data for training.

- There have to be enough `date` values. Since we are computing the hash and finding the modulo with respect to 10, we need at least 10 unique hash values. The more unique values we have, the better. To be safe, a rule of thumb is to shoot for 3–5× the denominator for the modulo, so in this case, we want 40 or so unique dates.

- The label has to be well distributed among the dates. If it turns out that all the delays happened on January 1 and there were no delays the rest of the year, this wouldn't work since the split datasets will be skewed. To be safe, look at a graph and make sure that all three splits have a similar distribution of labels. To be extra safe, ensure that the distributions of label by departure delay and other input values are similar across the three datasets.

 We can automate checking whether the label distributions are similar across the three datasets by using the Kolomogorov–Smirnov test: just plot the cumulative distribution functions of the label in the three datasets and find the maximum distance between each pair. The smaller the maximum distance, the better the split.

Trade-Offs and Alternatives

Let's look at a couple of variants of how we might do repeatable splitting and discuss the pros and cons of each. Let's also examine how to extend this idea to do repeatable sampling, not just splitting.

Single query

We don't need three separate queries to generate training, validation, and test splits. We can do it in a single query as follows:

```
CREATE OR REPLACE TABLE mydataset.mytable AS
SELECT
  airline,
  departure_airport,
```

```
    departure_schedule,
    arrival_airport,
    arrival_delay,
    CASE(ABS(MOD(FARM_FINGERPRINT(date), 10)))
        WHEN 9 THEN 'test'
        WHEN 8 THEN 'validation'
        ELSE 'training' END AS split_col
FROM
    `bigquery-samples`.airline_ontime_data.flights
```

We can then use the `split_col` column to decide which of three datasets any particular row falls in. Using a single query decreases computational time but requires creating a new table or modifying the source table to add the extra `split_col` column.

Random split

What if the rows are not correlated? In that case, we want a random, repeatable split but do not have a natural column to split by. We can hash the entire row of data by converting it to a string and hashing that string:

```
SELECT
    airline,
    departure_airport,
    departure_schedule,
    arrival_airport,
    arrival_delay
FROM
    `bigquery-samples`.airline_ontime_data.flights f
WHERE
    ABS(MOD(FARM_FINGERPRINT(TO_JSON_STRING(f), 10)) < 8
```

Note that if we have duplicate rows, they will always end up in the same split. This might be exactly what we desire. If not, then we will have to add a unique ID column to the SELECT query.

Split on multiple columns

We have talked about a single column that captures the correlation between rows. What if it is a combination of columns that capture when two rows are correlated? In such cases, simply concatenate the fields (this is a feature cross) before computing the hash. For example, suppose we only wish to ensure that flights from the same airport on the same day do not show up in different splits. In that case, we'd do the following:

```
SELECT
    airline,
    departure_airport,
    departure_schedule,
    arrival_airport,
    arrival_delay
```

```
FROM
  `bigquery-samples`.airline_ontime_data.flights
WHERE
  ABS(MOD(FARM_FINGERPRINT(CONCAT(date, arrival_airport)), 10)) < 8
```

If we split on a feature cross of multiple columns, we *can* use `arrival_airport` as one of the inputs to the model, since there will be examples of any particular airport in both the training and test sets. On the other hand, if we had split only on `arrival_airport`, then the training and test sets will have a mutually exclusive set of arrival airports and, therefore, `arrival_airport` cannot be an input to the model.

Repeatable sampling

The basic solution is good if we want 80% of the entire dataset as training, but what if we want to play around with a smaller dataset than what we have in BigQuery? This is common for local development. The flights dataset is 70 million rows, and perhaps what we want is a smaller dataset of one million flights. How would we pick 1 in 70 flights, and then 80% of those as training?

What we can *not* do is something along the lines of:

```
SELECT
    date,
    airline,
    departure_airport,
    departure_schedule,
    arrival_airport,
    arrival_delay
FROM
  `bigquery-samples`.airline_ontime_data.flights
WHERE
  ABS(MOD(FARM_FINGERPRINT(date), 70)) = 0
  AND ABS(MOD(FARM_FINGERPRINT(date), 10)) < 8
```

We cannot pick 1 in 70 rows and then pick 8 in 10. If we are picking numbers that are divisible by 70, of course they are also going to be divisible by 10! That second modulo operation is useless.

Here's a better solution:

```
SELECT
    date,
    airline,
    departure_airport,
    departure_schedule,
    arrival_airport,
    arrival_delay
FROM
  `bigquery-samples`.airline_ontime_data.flights
WHERE
```

```
ABS(MOD(FARM_FINGERPRINT(date), 70)) = 0
AND ABS(MOD(FARM_FINGERPRINT(date), 700)) < 560
```

In this query, the 700 is 70*10 and 560 is 70*8. The first modulo operation picks 1 in 70 rows and the second modulo operation picks 8 in 10 of those rows.

For validation data, you'd replace < 560 by the appropriate range:

```
ABS(MOD(FARM_FINGERPRINT(date), 70)) = 0
AND ABS(MOD(FARM_FINGERPRINT(date), 700)) BETWEEN 560 AND 629
```

In the preceding code, our one million flights come from only 1/70th of the days in the dataset. This may be precisely what we want—for example, we may be modeling the full spectrum of flights on a particular day when experimenting with the smaller dataset. However, if what we want is 1/70th of the flights on any particular day, we'd have to use RAND() and save the result as a new table for repeatability. From this smaller table, we can sample 80% of dates using FARM_FINGERPRINT(). Because this new table is only one million rows and only for experimentation, the duplication may be acceptable.

Sequential split

In the case of time-series models, a common approach is to use sequential splits of data. For example, to train a demand forecasting model where we train a model on the past 45 days of data to predict demand over the next 14 days, we'd train the model (full code (*https://github.com/GoogleCloudPlatform/bigquery-oreilly-book/ blob/master/blogs/bqml_arima/bqml_arima.ipynb*)) by pulling the necessary data:

```
CREATE OR REPLACE MODEL ch09eu.numrentals_forecast
OPTIONS(model_type='ARIMA',
        time_series_data_col='numrentals',
        time_series_timestamp_col='date') AS
SELECT
   CAST(EXTRACT(date from start_date) AS TIMESTAMP) AS date
   , COUNT(*) AS numrentals
FROM
   `bigquery-public-data`.london_bicycles.cycle_hire
GROUP BY date
HAVING date BETWEEN
DATE_SUB(CURRENT_DATE(), INTERVAL 45 DAY) AND CURRENT_DATE()
```

Such a sequential split of data is also necessary in fast-moving environments even if the goal is not to predict the future value of a time series. For example, in a fraud-detection model, bad actors adapt quickly to the fraud algorithm, and the model has to therefore be continually retrained on the latest data to predict future fraud. It is not sufficient to generate the evaluation data from a random split of the historical dataset because the goal is to predict behavior that the bad actors will exhibit in the future. The indirect goal is the same as that of a time-series model in that a good model will be able to train on historical data and predict future fraud. The data has to be split sequentially in terms of time to correctly evaluate this. For example (full code (*https://github.com/GoogleCloudPlatform/training-data-analyst/blob/master/blogs/ bigquery_datascience/bigquery_tensorflow.ipynb*)):

```
def read_dataset(client, row_restriction, batch_size=2048):
    ...
    bqsession = client.read_session(
        ...
        row_restriction=row_restriction)
    dataset = bqsession.parallel_read_rows()
    return (dataset.prefetch(1).map(features_and_labels)
            .shuffle(batch_size*10).batch(batch_size))

client = BigQueryClient()
train_df = read_dataset(client, 'Time <= 144803', 2048)
eval_df = read_dataset(client, 'Time > 144803', 2048)
```

Another instance where a sequential split of data is needed is when there are high correlations between successive times. For example, in weather forecasting, the weather on consecutive days is highly correlated. Therefore, it is not reasonable to put October 12 in the training dataset and October 13 in the testing dataset because there will be considerable leakage (imagine, for example, that there is a hurricane on October 12). Further, weather is highly seasonal, and so it is necessary to have days from all seasons in all three splits. One way to properly evaluate the performance of a forecasting model is to use a sequential split but take seasonality into account by using the first 20 days of every month in the training dataset, the next 5 days in the validation dataset, and the last 5 days in the testing dataset.

In all these instances, repeatable splitting requires only that we place the logic used to create the split into version control and ensure that the model version is updated whenever this logic is changed.

Stratified split

The example above of how weather patterns are different between different seasons is an example of a situation where the splitting needs to happen after the dataset is *stratified*. We needed to ensure that there were examples of all seasons in each split, and so we stratified the dataset in terms of months before carrying out the split. We used the first 20 days of every month in the training dataset, the next 5 days in the validation dataset, and the last 5 days in the testing dataset. Had we not been concerned with the correlation between successive days, we could have randomly split the dates within each month.

The larger the dataset, the less concerned we have to be with stratification. On very large datasets, the odds are very high that the feature values will be well distributed among all the splits. Therefore, in large-scale machine learning, the need to stratify happens quite commonly only in the case of skewed datasets. For example, in the flights dataset, less than 1% of flights take off before 6 a.m., and so the number of flights that meet this criterion may be quite small. If it is critical for our business use case to get the behavior of these flights correct, we should stratify the dataset based on departure hour and split each stratification evenly.

The departure time was an example of a skewed feature. In an imbalanced classification problem (such as fraud detection, where the number of fraud examples is quite small), we might want to stratify the dataset by the label and split each stratification evenly. This is also important if we have a multilabel problem and some of the labels are rarer than others. These are discussed in "Design Pattern 10: Rebalancing " on page 122 in Chapter 3.

Unstructured data

Although we have focused in this section on structured data, the same principles apply to unstructured data such as images, video, audio, or free-form text as well. Just use the metadata to carry out the split. For example, if videos taken on the same day are correlated, use a video's capture date from its metadata to split the videos among independent datasets. Similarly, if text reviews from the same person tend to be correlated, use the Farm Fingerprint of the user_id of the reviewer to repeatedly split reviews among the datasets. If the metadata is not available or there is no correlation between instances, encode the image or video using Base64 encoding and compute the fingerprint of the encoding.

A natural way to split text datasets might be to use the hash of the text itself for splitting. However, this is akin to a random split and does not address the problem of correlations between reviews. For example, if a person uses the word "stunning" a lot in their negative reviews or if a person rates all Star Wars movies as bad, their reviews are correlated. Similarly, a natural way to split image or audio datasets might be to use the hash of the filename for splitting, but it does not address the problem of correlations between images or videos. It is worth thinking carefully about the best way to split a dataset. In our experience, many problems with poor performance of ML can be addressed by designing the data split (and data collection) with potential correlations in mind.

When computing embeddings or pre-training autoencoders, we should make sure to first split the data and perform these pre-computations on the training dataset only. Because of this, splitting should not be done on the embeddings of the images, videos, or text unless these embeddings were created on a completely separate dataset.

Design Pattern 23: Bridged Schema

The Bridged Schema design pattern provides ways to adapt the data used to train a model from its older, original data schema to newer, better data. This pattern is useful because when an input provider makes improvements to their data feed, it often takes time for enough data of the improved schema to be collected for us to adequately train a replacement model. The Bridged Schema pattern allows us to use as much of the newer data as is available, but augment it with some of the older data to improve model accuracy.

Problem

Consider a point-of-sale application that suggests how much to tip a delivery person. The application might use a machine learning model that predicts the tip amount, taking into account the order amount, delivery time, delivery distance, and so on. Such a model would be trained on the actual tips added by customers.

Assume that one of the inputs to the model is the payment type. In the historical data, this has been recorded as "cash" or "card." However, let's say the payment system has been upgraded and it now provides more detail on the type of card (gift card, debit card, credit card) that was used. This is extremely useful information because the tipping behavior varies between the three types of cards.

At prediction time, the newer information will always be available since we are always predicting tip amounts on transactions conducted after the payment system upgrade. Because the new information is extremely valuable, and it is already available in production to the prediction system, we would like to use it in the model as soon as possible.

We cannot train a new model exclusively on the newer data because the quantity of new data will be quite small, limited as it is to transactions after the payment system upgrade. Because the quality of an ML model is highly dependent on the amount of data used to train it, it is likely that a model trained with only the new data is going to fare poorly.

Solution

The solution is to bridge the schema of the old data to match the new data. Then, we train an ML model using as much of the new data as is available and augment it with the older data. There are two questions to answer. First, how will we square the fact that the older data has only two categories for payment type, whereas the new data has four categories? Second, how will the augmentation be done to create datasets for training, validation, and testing?

Bridged schema

Consider the case where the older data has two categories (cash and card). In the new schema, the card category is now much more granular (gift card, debit card, credit card). What we do know is that a transaction coded as "card" in the old data would have been one of these types but the actual type was not recorded. It's possible to bridge the schema probabilistically or statically. The static method is what we recommend, but it is easier to understand if we walk through the probabilistic method first.

Probabilistic method. Imagine that we estimate from the newer training data that of the card transactions, 10% are gift cards, 30% are debit cards, and 60% are credit cards. Each time an older training example is loaded into the trainer program, we could choose the card type by generating a uniformly distributed random number in the range [0, 100) and choosing a gift card when the random number is less than 10, a debit card if it is in [10, 40), and a credit card otherwise. Provided we train for enough epochs, any training example would be presented as all three categories, but proportional to the actual frequency of occurrence. The newer training examples, of course, would always have the actually recorded category.

The justification for the probabilistic approach is that we treat each older example as having happened hundreds of times. As the trainer goes through the data, in each epoch, we simulate one of those instances. In the simulation, we expect that 10% of the time that a card was used, the transaction would have occurred with a gift card. That's why we pick "gift card" for the value of the categorical input 10% of the time. This is, of course, simplistic—just because gift cards are used 10% of the time overall, it is not the case that gift cards will be used 10% of the time for any specific transaction. As an extreme example, maybe taxi companies disallow use of gift cards on airport trips, and so a gift card is not even a legal value for some historical examples. However, in the absence of any extra information, we will assume that the frequency distribution is the same for all the historical examples.

Static method. Categorical variables are usually one-hot encoded. If we follow the probabilistic approach above and train long enough, the average one-hot encoded value presented to the training program of a "card" in the older data will be [0, 0.1, 0.3, 0.6]. The first 0 corresponds to the cash category. The second number is 0.1 because 10% of the time, on card transactions, this number will be 1 and it will be zero in all other cases. Similarly, we have 0.3 for debit cards and 0.6 for credit cards.

To bridge the older data into the newer schema, we can transform the older categorical data into this representation where we insert the a priori probability of the new classes as estimated from the training data. The newer data, on the other hand, will have [0, 0, 1, 0] for a transaction that is known to have been paid by a debit card.

We recommend the static method over the probabilistic method because it is effectively what happens if the probabilistic method runs for long enough. It is also much simpler to implement since every card payment from the old data will have the exact same value (the 4-element array [0, 0.1, 0.3, 0.6]). We can update the older data in one line of code, rather than writing a script to generate random numbers as in the probabilistic method. It is also computationally much less expensive.

Augmented data

In order to maximize use of the newer data, make sure to use only two splits of the data, which is discussed in "Design Pattern 12: Checkpoints" on page 149 in Chapter 4. Let's say that we have 1 million examples available with the old schema, but only 5,000 examples available with the new schema. How should we create the training and evaluation datasets?

Let's take the evaluation dataset first. It is important to realize that the purpose of training an ML model is to make predictions on unseen data. The unseen data in our case will be exclusively data that matches the new schema. Therefore, we need to set aside a sufficient number of examples from the new data to adequately evaluate generalization performance. Perhaps we need 2,000 examples in our evaluation dataset in order to be confident that the model will perform well in production. The evaluation dataset will not contain any older examples that have been bridged to match the newer schema.

How do we know whether we need 1,000 examples in the evaluation dataset or 2,000? To estimate this number, compute the evaluation metric of the current production model (which was trained on the old schema) on subsets of its evaluation dataset and determine how large the subset has to be before the evaluation metric is consistent.

Computing the evaluation metric on different subsets could be done as follows (as usual, the full code is on GitHub (*https://github.com/GoogleCloudPlatform/ml-design-patterns/blob/master/06_reproducibility/bridging_schema.ipynb*) in the code repository for this book):

```
for subset_size in range(100, 5000, 100):
    sizes.append(subset_size)
    # compute variability of the eval metric
    # at this subset size over 25 tries
    scores = []
    for x in range(1, 25):
        indices = np.random.choice(N_eval,
                        size=subset_size, replace=False)
        scores.append(
            model.score(df_eval[indices],
                    df_old.loc[N_train+indices, 'tip'])
        )
    score_mean.append(np.mean(scores))
    score_stddev.append(np.std(scores))
```

In the code above, we are trying out evaluation sizes of 100, 200, ..., 5,000. At each subset size, we are evaluating the model 25 times, each time on a different, randomly sampled subset of the full evaluation set. Because this is the evaluation set of the current production model (which we were able to train with one million examples), the evaluation dataset here might hold hundreds of thousands of examples. We can then compute the standard deviation of the evaluation metric over the 25 subsets, repeat this on different evaluation sizes, and graph this standard deviation against the evaluation size. The resulting graph will be something like Figure 6-3.

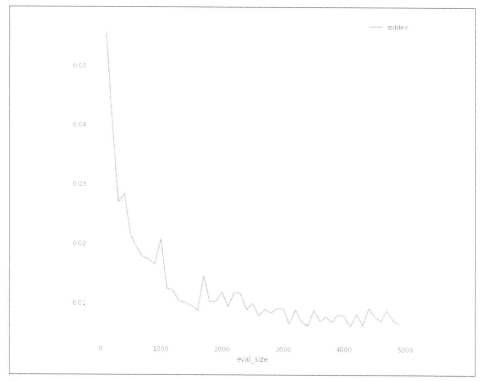

Figure 6-3. Determine the number of evaluation examples needed by evaluating the production model on subsets of varying sizes and tracking the variability of the evaluation metric by the size of the subset. Here, the standard deviation starts to plateau at around 2,000 examples.

From Figure 6-3, we see that the number of evaluation examples needs to be at least 2,000, and is ideally 3,000 or more. Let's assume for the rest of this discussion that we choose to evaluate on 2,500 examples.

The training set would contain the remaining 2,500 new examples (the amount of new data available after withholding 2,500 for evaluation) augmented by some number of older examples that have been bridged to match the new schema. How do we know how many older examples we need? We don't. This is a hyperparameter that we will have to tune. For example, on the tip problem, using grid search, we see from Figure 6-4 (the notebook on GitHub (*https://github.com/GoogleCloudPlatform/ml-design-patterns/blob/master/06_reproducibility/bridging_schema.ipynb*) has the full details) that the evaluation metric drops steeply until 20,000 examples and then starts to plateau.

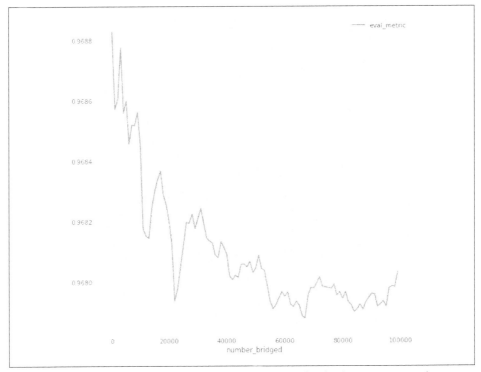

Figure 6-4. Determine the number of older examples to bridge by carrying out hyper-parameter tuning. In this case, it is apparent that there are diminishing returns after 20,000 bridged examples.

For best results, we should choose the smallest number of older examples that we can get away with—ideally, over time, as the number of new examples grows, we'll rely less and less on bridged examples. At some point, we'll be able to get rid of the older examples altogether.

It is worth noting that, on this problem, bridging does bring benefits because when we use no bridged examples, the evaluation metric is worse. If this is not the case, then the imputation method (the method of choosing the static value used for bridging) needs to be reexamined. We suggest an alternate imputation method (Cascade) in the next section.

It is extremely important to compare the performance of the newer model trained on bridged examples against the older, unchanged model on the evaluation dataset. It might be the case that the new information does not yet have adequate value.

Because we will be using the evaluation dataset to test whether or not the bridged model has value, it is critical that the evaluation dataset not be used during training or hyperparameter tuning. So, techniques like early stopping or checkpoint selection must be avoided. Instead, use regularization to control overfitting. The training loss will have to serve as the hyperparameter tuning metric. See the discussion of the Checkpoints design pattern in Chapter 4 for more details on how to conserve data by using only two splits.

Trade-Offs and Alternatives

Let's look at a commonly proposed approach that doesn't work, a complex alternative to bridging, and an extension of the solution to a similar problem.

Union schema

It can be tempting to simply create a union of the older and newer schemas. For example, we could define the schema for the payment type as having five possible values: cash, card, gift card, debit card, and credit card. This will make both the historical data and the newer data valid and is the approach that we would take in data warehousing to deal with changes like this. This way, the old data and the new data are valid as is and without any changes.

The backward-compatible, union-of-schemas approach doesn't work for machine learning though.

At prediction time, we will never get the value "card" for the payment type because the input providers have all been upgraded. Effectively, all those training instances will have been for nought. For reproducibility (this is the reason that this pattern is classified as a reproducibility pattern), we need to bridge the older schema into the newer schema and can't do a union of the two schemas.

Cascade method

Imputation in statistics is a set of techniques that can be used to replace missing data by some valid value. A common imputation technique is to replace a NULL value by the mean value of that column in the training data. Why do we choose the mean? Because, in the absence of any more information and assuming that the values are normally distributed, the most likely value is the mean.

The static method discussed in the main solution, of assigning a priori frequencies, is also an imputation method. We assume that the categorical variable is distributed according to a frequency chart (that we estimate from the training data) and impute the mean one-hot encoded value (according to that frequency distribution) to the "missing" categorical variable.

Do we know any other way to estimate unknown values given some examples? Of course! Machine learning. What we can do is to train a cascade of models (see "Design Pattern 8: Cascade " on page 108 in Chapter 3). The first model uses whatever new examples we have to train a machine learning model to predict the card type. If the original tips model had five inputs, this model will have four inputs. The fifth input (the payment type) will be the label for this model. Then, the output of the first model will be used to train the second model.

In practice, the Cascade pattern adds too much complexity for something that is meant to be a temporary workaround until you have enough new data. The static method is effectively the simplest machine learning model—it's the model we would get if we had uninformative inputs. We recommend the static approach and to use Cascade only if the static method doesn't do well enough.

Handling new features

Another situation where bridging might be needed is when the input provider adds extra information to the input feed. For example, in our taxi fare example, we may start receiving data on whether the taxi's wipers are on or whether the vehicle is moving. From this data, we can craft a feature on whether it was raining at the time the taxi trip started, the fraction of the trip time that the taxi is idle, and so on.

If we have new input features we want to start using immediately, we should bridge the older data (where this new feature will be missing) by imputing a value for the new feature. Recommended choices for the imputation value are:

- The mean value of the feature if the feature is numeric and normally distributed
- The median value of the feature if the feature is numeric and skewed or has lots of outliers
- The median value of the feature if the feature is categorical and sortable
- The mode of the feature if the feature is categorical and not sortable
- The frequency of the feature being true if it is boolean

If the feature is whether or not it was raining, it is boolean, and so the imputed value would be something like 0.02 if it rains 2% of the time in the training dataset. If the feature is the proportion of idle minutes, we could use the median value. The Cascade pattern approach remains viable for all these cases, but a static imputation is simpler and often sufficient.

Handling precision increases

When the input provider increases the precision of their data stream, follow the bridging approach to create a training dataset that consists of the higher-resolution data, augmented with some of the older data.

For floating-point values, it is not necessary to explicitly bridge the older data to match the newer data's precision. To see why, consider the case where some data was originally provided to one decimal place (e.g., 3.5 or 4.2) but is now being provided to two decimal places (e.g., 3.48 or 4.23). If we assume that 3.5 in the older data consists of values that would be uniformly distributed[1] in [3.45, 3.55] in the newer data, the statically imputed value would be 3.5, which is precisely the value that is stored in the older data.

For categorical values—for example, if the older data stored the location as a state or provincial code and the newer data provided the county or district code—use the frequency distribution of counties within states as described in the main solution to carry out static imputation.

Design Pattern 24: Windowed Inference

The Windowed Inference design pattern handles models that require an ongoing sequence of instances in order to run inference. This pattern works by externalizing the model state and invoking the model from a stream analytics pipeline. This pattern is also useful when a machine learning model requires features that need to be computed from aggregates over time windows. By externalizing the state to a stream pipeline, the Windowed Inference design pattern ensures that features calculated in a dynamic, time-dependent way can be correctly repeated between training and serving. It is a way of avoiding training–serving skew in the case of temporal aggregate features.

Problem

Take a look at the arrival delays at Dallas Fort Worth (DFW) airport depicted for a couple of days in May 2010 in Figure 6-5 (the full notebook (*https://github.com/ GoogleCloudPlatform/ml-design-patterns/blob/master/06_reproducibility/state ful_stream.ipynb*) is on GitHub).

---

[1] Note that the overall probability distribution function doesn't need to be uniform—all that we require is that the original bins are narrow enough for us to be able to approximate the probability distribution function by a staircase function. Where this assumption fails is when we have a highly skewed distribution that was inadequately sampled in the older data. In such cases, it is possible that 3.46 is more likely than 3.54, and this would need to be reflected in the bridged dataset.

Figure 6-5. Arrival delays at Dallas Fort Worth (DFW) airport on May 10–11, 2010. Abnormal arrival delays are marked with a dot.

The arrival delays exhibit considerable variability, but it is still possible to note unusually large arrival delays (marked by a dot). Note that the definition of "unusual" varies by context. Early in the morning (left corner of the plot), most flights are on time, so even the small spike is anomalous. By the middle of the day (after 12 p.m. on May 10), variability picks up and 25-minute delays are quite common, but a 75-minute delay is still unusual.

Whether or not a specific delay is anomalous depends on a time context, for example, on the arrival delays observed over the past two hours. To determine that a delay is anomalous requires that we first sort the dataframe based on the time (as in the graph in Figure 6-5 and shown below in pandas):

```
df = df.sort_values(by='scheduled_time').set_index('scheduled_time')
```

Then, we need to apply an anomaly detection function to sliding windows of two hours:

```
df['delay'].rolling('2h').apply(is_anomaly, raw=False)
```

The anomaly detection function, `is_anomaly`, can be quite sophisticated, but let's take the simple case of discarding extrema and calling a data value an anomaly if it is more than four standard deviations from the mean in the two-hour window:

```
def is_anomaly(d):
    outcome = d[-1] # the last item

    # discard min & max value & current (last) item
    xarr = d.drop(index=[d.idxmin(), d.idxmax(), d.index[-1]])
    prediction = xarr.mean()
    acceptable_deviation = 4 * xarr.std()
    return np.abs(outcome - prediction) > acceptable_deviation
```

This works on historical (training) data because the entire dataframe is at hand. Of course, when running inference on our production model, we will not have the entire dataframe. In production, we will be receiving flight arrival information one by one, as each flight arrives. So, all that we will have is a single delay value at a timestamp:

```
2010-02-03 08:45:00,19.0
```

Given that the flight above (at 08:45 on February 3) is 19 minutes late, is that unusual or not? Commonly, to carry out ML inference on a flight, we only need the features of that flight. In this case, however, the model requires information about all flights to DFW airport between 06:45 and 08:45:

```
2010-02-03 06:45:00,?
2010-02-03 06:?:00,?
...
2010-02-03 08:45:00,19.0
```

It is not possible to carry out inference one flight at a time. We need to somehow provide the model information about all the previous flights.

How do we carry out inference when the model requires not just one instance, but a sequence of instances?

Solution

The solution is to carry out stateful stream processing—that is, stream processing that keeps track of the model state through time:

- A sliding window is applied to flight arrival data. The sliding window will be over 2 hours, but the window can be closed more often, such as every 10 minutes. In such a case, aggregate values will be calculated every 10 minutes over the previous 2 hours.

- The internal model state (this could be the list of flights) is updated with flight information every time a new flight arrives, thus building a 2-hour historical record of flight data.

- Every time the window is closed (every 10 minutes in our example), a time-series ML model is trained on the 2-hour list of flights. This model is then used to predict future flight delays and the confidence bounds of such predictions.

- The time-series model parameters are externalized into a state variable. We could use a time-series model such as autoregressive integrated moving average (ARIMA) or long short-term memory (LSTMs), in which case, the model parameters would be the ARIMA model coefficients or the LSTM model weights. To keep the code understandable, we will use a zero-order regression model,[2] and so our model parameters will be the average flight delay and the variance of the flight delays over the two-hour window.

---

2 In other words, we are computing the average.

- When a flight arrives, its arrival delay can be classified as anomalous or not using the externalized model state—it is not necessary to have the full list of flights over the past 2 hours.

We can use Apache Beam for streaming pipelines because then, the same code will work on both the historical data and on newly arriving data. In Apache Beam, the sliding window is set up as follows (full code is on GitHub (*https://github.com/Google CloudPlatform/ml-design-patterns/blob/master/06_reproducibility/find_anoma lies_model.py*)):

```
windowed = (data
        | 'window' >> beam.WindowInto(
              beam.window.SlidingWindows(2 * 60 * 60, 10*60))
```

The model is updated by combining all the flight data collected over the past two hours and passing it to a function that we call ModelFn:

```
model_state = (windowed
        | 'model' >> beam.transforms.CombineGlobally(ModelFn()))
```

ModelFn updates the internal model state with flight information. Here, the internal model state will consist of a pandas dataframe that is updated with the flights in the window:

```
class ModelFn(beam.CombineFn):
    def create_accumulator(self):
        return pd.DataFrame()

    def add_input(self, df, window):
        return df.append(window, ignore_index=True)
```

Every time the window is closed, the output is extracted. The output here (we refer to it as externalized model state) consists of the model parameters:

```
    def extract_output(self, df):
        if len(df) < 1:
            return {}
        orig = df['delay'].values
        xarr = np.delete(orig, [np.argmin(orig), np.argmax(orig)])
        return {
            'prediction': np.mean(xarr),
            'acceptable_deviation': 4 * np.std(xarr)
        }
```

The externalized model state gets updated every 10 minutes based on a 2-hour rolling window:

Window close time	prediction	acceptable_deviation
2010-05-10T06:35:00	-2.8421052631578947	10.48412597725367
2010-05-10T06:45:00	-2.6818181818181817	12.083729926046008
2010-05-10T06:55:00	-2.9615384615384617	11.765962341537781

The code to extract the model parameters shown above is similar to that of the pandas case, but it is done within a Beam pipeline. This allows the code to work in streaming, but the model state is available only within the context of the sliding window. In order to carry out inference on every arriving flight, we need to externalize the model state (similar to how we export the model weights out to a file in the Stateless Serving Function pattern to decouple it from the context of the training program where these weights are computed):

```
model_external = beam.pvalue.AsSingleton(model_state)
```

This externalized state can be used to detect whether or not a given flight is an anomaly:

```
def is_anomaly(flight, model_external_state):
    result = flight.copy()
    error = flight['delay'] - model_external_state['prediction']
    tolerance = model_external_state['acceptable_deviation']
    result['is_anomaly'] = np.abs(error) > tolerance
    return result
```

The is_anomaly function is then applied to every item in the last pane of the sliding window:

```
anomalies = (windowed
        | 'latest_slice' >> beam.FlatMap(is_latest_slice)
        | 'find_anomaly' >> beam.Map(is_anomaly, model_external))
```

Trade-Offs and Alternatives

The solution suggested above is computationally efficient in the case of high-throughput data streams but can be improved further if the ML model parameters can be updated online. This pattern is also applicable to stateful ML models such as recurrent neural networks and when a stateless model requires stateful input features.

Reduce computational overhead

In the Problem section, we used the following pandas code:

```
dfw['delay'].rolling('2h').apply(is_anomaly, raw=False);
```

Whereas, in the Solution section, the Beam code was as follows:

```
windowed = (data
        | 'window' >> beam.WindowInto(
                beam.window.SlidingWindows(2 * 60 * 60, 10*60))
model_state = (windowed
        | 'model' >> beam.transforms.CombineGlobally(ModelFn()))
```

There are meaningful differences between the rolling window in pandas and the sliding window in Apache Beam because of how often the is_anomaly function is called and how often the model parameters (mean and standard deviation) need to be computed. These are discussed below.

Per element versus over a time interval.   In the pandas code, the is_anomaly function is being called on every instance in the dataset. The anomaly detection code computes the model parameters and applies it immediately to the last item in the window. In the Beam pipeline, the model state is also created on every sliding window, but the sliding window in this case is based on time. Therefore, the model parameters are computed just once every 10 minutes.

The anomaly detection itself is carried out on every instance:

```
anomalies = (windowed
        | 'latest_slice' >> beam.FlatMap(is_latest_slice)
        | 'find_anomaly' >> beam.Map(is_anomaly, model_external))
```

Notice that this carefully separates out computationally expensive training from computationally cheap inference. The computationally expensive part is carried out only once every 10 minutes while allowing every instance to be classified as being an anomaly or not.

High-throughput data streams.   Data volumes keep increasing, and much of that increase in data volume is due to real-time data. Consequently, this pattern has to be applied to high-throughput data streams—streams where the number of elements can be in excess of thousands of items per second. Think, for example, of clickstreams from websites or streams of machine activity from computers, wearable devices, or cars.

The suggested solution using a streaming pipeline is advantageous in that it avoids retraining the model at every instance, something that the pandas code in the Problem statement does. However, the suggested solution gives back those gains by creating an in-memory dataframe of all the records received. If we receive 5,000 items a second, then the in-memory dataframe over 10 minutes will contain 3 million records. Because there are 12 sliding windows that will need to be maintained at any point in time (10-minute windows, each over 2 hours), the memory requirements can become considerable.

Storing all the received records in order to compute the model parameters at the end of the window can become problematic. When the data stream is high throughput, it becomes important to be able to update the model parameters with each element. This can be done by changing the ModelFn as follows (full code is on GitHub (*https:// github.com/GoogleCloudPlatform/ml-design-patterns/blob/master/06_reproducibility/ find_anomalies_model.py*)):

```
class OnlineModelFn(beam.CombineFn):
    ...
    def add_input(self, inmem_state, input_dict):
        (sum, sumsq, count) = inmem_state
        input = input_dict['delay']
        return (sum + input, sumsq + input*input, count + 1)

    def extract_output(self, inmem_state):
        (sum, sumsq, count) = inmem_state
        ...
            mean = sum / count
            variance = (sumsq / count) - mean*mean
            stddev = np.sqrt(variance) if variance > 0 else 0
            return {
                'prediction': mean,
                'acceptable_deviation': 4 * stddev
            }
    ...
```

The key difference is that the only thing held in memory are three floating point numbers (sum, sum^2, count) required to extract the output model state, not the entire dataframe of received instances. Updating the model parameters one instance at a time is called an *online update* and is something that can be done only if the model training doesn't require iteration over the entire dataset. Therefore, in the above implementation, the variance is computed by maintaining a sum of x^2 so that we don't need a second pass through the data after computing the mean.

Streaming SQL

If our infrastructure consists of a high-performance SQL database that is capable of processing streaming data, it is possible to implement the Windowed Inference pattern in an alternative way by using an aggregation window (full code is on GitHub (*https://github.com/GoogleCloudPlatform/ml-design-patterns/blob/master/06_reprodu cibility/find_anomalies_model.py*)).

We pull out the flight data from BigQuery:

```
WITH data AS (
  SELECT
    PARSE_DATETIME('%Y-%m-%d-%H%M',
               CONCAT(CAST(date AS STRING),
               '-', FORMAT('%04d', arrival_schedule))
               ) AS scheduled_arrival_time,
      arrival_delay
  FROM `bigquery-samples.airline_ontime_data.flights`
```

```
    WHERE arrival_airport = 'DFW' AND SUBSTR(date, 0, 7) = '2010-05'
  ),
```

Then, we create the `model_state` by computing the model parameters over a time window specified as two hours preceding to one second preceding:

```
model_state AS (
  SELECT
    scheduled_arrival_time,
    arrival_delay,
    AVG(arrival_delay) OVER (time_window) AS prediction,
    4*STDDEV(arrival_delay) OVER (time_window) AS acceptable_deviation
  FROM data
  WINDOW time_window AS
    (ORDER BY UNIX_SECONDS(TIMESTAMP(scheduled_arrival_time))
     RANGE BETWEEN 7200 PRECEDING AND 1 PRECEDING)
)
```

Finally, we apply the anomaly detection algorithm to each instance:

```
SELECT
  *,
  (ABS(arrival_delay - prediction) > acceptable_deviation) AS is_anomaly
FROM model_state
```

The result looks like Table 6-1, with the arrival delay of 54 minutes marked as an anomaly given that all the previous flights arrived early.

Table 6-1. The result of a BigQuery query determining whether incoming flight data is an anomaly

scheduled_arrival_time	arrival_delay	prediction	acceptable_deviation	is_anomaly
2010-05-01T05:45:00	-18.0	-8.25	62.51399843235114	false
2010-05-01T06:00:00	-13.0	-10.2	56.878818553131005	false
2010-05-01T06:35:00	-1.0	-10.666	51.0790237442599	false
2010-05-01T06:45:00	-9.0	-9.28576	48.86521793473886	false
2010-05-01T07:00:00	54.0	-9.25	45.24220532707422	true

Unlike the Apache Beam solution, the efficiency of distributed SQL will allow us to calculate the 2-hour time window centered on each instance (instead of at a resolution of 10-minute windows). However, the drawback is that BigQuery tends to have relatively high latency (on the order of seconds), and so it cannot be used for real-time control applications.

Sequence models

The Windowed Inference pattern of passing a sliding window of previous instances to an inference function is useful beyond anomaly detection or even time-series models. Specifically, it is useful in any class of models, such as Sequence models, that require a historical state. For example, a translation model needs to see several successive words before it can carry out the translation so that the translation takes into account the context of the word. After all, the translation of the words "left," "Chicago," and "road" vary between the sentences "I left Chicago by road" and "Turn left on Chicago Road."

For performance reasons, the translation model will be set up to be stateless and require the user to provide the context. For example, if the model is stateless, instances of the model can be autoscaled in response to increased traffic, and can be invoked in parallel to obtain faster translations. Thus, the translation of the famous soliloquy from Shakespeare's Hamlet into German might follow these steps, picking off in the middle where the bolded word is the one to be translated:

Input (9 words, 4 on either side)	Output
The undiscovered country, from **whose** bourn No traveller returns	dessen
undiscovered country, from whose **bourn** No traveller returns, puzzles	Bourn
country, from whose bourn **No** traveller returns, puzzles the	Kein
from whose bourn No **traveller** returns, puzzles the will,	Reisender

The client, therefore, will need a streaming pipeline. The pipeline could take the input English text, tokenize it, send along nine tokens at a time, collect the outputs, and concatenate them into German sentences and paragraphs.

Most sequence models, such as recurrent neural networks and LSTMs, require streaming pipelines for high-performance inference.

Stateful features

The Windowed Inference pattern can be useful if an input feature to the model requires state, even if the model itself is stateless. For example, suppose we are training a model to predict arrival delays, and one of the inputs to the model is the departure delay. We might want to include, as an input to the model, the average departure delay of flights from that airport in the past two hours.

During training, we can create the dataset using a SQL window function:

```
WITH data AS (
  SELECT
    SAFE.PARSE_DATETIME('%Y-%m-%d-%H%M',
                CONCAT(CAST(date AS STRING), '-',
                FORMAT('%04d', departure_schedule))
                ) AS scheduled_depart_time,
    arrival_delay,
    departure_delay,
    departure_airport
  FROM `bigquery-samples.airline_ontime_data.flights`
  WHERE arrival_airport = 'DFW'
```

```
),
    SELECT
      * EXCEPT(scheduled_depart_time),
      EXTRACT(hour from scheduled_depart_time) AS hour_of_day,
      AVG(departure_delay) OVER (depart_time_window) AS avg_depart_delay
    FROM data
    WINDOW depart_time_window AS
      (PARTITION BY departure_airport ORDER BY
       UNIX_SECONDS(TIMESTAMP(scheduled_depart_time))
       RANGE BETWEEN 7200 PRECEDING AND 1 PRECEDING)
```

The training dataset now includes the average delay as just another feature:

Row	arrival_delay	departure_delay	departure_airport	hour_of_day	avg_depart_delay
1	-3.0	-7.0	LFT	8	**-4.0**
2	56.0	50.0	LFT	8	**41.0**
3	-14.0	-9.0	LFT	8	**5.0**
4	-3.0	0.0	LFT	8	**-2.0**

During inference, though, we will need a streaming pipeline to compute this average departure delay so that we can supply it to the model. To limit training–serving skew, it is preferable to use the same SQL in a tumbling window function in a streaming pipeline, rather than trying to translate the SQL into Scala, Python, or Java.

Batching prediction requests

Another scenario where we might want to use Windowed Inference even if the model is stateless is when the model is deployed on the cloud, but the client is embedded into a device or on-premises. In such cases, the network latency of sending inference requests one by one to a cloud-deployed model might be overwhelming. In this situation, "Design Pattern 19: Two-Phase Predictions" on page 232 from Chapter 5 can be used where the first phase uses a pipeline to collect a number of requests and the second phase sends it to the service in one batch.

This is suitable only for latency-tolerant use cases. If we are collecting input instances over five minutes, then the client will have to be tolerant of up to five minutes delay in getting back the predictions.

Design Pattern 25: Workflow Pipeline

In the Workflow Pipeline design pattern, we address the problem of creating an end-to-end reproducible pipeline by containerizing and orchestrating the steps in our machine learning process. The containerization might be done explicitly, or using a framework that simplifies the process.

Problem

An individual data scientist may be able to run data preprocessing, training, and model deployment steps from end to end (depicted in Figure 6-6) within a single script or notebook. However, as each step in an ML process becomes more complex, and more people within an organization want to contribute to this code base, running these steps from a single notebook will not scale.

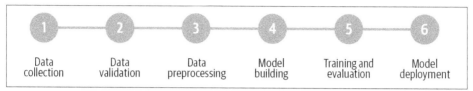

Figure 6-6. The steps in a typical end-to-end ML workflow. This is not meant to be all encompassing, but captures the most common steps in the ML development process.

In traditional programming, *monolithic applications* are described as those where all of the application's logic is handled by a single program. To test a small feature in a monolithic app, we must run the entire program. The same goes for deploying or debugging monolithic applications. Deploying a small bug fix for one piece of the program requires deploying the entire application, which can quickly become unwieldy. When the entire codebase is inextricably linked, it becomes difficult for individual developers to debug errors and work independently on different parts of the application. In recent years, monolithic apps have been replaced in favor of a *microservices* architecture where individual pieces of business logic are built and deployed as isolated (micro) packages of code. With microservices, a large application is split into smaller, more manageable parts so that developers can build, debug, and deploy pieces of an application independently.

This monolith-versus-microservice discussion provides a good analogy for scaling ML workflows, enabling collaboration, and ensuring ML steps are reproducible and reusable across different workflows. When someone is building an ML model on their own, a "monolithic" approach may be faster to iterate on. It also often works because one person is actively involved in developing and maintaining each piece: data gathering and preprocessing, model development, training, and deployment. However, when scaling this workflow, different people or groups in an organization might be responsible for different steps. To scale the ML workflow, we need a way for the team building out the model to run trials independently of the data preprocessing step. We'll also need to track the performance for each step of the pipeline and manage the output files generated by each part of the process.

Additionally, when initial development for each step is complete, we'll want to schedule operations like retraining, or create event-triggered pipeline runs that are invoked in response to changes in your environment, like new training data being added to a bucket. In such cases, it'll be necessary for the solution to allow us to run the entire workflow from end to end in one call while still being able to track output and trace errors from individual steps.

Solution

To handle the problems that come with scaling machine learning processes, we can make each step in our ML workflow a separate, containerized service. Containers guarantee that we'll be able to run the same code in different environments, and that we'll see consistent behavior between runs. These individual containerized steps together are then chained together to make a *pipeline* that can be run with a REST API call. Because pipeline steps run in containers, we can run them on a development laptop, with on-premises infrastructure, or with a hosted cloud service. This pipeline workflow allows team members to build out pipeline steps independently. Containers also provide a reproducible way to run an entire pipeline end to end, since they guarantee consistency among library dependency versions and runtime environments. Additionally, because containerizing pipeline steps allows for a separation of concerns, individual steps can use different runtimes and language versions.

There are many tools for creating pipelines with both on-premise and cloud options available, including Cloud AI Platform Pipelines (*https://oreil.ly/nJo1p*), TensorFlow Extended (*https://oreil.ly/OznI3*) (TFX), Kubeflow Pipelines (*https://oreil.ly/BoegQ*) (KFP), MLflow (*https://mlflow.org*), and Apache Airflow (*https://oreil.ly/63_GG*). To demonstrate the Workflow Pipeline design pattern here, we'll define our pipeline with TFX and run it on Cloud AI Platform Pipelines, a hosted service for running ML pipelines on Google Cloud using Google Kubernetes Engine (GKE) as the underlying container infrastructure.

Steps in TFX pipelines are known as *components*, and both pre-built and customizable components are available. Typically, the first component in a TFX pipeline is one that ingests data from an external source. This is referred to as an `ExampleGen` component where example refers to the machine learning terminology for a labeled instance used for training. `ExampleGen` (*https://oreil.ly/Sjx9F*) components allow you to source data from CSV files, TFRecords, BigQuery, or a custom source. The `BigQueryExampleGen` component, for example, lets us connect data stored in Big-Query to our pipeline by specifying a query that will fetch the data. Then it will store that data as TFRecords in a GCS bucket so that it can be used by the next component. This is a component we customize by passing it a query. These `ExampleGen` components address the data collection phase of an ML workflow outlined in Figure 6-6.

The next step of this workflow is data validation. Once we've ingested data, we can pass it to other components for transformation or analysis before training a model. The `StatisticsGen` (*https://oreil.ly/kX1QY*) component takes data ingested from an `ExampleGen` step and generates summary statistics on the provided data. The

---

SchemaGen (*https://oreil.ly/QpBlu*) outputs the inferred schema from our ingested data. Utilizing the output of SchemaGen, the ExampleValidator (*https://oreil.ly/UD7Uh*) performs anomaly detection on our dataset and checks for signs of data drift or potential training–serving skew.[3] The Transform (*https://oreil.ly/xsJYT*) component also takes output from SchemaGen and is where we perform feature engineering to transform our data input into the right format for our model. This could include converting free-form text inputs into embeddings, normalizing numeric inputs, and more.

Once our data is ready to be fed into a model, we can pass it to the Trainer (*https://oreil.ly/XFtR_*) component. When we set up our Trainer component, we point to a function that defines our model code, and we can specify where we'd like to train the model. Here, we'll show how to use Cloud AI Platform Training from this component. Finally, the Pusher (*https://oreil.ly/qP8GU*) component handles model deployment. There are many other (*https://oreil.ly/gHv_z*) pre-built components provided by TFX—we've only included a few here that we'll use in our sample pipeline.

For this example, we'll use the NOAA hurricane dataset in BigQuery to build a model that infers the SSHS code[4] for a hurricane. We'll keep the features, components, and model code relatively short in order to focus on the pipeline tooling. The steps of our pipeline are outlined below, and roughly follow the workflow outlined in Figure 6-6:

1. Data collection: run a query to get the hurricane data from BigQuery.
2. Data validation: use the ExampleValidator component to identify anomalies and check for data drift.
3. Data analysis and preprocessing: generate some statistics on the data and define the schema.
4. Model training: train a tf.keras model on AI Platform.
5. Model deployment: deploy the trained model to AI Platform Prediction.[5]

When our pipeline is complete, we'll be able to invoke the entire process outlined above with a single API call. Let's start by discussing the scaffolding for a typical TFX pipeline and the process for running it on AI Platform.

---

3 For more on data validation, see "Design Pattern 30: Fairness Lens" on page 343 in Chapter 7, *Responsible AI*.

4 SSHS stands for Saffir–Simpson Hurricane Scale (*https://oreil.ly/62kf3*), and is a scale from 1 to 5 used to measure the strength and severity of a hurricane. Note that the ML model does not forecast the severity of the hurricane at a later time. Instead, it simply learns the wind speed thresholds used in the Saffir–Simpson scale.

5 While deployment is the last step in our example pipeline, production pipelines often include more steps, such as storing the model in a shared repository or executing a separate serving pipeline that does CI/CD and testing.

Building the TFX pipeline

We'll use the `tfx` command-line tools to create and invoke our pipeline. New invocations of a pipeline are known as *runs*, which are distinct from updates we make to the pipeline itself, like adding a new component. We can do both with the TFX CLI. We can define the scaffolding for our pipeline in a single Python script, which has two key parts:

- An instance of tfx.orchestration.pipeline (*https://oreil.ly/62kf3*) where we define our pipeline and the components it includes.
- An instance of kubeflow_dag_runner (*https://oreil.ly/62kf3*) from the tfx (*https://oreil.ly/62kf3*) library. We'll use this to create and run our pipeline. In addition to the Kubeflow runner, there's also an API for running TFX pipelines with Apache Beam (*https://oreil.ly/hn0vF*), which we could use to run our pipeline locally.

Our pipeline (see full code in GitHub (*https://github.com/GoogleCloudPlatform/ml-design-patterns/tree/master/06_reproducibility/workflow_pipeline*)) will have the five steps or components defined above, and we can define our pipeline with the following:

```
pipeline.Pipeline(
    pipeline_name='hurricane_prediction',
    pipeline_root='path/to/pipeline/code',
    components=[
        bigquery_gen, statistics_gen, schema_gen, train, model_pusher
    ]
)
```

To use the `BigQueryExampleGen` component provided by TFX, we provide the query that will fetch our data. We can define this component in one line of code, where query is our BigQuery SQL query as a string:

```
bigquery_gen = BigQueryExampleGen(query=query)
```

Another benefit of using pipelines is that it provides tooling to keep track of the input, output artifacts, and logs for each component. The output of the `statistics_gen` component, for example, is a summary of our dataset, which we can see in Figure 6-7. `statistics_gen` (*https://oreil.ly/wvq9n*) is a pre-built component available in TFX that uses TF Data Validation to generate summary statistics on our dataset.

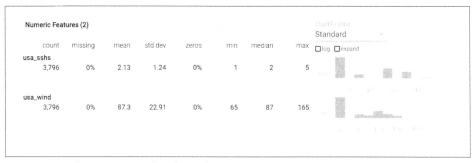

Numeric Features (2)								Chart to show
								Standard
	count	missing	mean	std dev	zeros	min	median	max	☐log ☐expand
usa_sshs								
	3,796	0%	2.13	1.24	0%	1	2	5	
usa_wind								
	3,796	0%	87.3	22.91	0%	65	87	165	

Figure 6-7. The output artifact from the statistics_gen component in a TFX pipeline.

Running the pipeline on Cloud AI Platform

We can run the TFX pipeline on Cloud AI Platform Pipelines, which will manage low-level details of the infrastructure for us. To deploy a pipeline to AI Platform, we package our pipeline code as a Docker container (*https://oreil.ly/rdXeb*) and host it on Google Container Registry (*https://oreil.ly/m5wqD*) (GCR).[6] Once our containerized pipeline code has been pushed to GCR, we'll create the pipeline using the TFX CLI:

```
tfx pipeline create  \
--pipeline-path=kubeflow_dag_runner.py \
--endpoint='your-pipelines-dashboard-url' \
--build-target-image='gcr.io/your-pipeline-container-url'
```

In the command above, endpoint corresponds with the URL of our AI Platform Pipelines dashboard. When that completes, we'll see the pipeline we just created in our pipelines dashboard. The `create` command creates a pipeline *resource* that we can invoke by creating a run:

```
tfx run create --pipeline-name='your-pipeline-name' --endpoint='pipeline-url'
```

After running this command, we'll be able to see a graph that updates in real time as our pipeline moves through each step. From the Pipelines dashboard, we can further examine individual steps to see any artifacts they generate, metadata, and more. We can see an example of the output for an individual step in Figure 6-8.

We could train our model directly in our containerized pipeline on GKE, but TFX provides a utility for using Cloud AI Platform Training as part of our process. TFX also has an extension for deploying our trained model to AI Platform Prediction. We'll utilize both of these integrations in our pipeline. AI Platform Training lets us take advantage of specialized hardware for training our models, such as GPUs or

---

6 Note that in order to run TFX pipelines on AI Platform, you currently need to host your code on GCR and can't use another container registry service like DockerHub.

TPUs, in a cost-effective way. It also provides an option to use distributed training, which can accelerate training time and minimize training cost. We can track individual training jobs and their output within the AI Platform console.

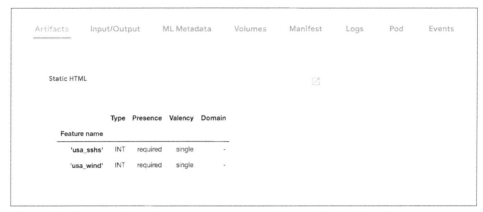

Figure 6-8. Output of the schema_gen component for an ML pipeline. The top menu bar shows the data available for each individual pipeline step.

 One advantage of building a pipeline with TFX or Kubeflow Pipelines is that we are not locked into Google Cloud. We can run the same code we're demonstrating here with Google's AI Platform Pipelines on Azure ML Pipelines (*https://oreil.ly/A5Rxe*), Amazon SageMaker (*https://oreil.ly/H3p3Y*), or on-premises.

To implement a training step in TFX, we'll use the `Trainer` component (*https://oreil.ly/TGKcP*) and pass it information on the training data to use as model input, along with our model training code. TFX provides an extension for running the training step on AI Platform that we can use by importing `tfx.extensions.google_cloud_ai_platform.trainer` and providing details on our AI Platform training configuration. This includes our project name, region, and GCR location of the container with training code.

Similarly, TFX also has an AI Platform `Pusher` (*https://oreil.ly/wS6lc*) component (*https://oreil.ly/bJavO*) for deploying trained models to AI Platform Prediction. In order to use the `Pusher` component with AI Platform, we provide details on the name and version of our model, along with a serving function that tells AI Platform the format of input data it should expect for our model. With that, we have a complete pipeline that ingests data, analyzes it, runs data transformation, and finally trains and deploys the model using AI Platform.

Why It Works

Without running our ML code as a pipeline, it would be difficult for others to reliably reproduce our work. They'd need to take our preprocessing, model development, training, and serving code and try to replicate the same environment where we ran it while taking into account library dependencies, authentication, and more. If there is logic controlling the selection of downstream components based on the output of upstream components, that logic will also have to be reliably replicated. The Workflow Pipeline design pattern lets others run and monitor our entire ML workflow from end to end in both on-premises and cloud environments, while still being able to debug the output of individual steps. Containerizing each step of the pipeline ensures that others will be able to reproduce both the environment we used to build it and the entire workflow captured in the pipeline. This also allows us to potentially reproduce the environment months later to support regulatory needs. With TFX and AI Platform Pipelines, the dashboard also gives us a UI for tracking the output artifacts produced from every pipeline execution. This is discussed further in "Trade-Offs and Alternatives" on page 315.

Additionally, with each pipeline component in its own container, different team members can build and test separate pieces of a pipeline in parallel. This allows for faster development and minimizes the risks associated with a more monolithic ML process where steps are inextricably linked to one another. The package dependencies and code required to build out the data preprocessing step, for example, may be significantly different than those for model deployment. By building these steps as part of a pipeline, each piece can be built in a separate container with its own dependencies and incorporated into a larger pipeline when completed.

To summarize, the Workflow Pipeline pattern gives us the benefits that come with a directed acyclic graph (DAG), along with the pre-built components that come with pipeline frameworks like TFX. Because the pipeline is a DAG, we have the option of executing individual steps or running an entire pipeline from end to end. This also gives us logging and monitoring for each step of the pipeline across different runs, and allows for tracking artifacts from each step and pipeline execution in a centralized place. Pre-built components provide standalone, ready-to-use steps for common components of ML workflows, including training, evaluation, and inference. These components run as individual containers wherever we choose to run our pipeline.

Trade-Offs and Alternatives

The main alternative to using a pipeline framework is to run the steps of our ML workflow using a makeshift approach for keeping track of the notebooks and output associated with each step. Of course, there is some overhead involved in converting the different pieces of our ML workflow into an organized pipeline. In this section, we'll look at some variations and extensions of the Workflow Pipeline design pattern:

creating containers manually, automating a pipeline with tools for continuous integration and continuous delivery (CI/CD), processes for moving from a development to production workflow pipeline, and alternative tools for building and orchestrating pipelines. We'll also explore how to use pipelines for metadata tracking.

Creating custom components

Instead of using pre-built or customizable TFX components to construct our pipeline, we can define our own containers to use as components, or convert a Python function to a component.

To use the container-based components (*https://oreil.ly/5ryEn*) provided by TFX, we use the `create_container_component` method, passing it the inputs and outputs for our component and a base Docker image along with any entrypoint commands for the container. For example, the following container-based component invokes the command-line tool bq to download a BigQuery dataset:

```
component = create_container_component(
    name='DownloadBQData',
    parameters={
        'dataset_name': string,
        'storage_location': string
    },
    image='google/cloud-sdk:278.0.0',
,
    command=[
        'bq', 'extract', '--compression=csv', '--field_delimiter=,',
        InputValuePlaceholder('dataset_name'),
        InputValuePlaceholder('storage_location'),
    ]
)
```

It's best to use a base image that already has most of the dependencies we need. We're using the Google Cloud SDK image, which provides us the bq command-line tool.

It is also possible to convert a custom Python function into a TFX component using the `@component` decorator. To demonstrate it, let's say we have a step for preparing resources used throughout our pipeline where we create a Cloud Storage bucket. We can define this custom step using the following code:

```
from google.cloud import storage
client = storage.Client(project="your-cloud-project")

@component
def CreateBucketComponent(
    bucket_name: Parameter[string] = 'your-bucket-name',
    ) -> OutputDict(bucket_info=string):
  client.create_bucket('gs://' + bucket_name)
  bucket_info = storage_client.get_bucket('gs://' + bucket_name)
```

```
    return {
      'bucket_info': bucket_info
    }
```

We can then add this component to our pipeline definition:

```
create_bucket = CreateBucketComponent(
    bucket_name='my-bucket')
```

Integrating CI/CD with pipelines

In addition to invoking pipelines via the dashboard or programmatically via the CLI or API, chances are we'll want to automate runs of our pipeline as we productionize the model. For example, we may want to invoke our pipeline whenever a certain amount of new training data is available. Or we might want to trigger a pipeline run when the source code for the pipeline changes. Adding CI/CD to our Workflow Pipeline can help connect trigger events to pipeline runs.

There are many managed services available for setting up triggers to run a pipeline when we want to retrain a model on new data. We could use a managed scheduling service to invoke our pipeline on a schedule. Alternatively, we could use a serverless event-based service like Cloud Functions (*https://oreil.ly/rVyzX*) to invoke our pipeline when new data is added to a storage location. In our function, we could specify conditions—like a threshold for the amount of new data added to necessitate retraining—for creating a new pipeline run. Once enough new training data is available, we can instantiate a pipeline run for retraining and redeploying the model as demonstrated in Figure 6-9.

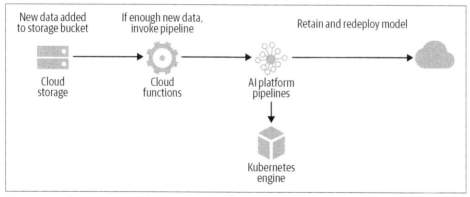

Figure 6-9. A CI/CD workflow using Cloud Functions to invoke a pipeline when enough new data is added to a storage location.

If we want to trigger our pipeline based on changes to source code, a managed CI/CD service like Cloud Build (*https://oreil.ly/kz8Aa*) can help. When Cloud Build executes our code, it is run as a series of containerized steps. This approach fits well within the context of pipelines. We can connect Cloud Build to GitHub Actions (*https://oreil.ly/G2Xwv*) or GitLab Triggers (*https://oreil.ly/m_dYr*) on the repository where our pipeline code is located. When the code is committed, Cloud Build will then build the containers associated with our pipeline based on the new code and create a run.

Apache Airflow and Kubeflow Pipelines

In addition to TFX, Apache Airflow (*https://oreil.ly/rQlqK*) and Kubeflow Pipelines (*https://oreil.ly/e_7zJ*) are both alternatives for implementing the Workflow Pipeline pattern. Like TFX, both Airflow and KFP treat pipelines as a DAG where the workflow for each step is defined in a Python script. They then take this script and provide APIs to handle scheduling and orchestrating the graph on the specified infrastructure. Both Airflow and KFP are open source and can therefore run on-premises or in the cloud.

It's common to use Airflow for data engineering, so it's worth considering for an organization's data ETL tasks. However, while Airflow provides robust tooling for running jobs, it was built as a general-purpose solution and wasn't designed with ML workloads in mind. KFP, on the other hand, was designed specifically for ML and operates at a lower level than TFX, providing more flexibility in how pipeline steps are defined. While TFX implements its own approach to orchestration, KFP lets us choose how to orchestrate our pipelines through its API. The relationship between TFX, KFP, and Kubeflow is summarized in Figure 6-10.

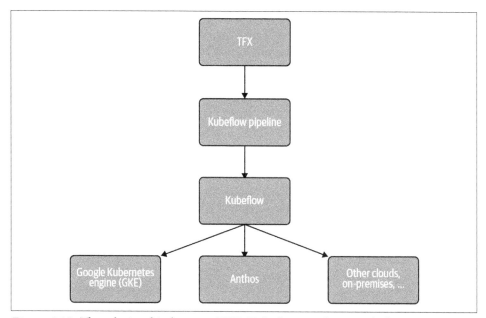

Figure 6-10. The relationship between TFX, Kubeflow Pipelines, Kubeflow, and under-lying infrastructure. TFX operates at the highest level on top of Kubeflow Pipelines, with pre-built components offering specific approaches to common workflow steps. Kubeflow Pipelines provides an API for defining and orchestrating an ML pipeline, providing more flexibility in how each step is implemented. Both TFX and KFP run on Kubeflow, a platform for running container-based ML workloads on Kubernetes. All of the tools in this diagram are open source, so the underlying infrastructure where pipelines run is up to the user—some options include GKE, Anthos, Azure, AWS, or on-premises.

Development versus production pipelines

The way a pipeline is invoked often changes as we move from development to production. We'll likely want to build and prototype our pipeline from a notebook, where we can re-invoke our pipeline by running a notebook cell, debug errors, and update code all from the same environment. Once we're ready to productionize, we can move our component code and pipeline definition to a single script. With our pipeline defined in a script, we'll be able to schedule runs and make it easier for others in our organization to invoke the pipeline in a reproducible way. One tool available for productionizing pipelines is Kale (*https://github.com/kubeflow-kale/kale*), which takes Jupyter notebook code and converts it into a script using the Kubeflow Pipelines API.

A production pipeline also allows for *orchestration* of an ML workflow. By orchestration, we mean adding logic to our pipeline to determine which steps will be executed, and what the outcome of those steps will be. For example, we might decide we only want to deploy models to production that have 95% accuracy or higher. When newly available data triggers a pipeline run and trains an updated model, we can add logic to check the output of our evaluation component to execute the deployment component if the accuracy is above our threshold, or end the pipeline run if not. Both Airflow and Kubeflow Pipelines, discussed previously in this section, provide APIs for pipeline orchestration.

Lineage tracking in ML pipelines

One additional feature of pipelines is using them for tracking model metadata and artifacts, also known as *lineage tracking*. Each time we invoke a pipeline, a series of artifacts is generated. These artifacts could include dataset summaries, exported models, model evaluation results, metadata on specific pipeline invocations, and more. Lineage tracking lets us visualize the history of our model versions along with other associated model artifacts. In AI Platform Pipelines, for example, we can use the pipelines dashboard to see which data a model version was trained on, broken down both by data schema and date. Figure 6-11 shows the Lineage Explorer dashboard for a TFX pipeline running on AI Platform. This allows us to track the input and output artifacts associated with a particular model.

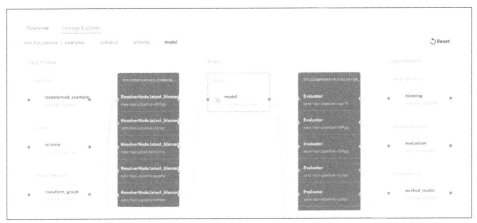

Figure 6-11. The Lineage Explorer section of the AI Platform Pipelines dashboard for a TFX pipeline.

One benefit of using lineage tracking to manage artifacts generated during our pipeline run is that it supports both cloud-based and on-premises environments. This gives us flexibility in where models are trained and deployed, and where model metadata is stored. Lineage tracking is also an important aspect of making ML pipelines reproducible, since it allows for comparisons between metadata and artifacts from different pipeline runs.

Design Pattern 26: Feature Store

The *Feature Store* design pattern simplifies the management and reuse of features across projects by decoupling the feature creation process from the development of models using those features.

Problem

Good feature engineering is crucial for the success of many machine learning solutions. However, it is also one of the most time-consuming parts of model development. Some features require significant domain knowledge to calculate correctly, and changes in the business strategy can affect how a feature should be computed. To ensure such features are computed in a consistent way, it's better for these features to be under the control of domain experts rather than ML engineers. Some input fields might allow for different choices of data representations (see Chapter 2) to make them more amenable for machine learning. An ML engineer or data scientist will typically experiment with multiple different transformations to determine which are helpful and which aren't, before deciding which features will be used in the final model. Many times, the data used for the ML model isn't drawn from a single source. Some data may come from a data warehouse, some data may sit in a storage bucket as unstructured data, and other data may be collected in real time through streaming. The structure of the data may also vary between each of these sources, requiring each input to have its own feature engineering steps before it can be fed into a model. This development is often done on a VM or personal machine, causing the feature creation to be tied to the software environment where the model is built, and the more complex the model gets, the more complicated these data pipelines become.

An ad hoc approach where features are created as needed by ML projects may work for one-off model development and training, but as organizations scale, this method of feature engineering becomes impractical and significant problems arise:

- Ad hoc features aren't easily reused. Features are re-created over and over again, either by individual users or within teams, or never leave the pipelines (or notebooks) in which they are created. This is particularly problematic for higher-level features that are complex to calculate. This could be because they are derived through expensive processes, such as pre-trained user or catalog item embeddings. Other times, it could be because the features are captured from upstream processes such as business priorities, availability of contracting, or market segmentations. Another source of complexity is when higher-level features, such as the number of orders by a customer in the past month, involve aggregations over time. Effort and time are wasted creating the same features from scratch for each new project.

- Data governance is made difficult if each ML project computes features from sensitive data differently.

- Ad hoc features aren't easily shared between teams or across projects. In many organizations, the same raw data is used by multiple teams, but separate teams may define features differently and there is no easy access to feature documentation. This also hinders effective cross-collaboration of teams, leading to siloed work and unnecessarily duplicated effort.

- Ad hoc features used for training and serving are inconsistent—i.e., training–serving skew. Training is typically done using historical data with batch features that are created offline. However, serving is typically carried out online. If the feature pipeline for training differs at all from the pipeline used in production for serving (for example, different libraries, preprocessing code, or languages), then we run the risk of training–serving skew.

- Productionizing features is difficult. When moving to production, there is no standardized framework to serve features for online ML models and to serve batch features for offline model training. Models are trained offline using features created in batch processes, but when served in production, these features are often created with an emphasis on low latency and less on high throughput. The framework for feature generation and storage is not flexible to handle both of these scenarios.

In short, the ad hoc approach to feature engineering slows model development and leads to duplicated effort and work stream inefficiency. Furthermore, feature creation is inconsistent between training and inference, running the risk of training–serving skew or data leakage by accidentally introducing label information into the model input pipeline.

Solution

The solution is to create a shared feature store, a centralized location to store and document feature datasets that will be used in building machine learning models and can be shared across projects and teams. The feature store acts as the interface between the data engineer's pipelines for feature creation and the data scientist's workflow building models using those features (Figure 6-12). This way, there is a central repository to house precomputed features, which speeds development time and aids in feature discovery. This also allows the basic software engineering principles of versioning, documentation, and access control to be applied to the features that are created.

A typical feature store is built with two key design characteristics: tooling to process large feature data sets quickly, and a way to store features that supports both low-latency access (for inference) and large batch access (for model training). There is also a metadata layer that simplifies documentation and versioning of different feature sets and an API that manages loading and retrieving feature data.

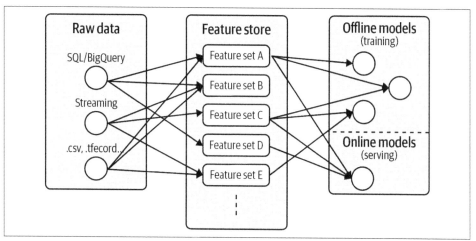

Figure 6-12. A feature store provides a bridge between raw data sources and model training and serving.

The typical workflow of a data or ML engineer is to read raw data (structured or streaming) from a data source, apply various transformations on the data using their favorite processing framework, and store the transformed feature within the feature store. Rather than creating feature pipelines to support a single ML model, the Feature Store pattern decouples feature engineering from model development. In particular, tools like Apache Beam, Flink, or Spark are often used when building a feature store since they can handle processing data in batch as well as streaming. This also reduces the incidence of training–serving skew, since the feature data is populated by the same feature creation pipelines.

After features are created, they are housed in a data store to be retrieved for training and serving. For serving feature retrieval, speed is optimized. A model in production backing some online application may need to produce real-time predictions within milliseconds, making low latency essential. However, for training, higher latency is not a problem. Instead the emphasis is on high throughput since historical features are pulled in large batches for training. A feature store addresses both these use cases by using different data stores for online and offline feature access. For example, a feature store may use Cassandra or Redis as a data store for online feature retrieval, and Hive or BigQuery for fetching historical, large batch feature sets.

In the end, a typical feature store will house many different feature sets containing features created from myriad raw data sources. The metadata layer is built in to document feature sets and provide a registry for easy feature discovery and cross collaboration among teams.

Feast

As an example of this pattern in action, consider Feast (*https://github.com/feast-dev*), which is an open source feature store for machine learning developed by Google Cloud and Gojek (*https://oreil.ly/PszIn*). It is built around Google Cloud services (*https://oreil.ly/ecJou*) using Big Query for offline model training and Redis for low-latency, online serving (Figure 6-13). Apache Beam is used for feature creation, which allows for consistent data pipelines for both batch and streaming.

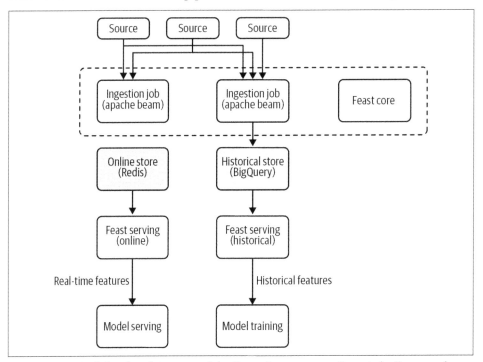

Figure 6-13. High-level architecture of the Feast feature store. Feast is built around Google BigQuery, Redis, and Apache Beam.

To see how this works in practice, we'll use a public BigQuery dataset containing information about taxi rides in New York City.[7] Each row of the table contains a timestamp of the pickup, the pickup latitude and longitude, the dropoff latitude and longitude, the number of passengers, and the cost of the taxi ride. The goal of the ML model will be to predict the cost of the taxi ride, denoted `fare_amount`, using these characteristics.

---

7 The data is available in the BigQuery table: *bigquery-public-data.new_york_taxi_trips.tlc_yellow_trips_2016*.

This model benefits from engineering additional features from the raw data. For example, since taxi rides are based on the distance and duration of the trip, pre-computing the distance between the pickup and dropoff is a useful feature. Once this feature is computed on the dataset, we can store it within a feature set for future use.

Adding feature data to Feast. Data is stored in Feast using `FeatureSets`. A `FeatureSet` contains the data schema and data source information, whether it is coming from a pandas dataframe or a streaming Kafka topic. `FeatureSets` are how Feast knows where to source the data it needs for a feature, how to ingest it, and some basic characteristics about the data types. Groups of features can be ingested and stored together, and feature sets provide efficient storage and logical namespacing of data within these stores.

Once our feature set is registered, Feast will start an Apache Beam job to populate the feature store with data from the source. A feature set is used to generate both offline and online feature stores, which ensures developers train and serve their model with the same data. Feast ensures that the source data complies with the expected schema of the feature set.

There are four steps to ingest feature data into Feast, as shown in Figure 6-14.

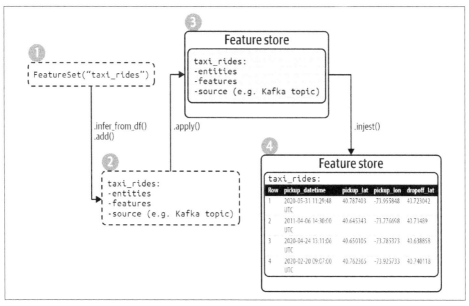

Figure 6-14. There are four steps to ingesting feature data into Feast: create a Feature-Set, add entities and features, register the FeatureSet, and ingest feature data into the FeatureSet.

The four steps are as follows:

1. Create a `FeatureSet`. The feature set specifies the entities, features, and source.

2. Add entities and features to the `FeatureSet`.

3. Register the `FeatureSet`. This creates a named feature set within Feast. The feature set contains no feature data.

4. Load feature data into the `FeatureSet`.

A notebook with the full code for this example can be found in the repository accompanying this book (*https://github.com/GoogleCloudPlatform/ml-design-patterns/blob/master/06_reproducibility/feature_store.ipynb*).

Creating a FeatureSet. We connect to a Feast deployment by setting up a client with the Python SDK:

```
from feast import Client, FeatureSet, Entity, ValueType

# Connect to an existing Feast deployment
client = Client(core_url='localhost:6565')
```

We can check that the client is connected by printing the existing feature sets with the command `client.list_feature_sets()`. If this is a new deployment, this will return an empty list. To create a new feature set, call the class `FeatureSet` and specify the feature set's name:

```
# Create a feature set
taxi_fs = FeatureSet("taxi_rides")
```

Adding entities and features to the FeatureSet. In the context of Feast, `FeatureSets` consist of entities and features. Entities are used as keys to look up feature values and are used to join features between different feature sets when creating datasets for training or serving. The entity serves as an identifier for whatever relevant characteristic you have in the dataset. It is an object that can be modeled and store information. In the context of a ride-hailing or food delivery service, a relevant entity could be `customer_id`, `order_id`, `driver_id`, or `restaurant_id`. In the context of a churn model, an entity could be a `customer_id` or `segment_id`. Here, the entity is the `taxi_id`, a unique identifier for the taxi vendor of each trip.

At this stage, the feature set we created called `taxi_rides` contains no entities or features. We can use the Feast core client to specify these from a pandas dataframe that contains the raw data inputs and entities as shown in Table 6-2.

Table 6-2. The taxi ride dataset contains information about taxi rides in New York. The entity is the taxi_id, a unique identifier for the taxi vendor of each trip

Row	pickup_datetime	pickup_lat	pickup_lon	dropoff_lat	dropoff_lon	num_pass	taxi_id	fare_amt
1	2020-05-31 11:29:48 UTC	40.787403	-73.955848	40.723042	-73.993106	2	0	15.3
2	2011-04-06 14:30:00 UTC	40.645343	-73.776698	40.71489	-73.987242	2	0	45.0
3	2020-04-24 13:11:06 UTC	40.650105	-73.785373	40.638858	-73.9678	2	2	32.1
4	2020-02-20 09:07:00 UTC	40.762365	-73.925733	40.740118	-73.986487	2	1	21.3

Defining Streaming Data Sources when Creating a Feature Set

Users can define streaming data sources when creating a feature set. Once a feature set is registered with a source, Feast will automatically start to populate its stores with data from this source. This is an example of a feature set with a user-provided source that retrieves streaming data from a Kafka topic:

```
feature_set = FeatureSet(
    name="stream_feature",
    entities=[
        Entity("taxi_id", ValueType.INT64)
    ],
    features=[
        Feature("traffic_last_5min", ValueType.INT64)
    ],
    source=KafkaSource(
        brokers="mybroker:9092",
        topic="my_feature_topic"
    )
)
```

The pickup_datetime timestamp here is important since it is necessary to retrieve batch features and is used to ensure time-correct joins for batch features. To create an additional feature, such as the Euclidean distance, load the dataset into a pandas dataframe and compute the feature:

```
# Load dataframe
taxi_df = pd.read_csv("taxi-train.csv")

# Engineer features, Euclidean distance
taxi_df['euclid_dist'] = taxi_df.apply(compute_dist, axis=1)
```

We can add entities and features to the feature set with .add(...). Alternatively, the method .infer_fields_from_df(...) will create the entities and features for our

FeatureSet directly from the pandas dataframe. We simply specify the column name that represents the entity. The schema and data types for the features of the Feature Set are then inferred from the dataframe:

```
# Infer the features of the feature set from the pandas DataFrame
    taxi_fs.infer_fields_from_df(taxi_df,
                entities=[Entity(name='taxi_id', dtype=ValueType.INT64)],
    replace_existing_features=True)
```

Registering the FeatureSet. Once the FeatureSet is created, we can register it with Feast using client.apply(taxi_fs). To confirm that the feature set was correctly registered or to explore the contents of another feature set, we can retrieve it using .get_feature_set(...):

```
print(client.get_feature_set("taxi_rides"))
```

This returns a JSON object containing the data schema for the taxi_rides feature set:

```
{
  "spec": {
    "name": "taxi_rides",
    "entities": [
      {
        "name": "key",
        "valueType": "INT64"
      }
    ],
    "features": [
      {
        "name": "dropoff_lon",
        "valueType": "DOUBLE"
      },
      {
        "name": "pickup_lon",
        "valueType": "DOUBLE"
      },
      ...
    ...
    ],
    }
}
```

Ingesting feature data into the FeatureSet. Once we are happy with our schema, we can ingest the dataframe feature data into Feast using .ingest(...). We'll specify the FeatureSet, called taxi_fs, and the dataframe from which to populate the feature data, called taxi_df.

```
# Load feature data into Feast for this specific feature set
client.ingest(taxi_fs, taxi_df)
```

Progress during this ingestion step is printed to the screen showing that we've ingested 28,247 rows into the `taxi_rides` feature set within Feast:

```
100%|██████████|28247/28247 [00:02<00:00, 2771.19rows/s]
Ingestion complete!

Ingestion statistics:
Success: 28247/28247 rows ingested
```

At this stage, calling `client.list_feature_sets()` will now list the feature set `taxi_rides` we just created and return [`default/taxi_rides`]. Here, `default` refers to the project scope of the feature set within Feast. This can be changed when instantiating the feature set to keep certain feature sets within project access.

 Datasets may change over time, causing feature sets to change as well. In Feast, once a feature set is created, there are only a few changes that can be made. For example, the following changes are allowed:

- Adding new features.
- Removing existing features. (Note that features are tombstoned and remain on record, so they are not removed completely. This will affect new features being able to take the names of previously deleted features.)
- Changing features' schemas.
- Changing the feature set's source or the `max_age` of the feature set examples.

The following changes are *not* allowed:

- Changes to the feature set name.
- Changes to entities.
- Changes to names of existing features.

Retrieving data from Feast

Once a feature set has been sourced with features, we can retrieve historical or online features. Users and production systems retrieve feature data through a Feast serving data access layer. Since Feast supports both offline and online store types, it's common to have Feast deployments for both, as shown in Figure 6-15. The same feature data is contained within the two feature stores, ensuring consistency between training and serving.

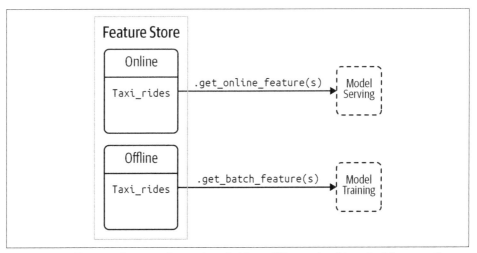

Figure 6-15. Feature data can be retrieved either offline, using historical features for model training, or online, for serving.

These deployments are accessed via a separate online and batch client:

```
_feast_online_client = Client(serving_url='localhost:6566')
_feast_batch_client = Client(serving_url='localhost:6567',
                             core_url='localhost:6565')
```

Batch serving. For training a model, historical feature retrieval is backed by BigQuery and accessed using `.get_batch_features(...)` with the batch serving client. In this case, we provide Feast with a pandas dataframe containing the entities and time-stamps that feature data will be joined to. This allows Feast to produce a point-in-time correct dataset based on the features that have been requested:

```
# Create a entity df of all entities and timestamps
entity_df = pd.DataFrame(
    {
        "datetime": taxi_df.datetime,
        "taxi_id": taxi_df.taxi_id,
    }
)
```

To retrieve historical features, the features in the feature set are referenced by the fea-ture set name and the feature name, separated by a colon—for example, `taxi_rides:pickup_lat`:

```
    FS_NAME = taxi_rides
model_features = ['pickup_lat',
                  'pickup_lon',
                  'dropoff_lat',
                  'dropoff_lon',
                  'num_pass',
```

```
                          'euclid_dist']
        label = 'fare_amt'

        features = model_features + [label]

    # Retrieve training dataset from Feast
    dataset = _feast_batch_client.get_batch_features(
        feature_refs=[FS_NAME + ":" + feature for feature in features],
        entity_rows=entity_df).to_dataframe()
```

The dataframe dataset now contains all features and the label for our model, pulled directly from the feature store.

Online serving.   For online serving, Feast only stores the latest entity values, as opposed to historical serving where all historical values are stored. Online serving with Feast is built to be very low latency, and Feast provides a gRPC API backed by Redis (*https://redis.io*). To retrieve online features, for example, when making online predictions with the trained model, we use .get_online_features(...) specifying the features we want to capture and the entity:

```
    # retrieve online features for a single taxi_id
    online_features = _feast_online_client.get_online_features(
        feature_refs=["taxi_rides:pickup_lat",
    "taxi_rides:pickup_lon",
        "taxi_rides:dropoff_lat",
    "taxi_rides:dropoff_lon",
                        "taxi_rides:num_pass",
    "taxi_rides:euclid_dist"],
        entity_rows=[
            GetOnlineFeaturesRequest.EntityRow(
                fields={
                    "taxi_id": Value(
                        int64_val=5)
                }
            )
        ]
    )
```

This saves online_features as a list of maps where each item in the list contains the latest feature values for the provided entity, here, taxi_id = 5:

```
    field_values {
      fields {
        key: "taxi_id"
        value {
          int64_val: 5
        }
      }
      fields {
        key: "taxi_rides:dropoff_lat"
        value {
```

```
      double_val: 40.78923797607422
    }
  }
  fields {
    key: "taxi_rides:dropoff_lon"
    value {
      double_val: -73.96871948242188
    }
  ...
```

To make an online prediction for this example, we pass the field values from the object returned in `online_features` as a pandas dataframe called `predict_df` to `model.predict`:

```
predict_df = pd.DataFrame.from_dict(online_features_dict)
model.predict(predict_df)
```

Why It Works

Feature stores work because they decouple feature engineering from feature usage, allowing feature development and creation to occur independently from the consumption of features during model development. As features are added to the feature store, they become available immediately for both training and serving and are stored in a single location. This ensures consistency between model training and serving.

For example, a model served as a customer-facing application may receive only 10 input values from a client, but those 10 inputs may need to be transformed into many more features via feature engineering before being sent to a model. Those engineered features are maintained within the feature store. It is crucial that the pipeline for retrieving features during development is the same as when serving the model. A feature store ensures that consistency (Figure 6-16).

Feast accomplishes this by using Beam on the backend for feature ingestion pipelines that write feature values into the feature sets, and uses Redis and BigQuery for online and offline (respectively) feature retrieval (Figure 6-17).[8] As with any feature store, the ingestion pipeline also handles partial failure or race conditions that might cause some data to be in one storage but not the other.

---

8 See the Gojek blog, "Feast: Bridging ML Models and Data (*https://oreil.ly/YVta5*)."

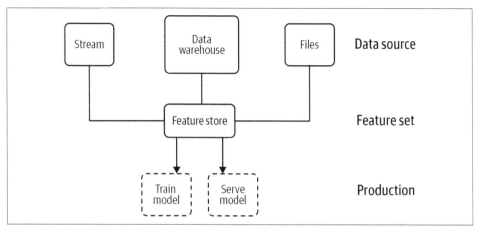

Figure 6-16. A feature store ensures the feature engineering pipelines are consistent between model training and serving. See also https://docs.feast.dev/.

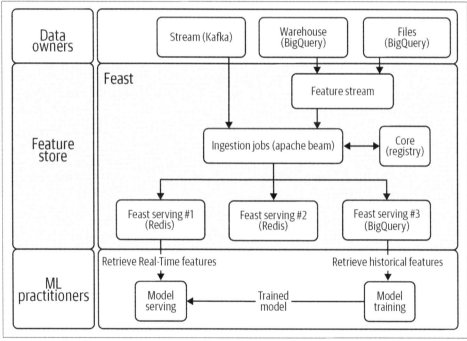

Figure 6-17. Feast uses Beam on the backend for feature ingestion and Redis and BigQuery for online and offline feature retrieval.

Different systems may produce data at different rates, and a feature store is flexible enough to handle those different cadences, both for ingestion and during retrieval (Figure 6-18). For example, sensor data could be produced in real time, arriving every second, or there could be a monthly file that is generated from an external system reporting a summary of the last month's transactions. Each of these need to be processed and ingested into the feature store. By the same token, there may be different time horizons for retrieving data from the feature store. For example, a user-facing online application may operate at very low latency using up-to-the-second features, whereas when training the model, features are pulled offline as a larger batch but with higher latency.

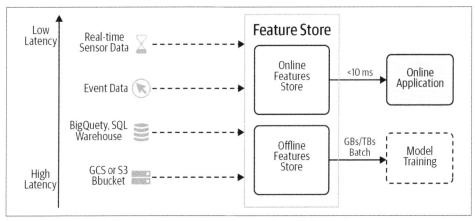

Figure 6-18. The Feature Store design pattern can handle both the requirements of data being highly scalable for large batches during training and extremely low latency for serving online applications.

There is no single database that can handle both scaling to potentially terabytes of data *and* extremely low latency on the order of milliseconds. The feature store achieves this with separate online and offline feature stores and ensures that features are handled in a consistent fashion in both scenarios.

Lastly, a feature store acts as a version-controlled repository for feature datasets, allowing the same CI/CD practices of code and model development to be applied to the feature engineering process. This means that new ML projects start with a process of feature selection from a catalog instead of having to do feature engineering from scratch, allowing organizations to achieve an economies-of-scale effect—as new features are created and added to the feature store, it becomes easier and faster to build new models that reuse those features.

Trade-Offs and Alternatives

The Feast framework that we discussed is built on Google BigQuery, Redis, and Apache Beam. However, there are feature stores that rely on other tools and tech stacks. And, although a feature store is the recommended way to manage features at scale, `tf.transform` provides an alternative solution that addresses the issue of

---

training–serving skew, but not feature reusability. There are also some alternative uses of a feature store that we have not yet detailed, such as how a feature store handles data from different sources and data arriving at different cadences.

Alternative implementations

Many large technology companies, like Uber, LinkedIn, Airbnb, Netflix, and Comcast, host their own version of a feature store, though the architectures and tools vary. Uber's Michelangelo Palette is built around Spark/Scala using Hive for offline feature creation and Cassandra for online features. Hopsworks provides another open source feature store alternative to Feast and is built around dataframes using Spark and pandas with Hive for offline and MySQL Cluster for online feature access. Airbnb built their own feature store as part of their production ML framework called Zipline. It uses Spark and Flink for feature engineering jobs and Hive for feature storage.

Whichever tech stack is used, the primary components of the feature store are the same:

- A tool to process large feature engineering jobs quickly, such as Spark, Flink or Beam.

- A storage component for housing the feature sets that are created, such as Hive, cloud storage (Amazon S3, Google Cloud Storage), BigQuery, Redis, BigTable, and/or Cassandra. The combination that Feast uses (BigQuery and Redis) is optimized for offline versus online (low-latency) feature retrieval.

- A metadata layer to record feature version information, documentation, and feature registry to simplify discovery and sharing of feature sets.

- An API for ingesting and retrieving features to/from the feature store.

Transform design pattern

If feature engineering code is not the same during training and inference, there is a risk that the two code sources will not be consistent. This leads to training–serving skew, and model predictions may not be reliable since the features may not be the same. Feature stores get around this problem by having their feature engineering jobs write feature data to both an online and an offline database. And, while a feature store itself doesn't perform the feature transformations, it provides a way to separate the upstream feature engineering steps from model serving and provide point in time correctness.

The Transform design pattern discussed in this chapter also provides a way to keep feature transformations separate and reproducible. For example, `tf.transform` can be used to preprocess data using exactly the same code for both training a model and serving predictions in production, thus eliminating training–serving skew. This ensures that training and serving feature engineering pipelines are consistent.

However, the feature store provides an added advantage of feature reusability that `tf.transform` does not have. Although `tf.transform` pipelines ensure reproducibility, the features are created and developed only for that model and are not easily shared or reused by other models and pipelines.

On the other hand, `tf.transform` takes special care to ensure that feature creation during serving is carried out on accelerated hardware, since it is part of the serving graph. Feature stores typically do not provide this capability today.

Design Pattern 27: Model Versioning

In the Model Versioning design pattern, backward compatibility is achieved by deploying a changed model as a microservice with a different REST endpoint. This is a necessary prerequisite for many of the other patterns discussed in this chapter.

Problem

As we've seen with *data drift* (introduced in Chapter 1), models can become stale over time and need to be updated regularly to make sure they reflect an organization's changing goals, and the environment associated with their training data. Deploying model updates to production will inevitably affect the way models behave on new data, which presents a challenge—we need an approach for keeping production models up to date while still ensuring backward compatibility for existing model users.

Updates to an existing model might include changing a model's architecture in order to improve accuracy, or retraining a model on more recent data to address drift. While these types of changes likely won't require a different model output format, they will affect the prediction results users get from a model. As an example, let's imagine we're building a model that predicts the genre of a book from its description and uses the predicted genres to make recommendations to users. We trained our initial model on a dataset of older classic books, but now have access to new data on thousands of more recent books to use for training. Training on this updated dataset improves our overall model accuracy, but slightly reduces accuracy on older "classic" books. To handle this, we'll need a solution that lets users choose an older version of our model if they prefer.

Alternatively, our model's end users might start to require more information on *how* the model is arriving at a specific prediction. In a medical use case, a doctor might need to see the regions in an x-ray that caused a model to predict the presence of disease rather than rely solely on the predicted label. In this case, the response from a deployed model would need to be updated to include these highlighted regions. This process is known as *explainability* and is discussed further in Chapter 7.

When we deploy updates to our model, we'll also likely want a way to track how the model is performing in production and compare this with previous iterations. We may also want a way to test a new model with only a subset of our users. Both performance monitoring and split testing, along with other possible model changes, will be difficult to solve by replacing a single production model each time we make updates. Doing this will break applications that are relying on our model output to match a specific format. To handle this, we'll need a solution that allows us to continuously update our model without breaking existing users.

Solution

To gracefully handle updates to a model, deploy multiple model versions with different REST endpoints. This ensures backward compatibility—by keeping multiple versions of a model deployed at a given time, those users relying on older versions will still be able to use the service. Versioning also allows for fine-grained performance monitoring and analytics tracking across versions. We can compare accuracy and usage statistics, and use this to determine when to take a particular version offline. If we have a model update that we want to test with only a small subset of users, the Model Versioning design pattern makes it possible to perform A/B testing.

Additionally, with model versioning, each deployed version of our model is a microservice—thus decoupling changes to our model from our application frontend. To add support for a new version, our team's application developers only need to change the name of the API endpoint pointing to the model. Of course, if a new model version introduces changes to the model's response format, we'll need to make changes to our app to accommodate this, but the model and application code are still separate. Data scientists or ML engineers can therefore deploy and test a new model version on our own without worrying about breaking our production app.

Types of model users

When we refer to "end users" of our model, this includes two different groups of people. If we're making our model API endpoint available to application developers outside our organization, these developers can be thought of as one type of model user. They are building applications that rely on our model for serving predictions to others. The backward compatibility benefit that comes with model versioning is most important for these users. If the format of our model's response changes, application developers may want to use an older model version until they've updated their application code to support the latest response format.

The other group of end users refers to those using an application that calls our deployed model. This could be a doctor relying on our model to predict the presence of disease in an image, someone using our book recommendation app, our organization's business unit analyzing the output of a revenue prediction model we built, and more. This group of users is less likely to run into backward compatibility issues, but may want the option to choose when to start using a new feature in our app. Also, if we can break users into distinct groups (i.e., based on their app usage), we can serve each group different model versions based on their preferences.

Model versioning with a managed service

To demonstrate versioning, we'll build a model that predicts flight delays and deploy this model to Cloud AI Platform Prediction (*https://oreil.ly/-GAVQ*). Because we looked at TensorFlow's SavedModel in previous chapters, we'll use an XGBoost model here.

Once we've trained our model, we can export it to get it ready for serving:

```
model.save_model('model.bst')
```

To deploy this model to AI Platform, we need to create a model version that will point to this `model.bst` in a Cloud Storage Bucket.

In AI Platform, a model resource can have many versions associated with it. To create a new version using the gcloud CLI, we'll run the following in a Terminal:

```
gcloud ai-platform versions create 'v1' \
  --model 'flight_delay_prediction' \
  --origin gs://your-gcs-bucket \
  --runtime-version=1.15 \
  --framework 'XGBOOST' \
  --python-version=3.7
```

With this model deployed, it's now accessible via the endpoint */models/ flight_delay_predictions/versions/v1* in an HTTPS URL tied to our project. Since this is the only version we've deployed so far, it's considered the *default*. This means that if we don't specify a version in our API request, the prediction service will use v1. Now we can make predictions to our deployed model by sending it examples in the format our model expects—in this case, a 110-element array of dummy-coded airport codes (for the full code, see the notebook on GitHub (*https://github.com/GoogleCloud Platform/ml-design-patterns/blob/master/06_reproducibility/model_version ing.ipynb*)). The model returns sigmoid output, a float value between 0 and 1 indicating the likelihood a given flight was delayed more than 30 minutes.

To make a prediction request to our deployed model, we'll use the following gcloud command, where *input.json* is a file with our newline delimited examples to send for prediction:

```
gcloud ai-platform predict --model 'flight_delay_prediction'
--version 'v1'
--json-request 'input.json'
```

If we send five examples for prediction, we'll get a five-element array back corresponding with the sigmoid output for each test example, like the following:

```
[0.019, 0.998, 0.213, 0.002, 0.004]
```

Now that we have a working model in production, let's imagine that our data science team decides to change the model from XGBoost to TensorFlow since it results in improved accuracy and gives them access to additional tooling in the TensorFlow ecosystem. The model has the same input and output format, but its architecture and exported asset format has changed. Instead of a *.bst* file, our model is now in the TensorFlow SavedModel format. Ideally we can keep our underlying model assets separate from our application frontend—this will allow application developers to focus on our application's functionality, rather than a change in model formatting that won't affect the way end users interact with the model. This is where model versioning can help. We'll deploy our TensorFlow model as a second version under the same `flight_delay_prediction` model resource. End users can upgrade to the new version for improved performance simply by changing the version name in the API endpoint.

To deploy our second version, we'll export the model and copy it to a new subdirectory in the bucket we used previously. We can use the same deploy command as above, replacing the version name with v2 and pointing to the Cloud Storage location of the new model. As shown in Figure 6-19, we're now able to see both deployed versions in our Cloud console.

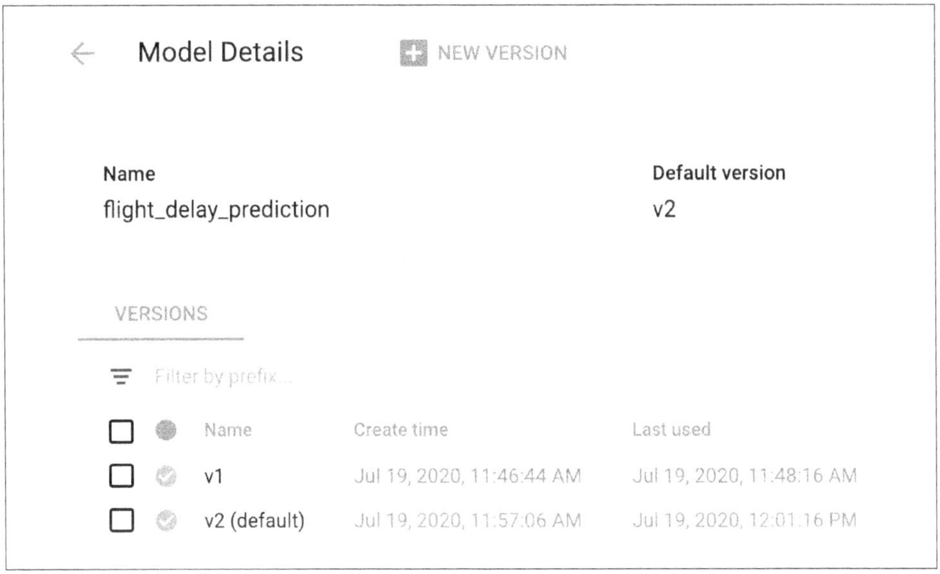

Figure 6-19. The dashboard for managing models and versions in the Cloud AI Platform console.

Notice that we've also set *v2* as the new default version, so that if users don't specify a version, they'll get a response from *v2*. Since the input and output format of our model are the same, clients can upgrade without worrying about breaking changes.

 Both Azure and AWS have similar model versioning services available. On Azure, model deployment and versioning is available with Azure Machine Learning (*https://oreil.ly/Q7NWh*). In AWS, these services are available in SageMaker (*https://oreil.ly/r98Ve*).

An ML engineer deploying a new version of a model as an ML model endpoint may want to use an API gateway such as Apigee that determines which model version to call. There are various reasons for doing this, including split testing a new version. For split testing, maybe they want to test a model update with a randomly selected group of 10% of application users to track how it affects their overall engagement with the app. The API gateway determines which deployed model version to call given a user's ID or IP address.

With multiple model versions deployed, AI Platform allows for performance monitoring and analytics across versions. This lets us trace errors to a specific version, monitor traffic, and combine this with additional data we're collecting in our application.

Versioning to Handle Newly Available Data

In addition to handling changes to our model itself, another reason to use versioning is when new training data becomes available. Assuming this new data follows the same schema used to train the original model, it's important to keep track of *when* the data was captured for each newly trained version. One approach to tracking this is to encode the timestamp range of each training dataset in the name of a model version. For example, if the latest version of a model is trained on data from 2019, we could name the version v20190101_20191231.

We can use this approach in combination with "Design Pattern 18: Continued Model Evaluation" on page 220 (discussed in Chapter 5) to determine when to take older model versions offline, or how far back training data should go. Continuous evaluation might help us determine that our model performs best when trained on data from the past two years. This could then inform the versions we decide to remove, and how much data to use when training newer versions.

Trade-Offs and Alternatives

While we recommend the Model Versioning design pattern over maintaining a single model version, there are a few implementation alternatives to the solution outlined above. Here, we'll look at other serverless and open source tooling for this pattern and the approach of creating multiple serving functions. We'll also discuss when to create an entirely new model resource instead of a version.

Other serverless versioning tools

We used a managed service specifically designed for versioning ML models, but we could achieve similar results with other serverless offerings. Under the hood, each model version is a stateless function with a specified input and output format, deployed behind a REST endpoint. We could therefore use a service like Cloud Run (*https://oreil.ly/KERBV*), for example, to build and deploy each version in a separate container. Each container has a unique URL and can be invoked by an API request. This approach gives us more flexibility in how to configure the deployed model environment, letting us add functionality like server-side preprocessing for model inputs. In our flight example above, we may not want to require clients to one-hot encode categorical values. Instead, we could let clients pass the categorical values as strings, and handle preprocessing in our container.

Why would we use a managed ML service like AI Platform Prediction instead of a more generalized serverless tool? Since AI Platform was built specifically for ML model deployment, it has built-in support for deploying models with GPUs optimized for ML. It also handles dependency management. When we deployed our XGBoost model above, we didn't need to worry about installing the correct XGBoost version or other library dependencies.

TensorFlow Serving

Instead of using Cloud AI Platform or another cloud-based serverless offering for model versioning, we could use an open source tool like TensorFlow Serving (*https://oreil.ly/NzDA9*). The recommended approach for implementing TensorFlow Serving is to use a Docker container via the latest `tensorflow/serving` (*https://oreil.ly/G0_Z7*) Docker image. With Docker, we could then serve the model using whichever hardware we'd like, including GPUs. The TensorFlow Serving API has built-in support for model versioning, following a similar approach to the one discussed in the Solution section. In addition to TensorFlow Serving, there are also other open source model serving options, including Seldon (*https://oreil.ly/Cddpi*) and MLFlow (*https://mlflow.org*).

Multiple serving functions

Another alternative to deploying multiple versions is to define multiple serving functions for a single version of an exported model. "Design Pattern 16: Stateless Serving Function" on page 201 (introduced in Chapter 5) explained how to export a trained model as a stateless function for serving in production. This is especially useful when model inputs require preprocessing to transform data sent by the client into the format the model expects.

To handle requirements for different groups of model end users, we can define multiple serving functions when we export our model. These serving functions are part of *one* exported model version, and this model is deployed to a single REST endpoint. In TensorFlow, serving functions are implemented using model *signatures*, which define the input and output format a model is expecting. We can define multiple serving functions using the @tf.function decorator and pass each function an input signature.

In the application code where we invoke our deployed model, we would determine which serving function to use based on the data sent from the client. For example, a request such as:

```
{"signature_name": "get_genre", "instances": … }
```

would be sent to the exported signature called get_genre, whereas a request like:

```
{"signature_name": "get_genre_with_explanation", "instances": … }
```

would be sent to the exported signature called get_genre_with_explanation.

Deploying multiple signatures can, therefore, solve the backward compatibility problem. However, there is a significant difference—there is only one model, and when that model is deployed, all the signatures are simultaneously updated. In our original example of changing the model from providing just one genre to providing multiple genres, the model architecture changed. The multiple-signature approach wouldn't work with that example since we have two different models. The multiple-signature solution is also not appropriate when we wish to keep different versions of the model separate and deprecate the older version over time.

Using multiple signatures is better than using multiple versions if you wish to maintain *both* model signatures going forward. In the scenario where there are some clients who simply want the best answer and other clients who want both the best answer and an explanation, there is an added benefit to updating all the signatures with a newer model instead of having to update versions one by one every time the model is retrained and redeployed.

What are some scenarios where we might want to maintain both versions of the model? With a text classification model, we may have some clients that need to send raw text to the model, and others that are able to transform raw text into matrices before getting a prediction. Based on the request data from the client, the model framework can determine which serving function to use. Passing text embedding matrices to a model is less expensive than preprocessing raw text, so this is an example where multiple serving functions could reduce server-side processing time. It's also worth noting that we can have multiple serving functions *with* multiple model versions, though there is a risk that this could create too much complexity.

New models versus new model versions

Sometimes it can be difficult to decide whether to create another model version or an entirely new model resource. We recommend creating a new model when a model's prediction task changes. A new prediction task typically results in a different model output format, and changing this could result in breaking existing clients. If we're unsure about whether to use a new version or model, we can think about whether we want existing clients to upgrade. If the answer is yes, chances are we have improved the model without changing the prediction task, and creating a new version will suffice. If we've changed the model in a way that would require users to decide whether they want the update, we'll likely want to create a new model resource.

To see this in practice, let's return to our flight prediction model to see an example. The current model has defined what it considers a delay (30+ minutes late), but our end users may have different opinions on this. Some users think just 15 minutes late counts as delayed, whereas others think a flight is only delayed if it's over an hour late. Let's imagine that we'd now like our users to be able to incorporate their own definition of delayed rather than use ours. In this case we'd use "Design Pattern 5: Reframing " on page 80 (discussed in Chapter 3) to change this to a regression model. The input format to this model is the same, but the output is now a numerical value representing the delay prediction.

The way our model users parse this response will obviously be different than the first version. With our latest regression model, app developers might choose to display the predicted delay when users search for flights, replacing something like "This flight is usually delayed more than 30 minutes" from the first version. In this scenario, the best solution is to create a new model *resource*, perhaps called `flight_model_regres sion`, to reflect the changes. This way, app developers can choose which to use, and we can continue to make performance updates to each model by deploying new versions.

Summary

This chapter focused on design patterns that address different aspects of reproducibility. Starting with the *Transform* design, we saw how this pattern is used to ensure reproducibility of the data preparation dependencies between the model training pipeline and the model serving pipeline. This is achieved by explicitly capturing the transformations applied to convert the model inputs into the model features. The *Repeatable Splitting* design pattern captures the way data is split among training, validation, and test datasets to ensure that an example used in training is never used for evaluation or testing even as the dataset grows.

The *Bridged Schema* design pattern looks at how to ensure reproducibility when a training dataset is a hybrid of newer data and older data with a different schema. This allows for combining two datasets with different schemas in a consistent way for training. Next, we discussed the *Windowed Inference* design pattern, which ensures that when features are calculated in a dynamic, time-dependent way, they can be correctly repeated between training and serving. This design pattern is particularly useful when machine learning models require features that are computed from aggregates over time windows.

The *Workflow Pipeline* design pattern addresses the problem of creating an end-to-end reproducible pipeline by containerizing and orchestrating the steps in our machine learning workflow. Next, we saw how the *Feature Store* design pattern can be used to address reproducibility and reusability of features across different machine learning jobs. Lastly, we looked at the *Model Versioning* design pattern, where backward compatibility is achieved by deploying a changed model as a microservice with a different REST endpoint.

In the next chapter, we look into design patterns that help carry out AI responsibly.

## CHAPTER 7

Responsible AI

Until this point, we've focused on patterns designed to help data and engineering teams prepare, build, train, and scale models for production use. These patterns mainly addressed teams directly involved in the ML model development process. Once a model is in production, its impact extends far beyond the teams who built it. In this chapter, we'll discuss the other *stakeholders* of a model, both those within and outside of an organization. Stakeholders could include executives whose business objectives dictate a model's goals, the end users of a model, auditors, and compliance regulators.

There are several groups of model stakeholders we'll be referring to in this chapter:

Model builders
> Data scientists and ML researchers directly involved in building ML models.

ML engineers
> Members of ML Ops teams directly involved in deploying ML models.

Business decision makers
> Decide whether or not to incorporate the ML model into their business processes or customer-facing applications and will need to evaluate whether the model is fit for this purpose.

End users of ML systems
> Make use of predictions from an ML model. There are many different types of model end users: customers, employees, and hybrids of these. Examples include a customer getting a movie recommendation from a model, an employee on a factory floor using a visual inspection model to determine whether a product is broken, or a medical practitioner using a model to aid in patient diagnosis.

Regulatory and compliance agencies
> People and organizations who need an executive-level summary of how a model is making decisions from a regulatory compliance perspective. This could include financial auditors, government agencies, or governance teams within an organization.

Throughout this chapter, we'll look at patterns that address a model's impact on individuals and groups outside the team and organization building a model. The *Heuristic Benchmark* design pattern provides a way of putting the model's performance in a context that end users and decision makers can understand. The *Explainable Predictions* pattern provides approaches to improving trust in ML systems by fostering an understanding of the signals a model is using to make predictions. The *Fairness Lens* design pattern aims to ensure that models behave equitably across different subsets of users and prediction scenarios.

Taken together, the patterns in this chapter fall under the practice of *Responsible AI* (*https://oreil.ly/MlJkM*). This is an area of active research and is concerned with the best ways to build fairness, interpretability, privacy, and security into AI systems. Recommended practices for responsible AI include employing a human-centered design approach by engaging with a diverse set of users and use-case scenarios throughout project development, understanding the limitations of datasets and models, and continuing to monitor and update ML systems after deployment. Responsible AI patterns are not limited to the three that we discuss in this chapter—many of the patterns in earlier chapters (like Continuous Evaluation, Repeatable Splitting, and Neutral Class, to name a few) provide methods to implement these recommended practices and attain the goal of building fairness, interpretability, privacy, and security into AI systems.

Design Pattern 28: Heuristic Benchmark

The Heuristic Benchmark pattern compares an ML model against a simple, easy-to-understand heuristic in order to explain the model's performance to business decision makers.

Problem

Suppose a bicycle rental agency wishes to use the expected duration of rentals to build a dynamic pricing solution. After training an ML model to predict the duration of a bicycle's rental period, they evaluate the model on a test dataset and determine that the mean absolute error (MAE) of the trained ML model is 1,200 seconds. When they present this model to the business decision makers, they will likely be asked: "Is an MAE of 1,200 seconds good or bad?" This is a question we need to be ready to handle whenever we develop a model and present it to business stakeholders. If we train an image classification model on items in a product catalog and the mean average precision (MAP) is 95%, we can expect to be asked: "Is a MAP of 95% good or bad?"

It is no good to wave our hands and say that this depends on the problem. Of course, it does. So, what is a good MAE for the bicycle rental problem in New York City? How about in London? What is a good MAP for the product catalog image classification task?

Model performance is typically stated in terms of cold, hard numbers that are difficult for end users to put into context. Explaining the formula for MAP, MAE, and so on does not provide the intuition that business decision makers are asking for.

Solution

If this is the second ML model being developed for a task, an easy answer is to compare the model's performance against the currently operational version. It is quite easy to say that the MAE is now 30 seconds lower or that the MAP is 1% higher. This works even if the current production workflow doesn't use ML. As long as this task is already being performed in production and evaluation metrics are being collected, we can compare the performance of our new ML model against the current production methodology.

But what if there is no current production methodology in place, and we are building the very first model for a green-field task? In such cases, the solution is to create a simple benchmark for the sole purpose of comparing against our newly developed ML model. We call this a *heuristic benchmark*.

A good heuristic benchmark should be intuitively easy to understand and relatively trivial to compute. If we find ourselves defending or debugging the algorithm used by the benchmark, we should search for a simpler, more understandable one. Good examples of a heuristic benchmark are constants, rules of thumb, or bulk statistics (such as the mean, median, or mode). Avoid the temptation to train even a simple machine learning model, such as a linear regression, on a dataset and use that as a benchmark—linear regression is likely not intuitive enough, especially once we start to include categorical variables, more than a handful of inputs, or engineered features.

 Do not use a heuristic benchmark if there is already an operational practice in place. Instead, we should compare our model against that existing standard. The existing operational practice does not need to use ML—it is simply whatever technique is currently being used to solve the problem.

Examples of good heuristic benchmarks and situations where we might employ them are shown in Table 7-1. Example code for the implementations of these heuristic benchmarks is in the GitHub repository (*https://github.com/GoogleCloudPlatform/ml-design-patterns/blob/master/07_responsible_ai/heuristic_benchmark.ipynb*) of this book.

Table 7-1. Heuristic benchmarks for a few selected scenarios (see code in GitHub (https://oreil.ly/WoESU))

Scenario	Heuristic benchmark	Example task	Implementation for example task
Regression problem where features and interactions between features are not well understood by the business.	Mean or median value of the label value over the training data. Choose the median if there are a lot of outliers.	Time interval before a question on Stack Overflow is answered.	Predict that it will take 2,120 seconds always. 2,120 seconds is the median time to first answer over the entire training dataset.
Binary classification problem where features and interactions between features are not well understood by the business.	Overall fraction of positives in the training data.	Whether or not an accepted answer in Stack Overflow will be edited.	Predict 0.36 as the output probability for all answers. 0.36 is the fraction of accepted answers overall that are edited.
Multilabel classification problem where features and interactions between features are not well understood by the business.	Distribution of the label value over the training data.	Country from which a Stack Overflow question will be answered.	Predict 0.03 for France, 0.08 for India, and so on. These are the fractions of answers written by people from France, India, and so on.
Regression problem where there is a single, very important, numeric feature.	Linear regression based on what is, intuitively, the single most important feature.	Predict taxi fare amount given pickup and dropoff locations. The distance between the two points is, intuitively, a key feature.	Fare = $4.64 per kilometer. The $4.64 is computed from the training data over all trips.
Regression problem with one or two important features. The features could be numeric or categorical but should be commonly used heuristics.	Lookup table where the rows and columns correspond to the key features (discretized if necessary) and the prediction for each cell is the average label in that cell estimated over the training data.	Predict duration of bicycle rental. Here, the two key features are the station that the bicycle is being rented from and whether or not it is peak hours for commuting.	Lookup table of average rental duration from each station based on peak hour versus nonpeak hour.
Classification problem with one or two important features. The features could be numeric or categorical.	As above, except that the prediction for each cell is the distribution of labels in that cell. If the goal is to predict a single class, compute the mode of the label in each cell.	Predict whether a Stack Overflow question will get answered within one day. The most important feature here is the primary tag.	For each tag, compute the fraction of questions that are answered within one day.

Scenario	Heuristic benchmark	Example task	Implementation for example task
Regression problem that involves predicting the future value of a time series.	Persistence or linear trend. Take seasonality into account. For annual data, compare against the same day/week/quarter of previous year.	Predict weekly sales volume	Predict that next week's sales $= s_0$ where s_0 is the sales this week. (or) Next week's sales $= s_0 + (s_0 - s_{-1})$ where s_{-1} is last week's sales. (or) Next week's sales $= s_{-1y}$ where s_{-1y} is the sales of the corresponding week last year. Avoid the temptation to combine the three options since the value of the relative weights is not intuitive.
Classification problem currently being solved by human experts. This is common for image, video, and text tasks and includes scenarios where it is cost-prohibitive to routinely solve the problem with human experts.	Performance of human experts.	Detecting eye disease from retinal scans.	Have three or more physicians examine each image. Treat the decision of a majority of physicians as being correct, and look at the percentile ranking of the ML model among human experts.
Preventive or predictive maintenance.	Perform maintenance on a fixed schedule.	Preventive maintenance of a car.	Bring cars in for maintenance once every three months. The three months is the median time to failure of cars from the last service date.
Anomaly detection.	99th percentile value estimated from the training dataset.	Identify a denial of service (DoS) attack from network traffic.	Find the 99th percentile of the number of requests per minute in the historical data. If over any one-minute period, the number of requests exceeds this number, flag it as a DoS attack.
Recommendation model.	Recommend the most popular item in the category of the customer's last purchase.	Recommend movies to users.	If a user just saw (and liked) *Inception* (a sci-fi movie), recommend *Icarus* to them (the most popular sci-fi movie they haven't yet watched).

Many of the scenarios in Table 7-1 refer to "important features." These are important features in the sense that they are widely accepted within the business as having a well-understood impact on the prediction problem. In particular, these are not features ascertained using feature importance methods on your training dataset. As an example, it's well accepted within the taxicab industry that the most important determinant of a taxi fare is distance, and that longer trips cost more. That's what makes distance an important feature, not the outcome of a feature importance study.

Trade-Offs and Alternatives

We will often find that a heuristic benchmark is useful beyond the primary purpose of explaining model performance. In some cases, the heuristic benchmark might require special data collection. Finally, there are instances where a heuristic benchmark may be insufficient because the comparison itself needs context.

Development check

It is often the case that a heuristic benchmark proves useful beyond explaining the performance of ML models. During development, it can also help with diagnosing problems with a particular model approach.

For example, say that we are building a model to predict the duration of rentals and our benchmark is a lookup table of average rental duration given the station name and whether or not it is peak commute hour:

```
CREATE TEMPORARY FUNCTION is_peak_hour(start_date TIMESTAMP) AS
    EXTRACT(DAYOFWEEK FROM start_date) BETWEEN 2 AND 6 -- weekday
    AND (
        EXTRACT(HOUR FROM start_date) BETWEEN 6 AND 10
        OR
        EXTRACT(HOUR FROM start_date) BETWEEN 15 AND 18)
;

SELECT
    start_station_name,
    is_peak_hour(start_date) AS is_peak,
    AVG(duration) AS predicted_duration,
FROM `bigquery-public-data.london_bicycles.cycle_hire`
GROUP BY 1, 2
```

As we develop our model, it is a good idea to compare the performance of our ML model against this benchmark. In order to do this, we will be evaluating model performance on different stratifications of the evaluation dataset. Here, the evaluation dataset will be stratified by start_station_name and is_peak. By doing so, we can easily diagnose whether our model is overemphasizing the busy, popular stations and ignoring infrequent stations in the training data. If that is happening, we can experiment with increasing model complexity or balancing the dataset to overweight less popular stations.

Human experts

We recommended that in classification problems like diagnosing eye disease—where the work is carried out by human experts—that the benchmark would involve a panel of such experts. By having three or more physicians examine each image, it is possible to identify the extent to which human physicians make errors and compare the error rate of the model against that of human experts. In the case of such image classification problems, this is a natural extension of the labeling phase because the labels for eye disease are created through human labeling.

It is sometimes advantageous to use human experts even if we have actual ground truth. For example, when building a model to predict the cost of auto repair after an accident, we can look at historical data and find the actual cost of the repair. We will not typically use human experts for this problem because the ground truth is directly available from the historical dataset. However, for the purposes of communicating the benchmark, it can be helpful to have insurance agents assess the cars for a damage estimate, and compare our model's estimates to those of the agents.

Using human experts need not be limited to unstructured data as with eye disease or damage cost estimation. For example, if we are building a model to predict whether or not a loan will get refinanced within a year, the data will be tabular and the ground truth will be available in the historical data. However, even in this case, we might ask human experts to identify loans that will get refinanced for the purposes of communicating how often loan agents in the field would get it right.

Utility value

Even if we have an operational model or excellent heuristic to compare against, we will still have to explain the impact of the improvement that our model offers. Communicating that the MAE is 30 seconds lower or that the MAP is 1% higher might not be enough. The next question might very well be, "Is a 1% improvement good? Is it worth the hassle of putting an ML model into production rather than the simple heuristic rule?"

If you can, it is important to translate the improvement in model performance into the model's utility value. This value could be monetary, but it could also correspond with other measures of utility, like better search results, earlier disease detection, or less waste resulting from improved manufacturing efficiency. This utility value is useful in deciding whether or not to deploy this model, since deploying or changing a production model always carries a certain cost in terms of reliability and error budgets. For example, if the image classification model is used to pre-fill an order form, we can calculate that a 1% improvement will translate to 20 fewer abandoned orders per day, and is therefore worth a certain amount of money. If this is more than the threshold set by our Site Reliability Engineering team, we'd deploy the model.

In our bicycle rental problem, it might be possible to measure the impact on the business by using this model. For example, we might be able to calculate the increased availability of bicycles or the increased profits based on using the model in a dynamic pricing solution.

Design Pattern 29: Explainable Predictions

The Explainable Predictions design pattern increases user trust in ML systems by providing users with an understanding of how and why models make certain predictions. While models such as decision trees are interpretable by design, the architecture of deep neural networks makes them inherently difficult to explain. For all models, it is useful to be able to interpret predictions in order to understand the combinations of features influencing model behavior.

Problem

When evaluating a machine learning model to determine whether it's ready for production, metrics like accuracy, precision, recall, and mean squared error only tell one piece of the story. They provide data on how *correct* a model's predictions are relative to ground truth values in the test set, but they carry no insight on *why* a model arrived at those predictions. In many ML scenarios, users may be hesitant to accept a model's prediction at face value.

To understand this, let's look at a model (*https://oreil.ly/5W-2n*) that predicts the severity of diabetic retinopathy (DR) from an image of a retina.[1] The model returns a softmax output, indicating the probability that an individual image belongs to 1 of 5 categories denoting the severity of DR in the image—ranging from 1 (no DR present) to 5 (proliferative DR, the worst form). Let's imagine that for a given image, the model returns 95% confidence that the image contains proliferative DR. This may seem like a high-confidence, accurate result, but if a medical professional is relying solely on this model output to provide a patient diagnosis, they still have no insight into *how* the model arrived at this prediction. Maybe the model identified the correct regions in the image that are indicative of DR, but there's also a chance the model's prediction is based on pixels in the image that show no indication of the disease. As an example, maybe some images in the dataset contain doctor notes or annotations. The model could be incorrectly using the presence of an annotation to make its prediction, rather than the diseased areas in the image.[2] In the model's current form, there is no way to attribute the prediction to regions in an image, making it difficult for the doctor to trust the model.

Medical imaging is just one example—there are many industries, scenarios, and model types where a lack of insight into a model's decision-making process can lead to problems with user trust. If an ML model is used to predict an individual's credit

---

1 DR is an eye condition affecting millions of people around the world. It can lead to blindness, but if caught early, it can be successfully treated. To learn more and find the dataset, see here (*https://oreil.ly/ix21h*).

2 Explanations were used to identify and correct for annotations present in radiology images in this study (*https://oreil.ly/qowNO*).

---

score or other financial health metric, people will likely want to know why they received a particular score. Was it a late payment? Too many lines of credit? Short credit history? Maybe the model is relying solely on demographic data to make its predictions, and subsequently introducing bias into the model without our knowledge. With only the score, there is no way to know how the model arrived at its prediction.

In addition to model end users, another group of stakeholders are those involved with regulatory and compliance standards for ML models, since models in certain industries may require auditing or additional transparency. Stakeholders involved in auditing models will likely need a higher-level summary of how the model is arriving at its predictions in order to justify its use and impact. Metrics like accuracy are not useful in this case—without insight into *why* a model makes the predictions it does, its use may become problematic.

Finally, as data scientists and ML engineers, we can only improve our model quality to a certain degree without an understanding of the features it's relying on to make predictions. We need a way to verify that models are performing in the way we expect. For example, let's say we are training a model on tabular data to predict whether a flight will be delayed. The model is trained on 20 features. Under the hood, maybe it's relying only on 2 of those 20 features, and if we removed the rest, we could significantly improve our system's performance. Or maybe each of those 20 features is necessary to achieve the degree of accuracy we need. Without more details on what the model is using, it's difficult to know.

Solution

To handle the inherent unknowns in ML, we need a way to understand how models work under the hood. Techniques for understanding and communicating how and why an ML model makes predictions is an area of active research. Also called interpretability or model understanding, explainability is a new and rapidly evolving field within ML, and can take a variety of forms depending on a model's architecture and the type of data it is trained on. Explainability can also help reveal bias in ML models, which we cover when discussing the Fairness Lens pattern in this chapter. Here, we'll focus on explaining deep neural networks using feature attributions. To understand this in context, first we'll look at explainability for models with less complex architectures.

Simpler models like decision trees are more straightforward to explain than deep models since they are often *interpretable by design*. This means that their learned weights provide direct insight into how the model is making predictions. If we have a linear regression model with independent, numeric input features, the weights may sometimes be interpretable. Take for example a linear regression model that predicts fuel efficiency of a car.[3] In scikit-learn (*https://oreil.ly/V9GT5*), we can get the learned coefficients of a linear regression model with the following:

```
model = LinearRegression().fit(x_train, y_train)
coefficients = model.coef_
```

The resulting coefficients for each feature in our model are shown in Figure 7-1.

	Learned coefficients
cylinders	-0.926610
displacement	0.037055
horsepower	-0.017953
weight	-0.007286
acceleration	0.164976
model year	0.723584
origin_1	-1.779775
origin_2	0.781041
origin_3	0.998735

Figure 7-1. The learned coefficients from our linear regression fuel efficiency model, which predicts a car's miles per gallon. We used get_dummies() from pandas to convert the origin feature to a boolean column since it is categorical.

The coefficients show us the relationship between each feature and the model's output, predicted miles per gallon (MPG). For example, from these coefficients, we can conclude that for each additional cylinder in a car, our model's predicted MPG will decrease. Our model has also learned that as new cars are introduced (denoted by the "model year" feature), they often have higher fuel efficiency. We can learn much more about the relationships between our model's features and output from these coefficients than we could from the learned weights of a hidden layer in a deep neural network. This is why models like the one demonstrated above are often referred to as *interpretable by design*.

---

3 The model discussed here is trained on a public UCI dataset (*https://oreil.ly/cNixp*).

While it's tempting to assign significant meaning to the learned weights in linear regression or decision tree models, we must be extremely cautious when doing so. The conclusions we drew earlier are still correct (i.e., inverse relationship between number of cylinders and fuel efficiency), but we cannot conclude from the magnitude of coefficients, for example, that the categorical origin feature or the number of cylinders are more important to our model than horsepower or weight. First, each of these features is represented in a different unit. One cylinder bears no equivalence to one pound— the cars in this dataset have a maximum of 8 cylinders, but weigh over 3,000 pounds. Additionally, origin is a categorical feature represented with dummy values, so each origin value can only be 0 or 1. The coefficients also don't tell us anything about the relationship *between* features in our model. More cylinders are often correlated with more horsepower, but we can't conclude this from the learned weights.[4]

When models are more complex, we use *post hoc* explainability methods to approximate the relationships between a model's features and its output. Typically, post hoc methods perform this analysis without relying on model internals like learned weights. This is an area of ongoing research, and there are a variety of proposed explanation methods, along with tooling for adding these methods to your ML workflow. The type of explanation methods we'll discuss are known as *feature attributions*. These methods aim to attribute a model's output—whether it be an image, classification, or numerical value—to its features, by assigning attribution values to each feature indicating how much that feature contributed to the output. There are two types of feature attributions:

Instance-level

Feature attributions that explain a model's output for an individual prediction. For example, in a model predicting whether someone should be approved for a line of credit, an instance-level feature attribution would provide insight into why a specific person's application was denied. In an image model, an instance-level attribution might highlight the pixels in an image that caused it to predict it contained a cat.

Global

Global feature attributions analyze the model's behavior across an aggregate to draw conclusions about how the model is behaving as a whole. Typically this is done by averaging instance-level feature attributions from a test dataset. In a model predicting whether a flight will be delayed, global attributions might tell

---

4 The scikit-learn documentation (*https://oreil.ly/DAmIm*) goes into more detail on how to correctly interpret the learned weights in linear models.

us that overall, extreme weather is the most significant feature when predicting delays.

The two feature attribution methods we'll explore[5] are outlined in Table 7-2 and provide different approaches that can be used for both instance-level and global explanations.

Table 7-2. Descriptions of different explanation methods and links to their research papers

Name	Description	Paper
Sampled Shapley	Based on the concept of Shapley Value,[a] this approach determines a feature's marginal contribution by calculating how much adding and removing that feature affects a prediction, analyzed over multiple combinations of feature values.	*https://oreil.ly/ubEjW*
Integrated Gradients (IG)	Using a predefined model baseline, IG calculates the derivatives (gradients) along the path from this baseline to a specific input.	*https://oreil.ly/sy8f8*

[a] The Shapley Value was introduced in a paper by Lloyd Shapley (*https://oreil.ly/xCrqU*) in 1951, and is based on concepts from game theory.

While we could implement these approaches from scratch, there is tooling designed to simplify the process of getting feature attributions. The available open source and cloud-based explainability tools let us focus on debugging, improving, and summarizing our models.

Model baseline

In order to use these tools, we first need to understand the concept of a *baseline* as it applies to explaining models with feature attributions. The goal of any explainability method is to answer the question, "Why did the model predict X?" Feature attributions attempt to do this by providing numerical values for each feature indicating how much that feature contributed to the final output. Take for example a model predicting whether a patient has heart disease given some demographic and health data. For a single example in our test dataset, let's imagine that the attribution value for a patient's cholesterol feature is 0.4, and the attribution for their blood pressure is −0.2. Without context, these attribution values don't mean much, and our first question will likely be, "0.4 and −0.2 relative to what?" That "what" is the model's *baseline*.

Whenever we get feature attribution values, they are all relative to a predefined baseline prediction value for our model. Baseline predictions can either be *informative* or *uninformative*. Uninformative baselines typically compare against some average case

---

5 We're focusing on these two explainability methods since they are widely used and cover a variety of model types, but there are many other methods and frameworks not included in this analysis, such as LIME (*https://oreil.ly/0c4uB*) and ELI5 (*https://github.com/TeamHG-Memex/eli5*).

across a training dataset. In an image model, an uninformative baseline could be a solid black or white image. In a text model, an uninformative baseline could be 0 values for the model's embedding matrices or stop words like "the," "is," or "and." In a model with numerical inputs, a common approach to choosing a baseline is to generate a prediction using the median value for each feature in the model.

Determining Baselines

The way we think about a baseline will differ depending on whether our model is performing a regression or classification task. For a regression task, a model will have *exactly one* numerical baseline prediction value. In our car mileage example, let's imagine we decide to use the median approach for calculating our baseline. The median for the eight features in our dataset is the following array:

    [151.0, 93.5, 2803.5, 15.5, 76.0, 1.0, 0.0, 0.0]

When we send this to our model, the predicted MPG is 22.9. Consequently, for every prediction we make to this model, we'll use 22.9 MPG as the baseline to compare predictions.

Let's now imagine that we follow the Reframing pattern to change this from a regression to a classification problem. To do this, we'll define "low," "medium," and "high" buckets for fuel efficiency, and our model will therefore output a three-element softmax array indicating the probability a given car corresponds with each class. Taking the same median baseline input as above, our classification model now returns the following as our baseline prediction:

    [0.1, 0.7, 0.2]

With this, we now have a *different* baseline prediction value for each class. Let's say we generate a new prediction on an example from our test set, and our model outputs the following array, predicting a 90% probability that this car has "low" fuel efficiency:

    [0.9, 0.06, 0.04]

The resulting feature attribution values should explain why the model predicted 0.9 compared to the baseline prediction value of 0.1 for the "low" class. We can also look at feature attribution values for the other classes to understand, for example, why our model predicted the same car had a 6% chance of belonging to our "medium" fuel efficiency class.

Figure 7-2 shows instance-level feature attributions for a model that predicts the duration of a bike trip. The uninformative baseline for this model is a trip duration of 13.6 minutes, which we get by generating a prediction using the median value for each feature in our dataset. When a model's prediction is *less than* the baseline prediction value, we should expect most attribution values to be negative, and vice versa.

In this example, we get a predicted duration of 10.71, which is less than the model's baseline, and explains why many of the attribution values are negative. We can determine the most important features by taking the absolute value of the feature attributions. In this example, the trip's distance was the most important feature, causing our model's prediction to decrease 2.4 minutes from the baseline. Additionally, as a sanity check, we should ensure that the feature attribution values roughly add up to the difference between the current prediction and the baseline prediction.

```
Baseline prediction:  13.61
Predicted duration:  10.71

Name                Feature value    Attribution value
------------        ---------------  --------------------
distance              1395.51             -2.44478
start_hr                18               -1.29039
max_temp               20.7239             0.690506
temp                   16.168              0.12629
dew_point               7.83396            0.0110318
prcp                    0.03              -0.00134132
weekday                 1                  0
wdsp                    0                  0
rain_drizzle            0                  0
```

Figure 7-2. The feature attribution values for a single example in a model predicting bike trip duration. The model's baseline, calculated using the median of each feature value, is 13.6 minutes, and the attribution values show how much each feature influenced the prediction.

Informative baselines, on the other hand, compare a model's prediction with a specific alternative scenario. In a model identifying fraudulent transactions, an informative baseline might answer the question, "Why was this transaction flagged as fraud instead of nonfraudulent?" Instead of using the median feature values across the entire training dataset to calculate the baseline, we would take the median of only the nonfraudulent values. In an image model, maybe the training images contain a significant portion of solid black and white pixels, and using these as a baseline would result in inaccurate predictions. In this case, we'd need to come up with a different *informative* baseline image.

SHAP

The open source library SHAP (*https://github.com/slundberg/shap*) provides a Python API for getting feature attributions on many types of models, and is based on the concept of Shapley Value introduced in Table 7-2. To determine feature attribution values, SHAP calculates how much adding or removing each feature contributes to a model's prediction output. It performs this analysis across many different combinations of feature values and model output.

SHAP is framework-agnostic and works with models trained on image, text, or tabular data. To see how SHAP works in practice, we'll use the fuel efficiency dataset referenced previously. This time, we'll build a deep model with the Keras `Sequential` API:

```
model = tf.keras.Sequential([
  tf.keras.layers.Dense(16, input_shape=(len(x_train.iloc[0])),
  tf.keras.layers.Dense(16, activation='relu'),
  tf.keras.layers.Dense(1)
])
```

To use SHAP, we'll first create a `DeepExplainer` object by passing it our model and a subset of examples from our training set. Then we'll get the attribution values for the first 10 examples in our test set:

```
import shap
explainer = shap.DeepExplainer(model, x_train[:100])
attribution_values = explainer.shap_values(x_test.values[:10])
```

SHAP has some built-in visualization methods that make it easier to understand the resulting attribution values. We'll use SHAP's `force_plot()` method to plot the attribution values for the first example in our test set with the following code:

```
shap.force_plot(
    explainer.expected_value[0],
    shap_values[0][0,:],
    x_test.iloc[0,:]
)
```

In the code above, `explainer.expected_value` is our model's baseline. SHAP calculates the baseline as the mean of the model's output across the dataset we passed when we created the explainer (in this case, `x_train[:100]`), though we could also pass our own baseline value to `force_plot`. The ground truth value for this example is 14 miles per gallon, and our model predicts 13.16. Our explanation will therefore explain our model's prediction of 13.16 with feature attribution values. In this case, the attribution values are relative to the model's baseline of 24.16 MPG. The attribution values should therefore add up to roughly 11, the difference between the model's baseline and the prediction for this example. We can identify the most important features by looking at the ones with the highest absolute value. Figure 7-3 shows the resulting plot for this example's attribution values.

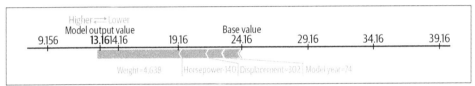

Figure 7-3. The feature attribution values for one example from our fuel efficiency prediction model. In this case, the car's weight is the most significant indicator of MPG with a feature attribution value of roughly 6. Had our model's prediction been above the baseline of 24.16, we would instead see mostly negative attribution values.

For this example, the most important indicator of fuel efficiency is weight, pushing our model's prediction down by about 6 MPG from the baseline. This is followed by horsepower, displacement, and then the car's model year. We can get a summary (or global explanation) of the feature attribution values for the first 10 examples from our test set with the following:

```
shap.summary_plot(
    shap_values,
    feature_names=data.columns.tolist(),
    class_names=['MPG']
)
```

This results in the summary plot shown in Figure 7-4.

In practice, we'd have a larger dataset and would want to calculate global-level attributions on more examples. We could then use this analysis to summarize the behavior on our model to other stakeholders within and outside our organization.

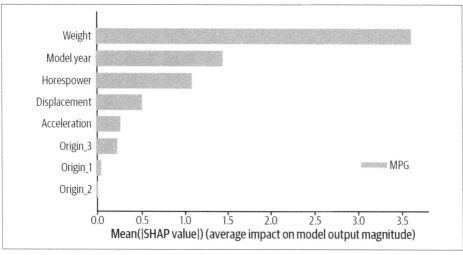

Figure 7-4. An example of global-level feature attributions for the fuel efficiency model, calculated on the first 10 examples from the test dataset.

Explanations from deployed models

SHAP provides an intuitive API for getting attributions in Python, typically used in a script or notebook environment. This works well during model development, but there are scenarios where you'd want to get explanations on a deployed model in addition to the model's prediction output. In this case, cloud-based explainability tools are the best option. Here, we'll demonstrate how to get feature attributions on a deployed model using Google Cloud's Explainable AI (*https://oreil.ly/lDocn*). At the time of this writing, Explainable AI works with custom TensorFlow models and tabular data models built with AutoML.

We'll deploy an image model to AI Platform to show explanations, but we could also use Explainable AI with TensorFlow models trained on tabular or text data. To start, we'll deploy a TensorFlow Hub (*https://oreil.ly/Ws8jx*) model trained on the Image-Net dataset. So that we can focus on the task of getting explanations, we won't do any transfer learning on the model and will use ImageNet's original 1,000 label classes:

```
model = tf.keras.Sequential([
    hub.KerasLayer(".../mobilenet_v2/classification/2",
            input_shape=(224,224,3)),
    tf.keras.layers.Softmax()
])
```

To deploy a model to AI Platform with explanations, we first need to create a metadata file that will be used by the explanation service to calculate feature attributions. This metadata is provided in a JSON file and includes information on the baseline we'd like to use and the parts of the model we want to explain. To simplify this process, Explainable AI provides an SDK that will generate metadata via the following code:

```
from explainable_ai_sdk.metadata.tf.v2 import SavedModelMetadataBuilder

model_dir = 'path/to/savedmodel/dir'

model_builder = SavedModelMetadataBuilder(model_dir)
model_builder.set_image_metadata('input_tensor_name')
model_builder.save_metadata(model_dir)
```

This code didn't specify a model baseline, which means it'll use the default (for image models, this is a black and white image). We can optionally add an `input_baselines` parameter to `set_image_metadata` to specify a custom baseline. Running the `save_metadata` method above creates an *explanation_metadata.json* file in a model directory (the full code is in the GitHub repository (*https://github.com/Google CloudPlatform/ml-design-patterns/blob/master/07_stakeholder_management/explaina bility.ipynb*)).

When using this SDK via AI Platform Notebooks, we also have the option to generate explanations locally within a notebook instance without deploying our model to the cloud. We can do this via the `load_model_from_local_path` method.

With our exported model and the *explanation_metadata.json* file in a Storage bucket, we're ready to create a new model version. When we do this, we specify the explanation method we'd like to use.

To deploy our model to AI Platform, we can copy our model directory to a Cloud Storage bucket and use the gcloud CLI to create a model version. AI Platform has three possible explanation methods to choose from:

Integrated Gradients (IG)
This implements the method introduced in the IG paper (*https://oreil.ly/FJhMd*) and works with any differentiable TensorFlow model—image, text, or tabular. For image models deployed on AI Platform, IG returns an image with highlighted pixels, indicating the regions that signaled the models prediction.

Sampled Shapley
Based on the Sampled Shapley paper (*https://oreil.ly/EAS8T*), this uses an approach similar to the open source SHAP library. On AI Platform, we can use this method with tabular and text TensorFlow models. Because IG works only with differentiable models, AutoML Tables uses Sampled Shapley to calculate feature attributions for all models.

XRAI

> This approach (*https://oreil.ly/niGVQ*) is built upon IG and applies smoothing to return region-based attributions. XRAI works only with image models deployed on AI Platform.

In our gcloud command, we specify the explanation method we'd like to use along with the number of integral steps or paths we want the method to use when computing attribution values.[6] The `steps parameter` refers to the number of feature combinations sampled for each output. In general, increasing this number will improve explanation accuracy:

```
!gcloud beta ai-platform versions create $VERSION_NAME \
--model $MODEL_NAME \
--origin $GCS_VERSION_LOCATION \
--runtime-version 2.1 \
--framework TENSORFLOW \
--python-version 3.7 \
--machine-type n1-standard-4 \
--explanation-method xrai \
--num-integral-steps 25
```

Once the model is deployed, we can get explanations using the Explainable AI SDK:

```
model = explainable_ai_sdk.load_model_from_ai_platform(
  GCP_PROJECT,
  MODEL_NAME,
  VERSION_NAME
)
request = model.explain([test_img])

# Print image with pixel attributions
request[0].visualize_attributions()
```

In Figure 7-5, we can see a comparison of the IG and XRAI explanations returned from Explainable AI for our ImageNet model. The highlighted pixel regions show the pixels that contributed most to our model's prediction of "husky."

Typically, IG is recommended for "non-natural" images like those taken in a medical, factory, or lab environment. XRAI usually works best for images taken in natural environments like the one of this husky. To understand why IG is preferred for non-natural images, see the IG attributions for the diabetic retinopathy image in Figure 7-6. In cases like this medical one, it helps to see attributions at a fine-grained, pixel level. In the dog image, on the other hand, knowing the exact pixels that caused our model to predict "husky" is less important, and XRAI gives us a higher-level summary of the important regions.

---

6 For more details on these explanation methods and their implementation, see the Explainable AI (*https://oreil.ly/PYn8P*) whitepaper.

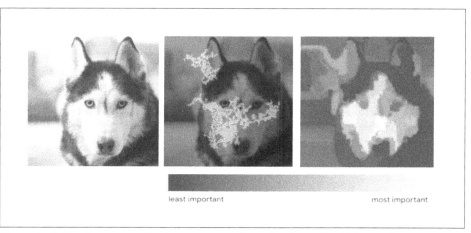

Figure 7-5. The feature attributions returned from Explainable AI for an ImageNet model deployed to AI Platform. On the left is the original image. The IG attributions are shown in the middle, and the XRAI attributions are shown on the right. The key below shows what the regions in XRAI correspond to—lighter regions are the most important, and darker areas represent the least important regions.

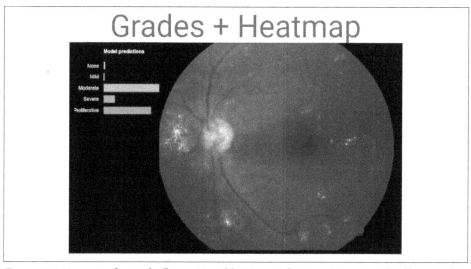

Figure 7-6. As part of a study (https://oreil.ly/Xp_vp) by Rory Sayres and colleagues in 2019, different groups of ophthalmologists were asked to evaluate the degree of DR on an image in three scenarios: the image by itself without model predictions, the image with model predictions, and the image with predictions and pixel attributions (shown here). We can see how pixel attributions can help increase confidence in the model's prediction.

Explainable AI also works in AutoML Tables (*https://oreil.ly/ CSQly*), a tool for training and deploying tabular data models. AutoML Tables handles data preprocessing and selects the best model for our data, which means we don't need to write any model code. Feature attributions through Explainable AI are enabled by default for models trained in AutoML Tables, and both global and instance-level explanations are provided.

Trade-Offs and Alternatives

While explanations provide important insight into how a model is making decisions, they are only as good as the model's training data, the quality of your model, and the chosen baseline. In this section, we'll discuss some limitations of explainability, along with some alternatives to feature attributions.

Data selection bias

It's often said that machine learning is "garbage in, garbage out." In other words, a model is only as good as the data used to train it. If we train an image model to identify 10 different cat breeds, those 10 cat breeds are all it knows. If we show the model an image of a dog, all it can do is try to classify the dog into 1 of the 10 cat categories it's been trained on. It might even do so with high confidence. That is to say, models are a direct representation of their training data.

If we don't catch data imbalances before training a model, explainability methods like feature attributions can help bring data selection bias to light. As an example, say we're building a model to predict the type of boat present in an image. Let's say it correctly labels an image from our test set as "kayak," but using feature attributions, we find that the model is relying on the boat's paddle to predict "kayak" rather than the shape of the boat. This is a signal that our dataset might not have enough variation in training images for each class—we'll likely need to go back and add more images of kayaks at different angles, both with and without paddles.

Counterfactual analysis and example-based explanations

In addition to feature attributions—described in the Solution section—there are many other approaches to explaining the output of ML models. This section is not meant to provide an exhaustive list of all explainability techniques, as this area is quickly evolving. Here, we will briefly describe two other approaches: counterfactual analysis and example-based explanations.

Counterfactual analysis is an instance-level explainability technique that refers to finding examples from our dataset with similar features that resulted in different predictions from our model. One way to do this is through the What-If Tool (*https:// oreil.ly/Vf3D-*), an open source tool for evaluating and visualizing the output of ML models. We'll provide a more in-depth overview of the What-If Tool in the Fairness Lens design pattern—here, we'll focus specifically on its counterfactual analysis functionality. When visualizing data points from our test set in the What-If Tool, we have the option to show the nearest counterfactual data point to the one we're selecting. Doing this will let us compare feature values and model predictions for these two data points, which can help us better understand the features our model is relying on most. In Figure 7-7, we see a counterfactual comparison for two data points from a mortgage application dataset. In bold, we see the features where these two data points are different, and at the bottom, we can see the model output for each.

Example-based explanations compare new examples and their corresponding predictions to similar examples from our training dataset. This type of explanation is especially useful for understanding how our training dataset affects model behavior. Example-based explanations work best on image or text data, and can be more intuitive than feature attributions or counterfactual analysis since they map a model's prediction directly to the data used for training.

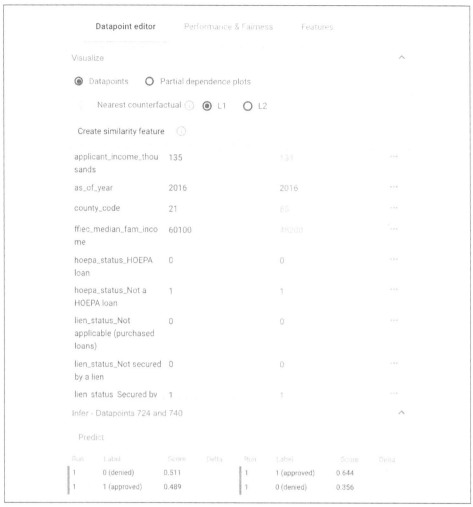

Figure 7-7. Counterfactual analysis in the What-If Tool for two data points from a US mortgage application dataset. Differences between the two data points are bolded. More information on this dataset can be found in the discussion of the Fairness Lens pattern in this chapter.

To better understand this approach, let's look at the game Quick, Draw (*https:// oreil.ly/-QsHl*)![7] The game asks players to draw an item, and guesses what they are drawing in real time, using a deep neural network trained on thousands of drawings by others. After players finish a drawing, they can see how the neural network arrived

---

7 For more details on Quick, Draw! and example-based explanations, see this paper (*https://oreil.ly/Yvexy*).

at its prediction by looking at examples from the training dataset. In Figure 7-8, we can see the example-based explanations for a drawing of french fries that the model successfully recognized.

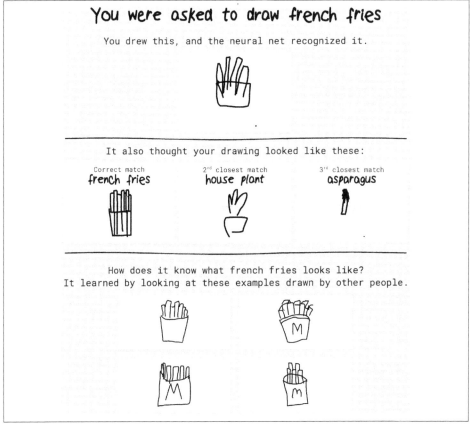

Figure 7-8. Example-based explanations from the game Quick, Draw! showing how the model correctly predicted "french fries" for the given drawing through examples from the training dataset.

Limitations of explanations

Explainability represents a significant improvement in understanding and interpreting models, but we should be cautious about placing too much trust in our model's explanations, or assuming they provide perfect insight into a model. Explanations in any form are a direct reflection of our training data, model, and selected baseline. That is to say, we can't expect our explanations to be high quality if our training dataset is an inaccurate representation of the groups reflected by our model, or if the baseline we've chosen doesn't work well for the problem we're solving.

Additionally, the relationship that explanations can identify between a model's features and output is representative only of our data and model, and not necessarily the environment outside this context. As an example, let's say we train a model to identify fraudulent credit card transactions and it finds, as a global-level feature attribution, that a transaction's amount is the feature most indicative of fraud. Following this, it would be incorrect to conclude that amount is *always* the biggest indicator of credit card fraud—this is only the case within the context of our training dataset, model, and specified baseline value.

We can think of explanations as an important addition to accuracy, error, and other metrics used to evaluate ML models. They provide useful insight into a model's quality and potential bias, but should not be the sole determinant of a high-quality model. We recommend using explanations as one piece of model evaluation criteria in addition to data and model evaluation, and many of the other patterns outlined in this and previous chapters.

Design Pattern 30: Fairness Lens

The Fairness Lens design pattern suggests the use of preprocessing and postprocessing techniques to ensure that model predictions are fair and equitable for different groups of users and scenarios. Fairness in machine learning is a continuously evolving area of research, and there is no single catch-all solution or definition to making a model "fair." Evaluating an entire end-to-end ML workflow—from data collection to model deployment—through a fairness lens is essential to building successful, high-quality models.

Problem

With the word "machine" in its name, it's easy to assume that ML models can't be biased. After all, models are the result of patterns learned by a computer, right? The problem with this thinking is that the datasets models learn from are created by *humans*, not machines, and humans are full of bias. This inherent human bias is inevitable, but is not necessarily always bad. Take for example a dataset used to train a financial fraud detection model—this data will likely be heavily imbalanced with very few fraudulent examples, since fraud is relatively rare in most cases. This is an example of naturally occurring bias, as it is a reflection of the statistical properties of the original dataset. Bias becomes *harmful* when it affects different groups of people differently. This is known as *problematic bias*, and it's what we'll be focusing on throughout this section. If this type of bias is not accounted for, it can find its way into models, creating adverse effects as production models directly reflect the bias present in the data.

Problematic bias is present even in situations where you may not expect it. As an example, imagine we're building a model to identify different types of clothing and accessories. We've been tasked with collecting all of the shoe images for the training dataset. When we think about shoes, we take note of the first thing that comes to mind. Is it a tennis shoe? Loafer? Flip flop? What about a stiletto? Let's imagine that we live in a climate that is warm year-round and most of the people we know wear sandals all the time. When we think of a shoe, a sandal is the first thing that comes to mind. As a result, we collect a diverse representation of sandal images with different types of straps, sole thicknesses, colors, and more. We contribute these to the larger clothing dataset, and when we test the model on a test set of images of our friend's shoes, it reaches 95% accuracy on the "shoe" label. The model looks promising, but problems arise when our colleagues from different locations test the model on images of their heels and sneakers. For their images, the label "shoe" is not returned at all.

This shoe example demonstrates bias in the training data distribution, and although it may seem oversimplified, this type of bias occurs frequently in production settings. Data distribution bias happens when the data we collect doesn't accurately reflect the entire population who will use our model. If our dataset is human-centered, this type of bias can be especially evident if our dataset fails to include an equal representation of ages, races, genders, religions, sexual orientations, and other identity characteristics.[8]

Even when our dataset does appear balanced with respect to these identity characteristics, it is still subject to bias in the way these groups are represented in the data. Suppose we are training a sentiment analysis model to classify restaurant reviews on a scale of 1 (extremely negative) to 5 (extremely positive). We've taken care to get a balanced representation of different types of restaurants in the data. However, it turns out that the majority of reviews for seafood restaurants are positive, whereas most of the vegetarian restaurant reviews are negative. This data representation bias will be directly represented by our model. Whenever new reviews are added for vegetarian restaurants, they'll have a much higher chance of being classified as negative, which could then influence someone's likelihood to visit one of these restaurants in the future. This is also known as *reporting bias*, since the dataset (here, the "reported" data) doesn't accurately reflect the real world.

A common fallacy when dealing with data bias issues is that removing the areas of bias from a dataset will fix the problem. Let's say we're building a model to predict the likelihood someone will default on a loan. If we find the model is treating people of different races unfairly, we might assume this could be fixed by simply removing race as a feature from the dataset. The problem with this is that, due to systemic bias, characteristics like race and gender are often reflected implicitly in other features like

---

8 For a more detailed look on how race and gender bias can find their way into image classification models, see Joy Buolamwini and Timmit Gebru, "Gender Shades: Intersectional Accuracy Disparities in Commercial Gender Classification" (*https://oreil.ly/1zw3e*), *Proceedings of Machine Learning Research* 81 (2018): 1-15.

zip code or income. This is known as *implicit* or *proxy bias*. Removing obvious features with potential bias like race and gender can often be worse than leaving them in, since it makes it harder to identify and correct instances of bias in the model.

When collecting and preparing data, another area where bias can be introduced is in the way the data is labeled. Teams often outsource labeling of large datasets, but it's important to take care in understanding how labelers can introduce bias to a dataset, especially if the labeling is subjective. This is known as *experimenter bias*. Imagine we're building a sentiment analysis model, and we have outsourced the labeling to a group of 20 people—it's their job to label each piece of text on a scale from 1 (negative) to 5 (positive). This type of analysis is extremely subjective and can be influenced by one's culture, upbringing, and many other factors. Before using this data to train our model, we should ensure this group of 20 labelers reflects a diverse population.

In addition to data, bias can also be introduced during model training by the objective function we choose. For example, if we optimize our model for overall accuracy, this may not accurately reflect model performance across all slices of data. In cases where datasets are inherently imbalanced, using accuracy as our only metric may miss cases where our model is underperforming or making unfair decisions on minority classes in our data.

Throughout this book, we've seen that ML has the power to improve productivity, add business value, and automate tasks that were previously manual. As data scientists and ML engineers, we have a shared responsibility to ensure the models we build don't have adverse effects on the populations that use them.

Solution

To handle problematic bias in machine learning, we need solutions both for identifying areas of harmful bias in data before training a model, and evaluating our trained model through a fairness lens. The Fairness Lens design pattern provides approaches for building datasets and models that treat all groups of users equally. We'll demonstrate techniques for both types of analysis using the What-If Tool (*https://oreil.ly/Sk36z*), an open source tool for dataset and model evaluation that can be run from many Python notebook environments.

Before proceeding with the tools outlined in this section, it's worth analyzing both the dataset and prediction task to determine whether there is potential for problematic bias. This requires looking closer at *who* will be impacted by a model, and *how* those groups will be impacted. If problematic bias seems likely, the technical approaches outlined in this section provide a good starting point for mitigating this type of bias. If, on the other hand, the skew in the dataset contains naturally occurring bias that will not have adverse effects on different groups of people, "Design Pattern 10: Rebalancing " on page 122 in Chapter 3 provides solutions for handling data that is inherently imbalanced.

Throughout this section, we'll be referencing a public dataset (*https://oreil.ly/azFUV*) of US mortgage applications. Loan agencies in the US are required to report information on an individual application, like the type of loan, the applicant's income, the agency handling the loan, and the status of the application. We will train a loan application approval model on this dataset in order to demonstrate different aspects of fairness. To our knowledge, this dataset is not used as is by any loan agency to train ML models, and so the fairness red flags we raise are only hypothetical.

We've created a subset of this dataset and done some preprocessing to turn this into a binary classification problem—whether an application was approved or denied. In Figure 7-9, we can see a preview of the dataset.

as_of_year	agency_code	loan_type	property_type	loan_purpose	occupancy	loan_amt_thousands	preapproval	county_code	applicant_income_thousands	purchaser_type	hoepa_status	lien_status	population
2016	Consumer Financial Protection Bureau (CFPB)	Conventional (any loan other than FHA, VA, FSA	One to four family (other than manufactured ho	Refinancing	1	110.0	Not applicable	119.0	65.0	Freddie Mac (FHLMC)	Not a HOEPA loan	Secured by a first lien	4930.0
2016	Department of Housing and Urban Development (HUD)	Conventional (any loan other than FHA, VA, FSA	One to four family (other than manufactured ho	Home purchase	1	480.0	Not applicable	33.0	270.0	Loan was not originated or was not sold in ca	Not a HOEPA loan	Secured by a first lien	4791.0
2016	Federal Deposit Insurance Corporation (FDIC)	Conventional (any loan other than FHA, VA, FSA	One to four family (other than manufactured ho	Refinancing	2	240.0	Not applicable	59.0	96.0	Commercial bank, savings bank or savings assoc	Not a HOEPA loan	Secured by a first lien	3439.0
2016	Office of the Comptroller of the Currency (OCC)	Conventional (any loan other than FHA, VA, FSA	One to four family (other than manufactured ho	Refinancing	1	76.0	Not applicable	65.0	85.0	Loan was not originated or was not sold in ca	Not a HOEPA loan	Secured by a subordinate lien	3952.0

Figure 7-9. A preview of a few columns from the US mortgage application dataset referenced throughout this section.

Before training

Because ML models are a direct representation of the data used to train them, it's possible to mitigate a significant amount of bias *before* building or training a model by performing thorough data analysis, and using the results of this analysis to adjust our data. In this phase, focus on identifying data collection or data representation bias, outlined in the Problem section. Table 7-3 shows some questions to consider for each type of bias depending on data type.

Table 7-3. Descriptions of different types of data bias

	Definition	Considerations for analysis
Data distribution bias	Data that doesn't contain an equal representation of all possible groups that will use the model in production	• Does the data contain a balanced set of examples across all relevant demographic slices (gender, age, race, religion, etc.)? • Does each label in the data contain a balanced split of all possible variations of this label? (E.g., the shoe example in the Problem section.)
Data representation bias	Data that is well balanced, but doesn't represent different slices of data equally	• For classification models, are *labels* balanced across relevant features? For example, in a dataset intended for credit worthiness prediction, does the data contain an equal representation across gender, race, and other identity characteristics of people marked as unlikely to pay back a loan? • Is there bias in the way different demographic groups are represented in the data? This is especially relevant for models predicting sentiment or a rating value. • Is there subjective bias introduced by data labelers?

Once we've examined our data and corrected for bias, we should take these same considerations into account when splitting our data into training, test, and validation sets. That is to say, once our full dataset is balanced, it's essential that our train, test, and validation splits maintain the same balance. Returning to our shoe image example, let's imagine we've improved our dataset to include varied images of 10 types of shoes. The training set should contain a similar percentage of each type of shoe as the test and validation sets. This will ensure that our model reflects and is being evaluated on real-world scenarios.

To see what this dataset analysis looks like in practice, we'll use the What-If Tool on the mortgage dataset introduced above. This will let us visualize the current balance of our data across various slices. The What-If Tool works both with and without a model. Since we haven't built our model yet, we can initialize the What-If Tool widget by passing it only our data:

```
config_builder = WitConfigBuilder(test_examples, column_names)
WitWidget(config_builder)
```

In Figure 7-10, we can see what the tool looks like when it loads when passed 1,000 examples from our dataset. The first tab is called the "Datapoint editor," which provides an overview of our data and lets us inspect individual examples. In this visualization, our data points are colored by the label—whether or not a mortgage application was approved. An individual example is also highlighted, and we can see the feature values associated with it.

Figure 7-10. The What-If Tool's "Datapoint editor," where we can see how our data is split by label class and inspect features for individual examples from our dataset.

There are many options for customizing the visualization in the Datapoint editor, and doing this can help us understand how our dataset is split across different slices. Keeping the same color-coding by label, if we select the `agency_code` column from the Binning | Y-Axis drop-down, the tool now shows a chart of how balanced our data is with regard to the agency underwriting each application's loan. This is shown in Figure 7-11. Assuming these 1,000 datapoints are a good representation of the rest of our dataset, there are a few instances of potential bias revealed in Figure 7-11:

Data representation bias

> The percentage of HUD applications *not* approved is higher than other agencies represented in our data. A model will likely learn this, causing it to predict "not approved" more frequently for applications originating through HUD.

Data collection bias

> We may not have enough data on loans originating from FRS, OCC, FDIC, or NCUA to accurately use `agency_code` as a feature in our model. We should make sure the percentage of applications for each agency in our dataset reflects real-world trends. For example, if a similar number of loans go through FRS and HUD, we should have an equal number of examples for each of those agencies in our dataset.

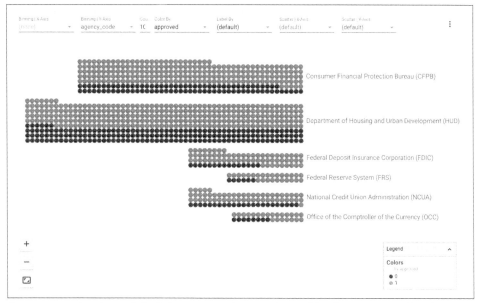

Figure 7-11. A subset of the US mortgage dataset, binned by the agency_code column in the dataset.

We can repeat this analysis across other columns in our data and use our conclusions to add examples and improve our data. There are many other options for creating custom visualizations in the What-If Tool—see the full code (*https://github.com/ GoogleCloudPlatform/ml-design-patterns/blob/master/07_responsible_ai/fair ness.ipynb*) on GitHub for more ideas.

Another way to understand our data using the What-If Tool is through the Features tab, shown in Figure 7-12. This shows how our data is balanced across each column in our dataset. From this we can see where we need to add or remove data, or change our prediction task.[9] For example, maybe we want to limit our model to making predictions only on refinancing or home purchase loans since there may not be enough data available for other possible values in the loan_purpose column.

---

9 To learn more about changing a prediction task, see "Design Pattern 5: Reframing " on page 80 and "Design Pattern 9: Neutral Class " on page 117 in Chapter 3.

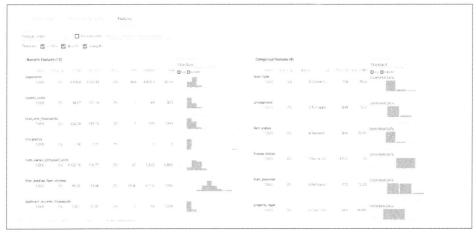

Figure 7-12. The Features tab in the What-If Tool, which shows histograms of how a dataset is balanced for each column.

Once we've refined our dataset and prediction task, we can consider anything else we might want to optimize during model training. For example, maybe we care most about our model's accuracy on applications it predicts as "approved." During model training, we'd want to optimize for AUC (or another metric) on the "approved" class in this binary classification model.

> If we've done all we can to eliminate data collection bias and find that there is not enough data available for a specific class, we can follow "Design Pattern 10: Rebalancing" on page 122 in Chapter 3. This pattern discusses techniques for building models to handle imbalanced data.

Bias in Other Forms of Data

Although we've shown a tabular dataset here, bias is equally common in other types of data. The Civil Comments dataset (*https://oreil.ly/xaocx*) provided by Jigsaw provides a good example of areas where we might find bias in text data. This dataset labels comments according to their toxicity (ranging from 0 to 1), and has been used to build models for flagging toxic online comments. Each comment in the dataset is tagged as to whether one of a collection of identity attributes is present, like the mention of a religion, race, or sexual orientation. If we plan to use this data to train a model, it's important that we look out for data representation bias. That is to say, the identity terms in a comment should *not* influence that comment's toxicity, and any such bias should be accounted for before training a model.

Take the following made-up comment as an example: "Mint chip is their best ice cream flavor, hands down." If we were to replace "Mint chip" with "Rocky road," the comment should be labeled with the same toxicity score (ideally 0). Similarly, if the comment were instead, "Mint chip is the worst. If you like this flavor you're an idiot," we'd expect a higher toxicity score, and that score should be the same any time we replace "Mint chip" with a different flavor name. We've used ice cream in this example, but it's easy to imagine how this would play out with more controversial identity terms, especially in a human-centered dataset—a concept known as counterfactual fairness.

After training

Even with rigorous data analysis, bias may find its way into a trained model. This can happen as a result of a model's architecture, optimization metrics, or data bias that wasn't identified before training. To solve for this, it's important to evaluate our model from a fairness perspective and dig deeper into metrics other than overall model accuracy. The goal of this post-training analysis is to understand the trade-offs between model accuracy and the effects a model's predictions will have on different groups.

The What-If Tool is one such option for post-model analysis. To demonstrate how to use it on a trained model, we'll build on our mortgage dataset example. Based on our previous analysis, we've refined the dataset to only include loans for the purpose of refinancing or home purchases,[10] and trained an XGBoost model to predict whether or not an application will be approved. Because we're using XGBoost, we converted all categorical features into boolean columns using the pandas `get_dummies()` method.

We'll make a few additions to our What-If Tool initialization code above, this time passing in a function that calls our trained model, along with configs specifying our label column and the name for each label:

```
def custom_fn(examples):
  df = pd.DataFrame(examples, columns=columns)
  preds = bst.predict_proba(df)
  return preds

config_builder = (WitConfigBuilder(test_examples, columns)
  .set_custom_predict_fn(custom_fn)
  .set_target_feature('mortgage_status')
  .set_label_vocab(['denied', 'approved']))
WitWidget(config_builder, height=800)
```

---

10 There are many more pre-training optimizations that could be made on this dataset. We've chosen just one here as a demo of what's possible.

Now that we've passed the tool our model, the resulting visualization shown in Figure 7-13 plots our test datapoints according to our model's prediction confidence indicated on the y-axis.

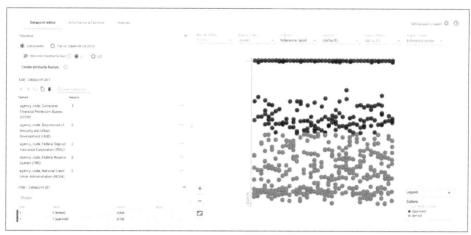

Figure 7-13. The What-If Tool's Datapoint editor for a binary classification model. The y-axis is the model's prediction output for each datapoint, ranging from 0 (denied) to 1 (approved).

The What-If Tool's Performance & Fairness tab lets us evaluate our model's fairness across different data slices. By selecting one of our model's features to "Slice by," we can compare our model's results for different values of this feature. In Figure 7-14, we've sliced by the agency_code_HUD feature—a boolean value indicating whether an application was underwritten by HUD (0 for non-HUD loans, 1 for HUD loans).

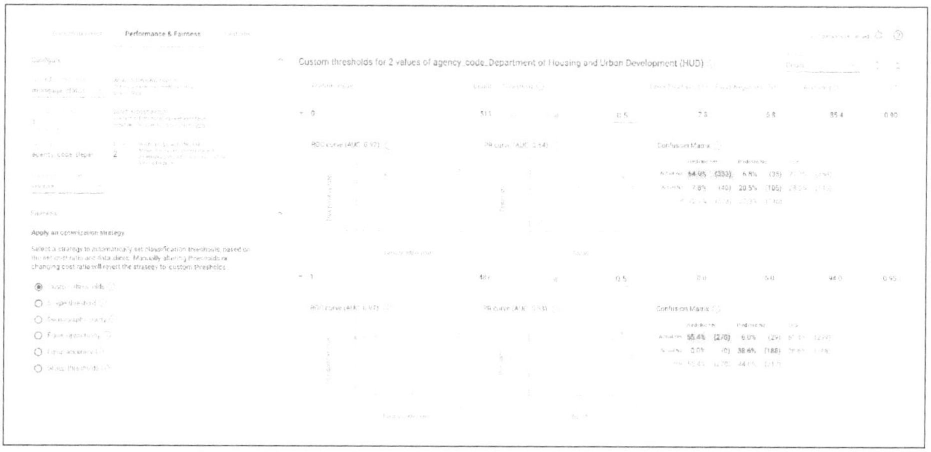

Figure 7-14. The What-If Tool Performance & Fairness tab, showing our XGBoost model performance across different feature values.

From these Performance & Fairness charts, we can see:

- Our model's accuracy on loans supervised by HUD is significantly higher—94% compared to 85%.

- According to the confusion matrix, non-HUD loans are approved at a higher rate—72% compared to 55%. This is likely due to the data representation bias identified in the previous section (we purposely left the dataset this way to show how models can amplify data bias).

There are a few ways to act on these insights, as shown in the "Optimization strategy" box in Figure 7-14. These optimization methods involve changing our model's *classification threshold*—the threshold at which a model will output a positive classification. In the context of this model, what confidence threshold are we OK with to mark an application as "approved"? If our model is more than 60% confident an application should be approved, should we approve it? Or are we only OK approving applications when our model is more than 98% confident? This decision is largely dependent on a model's context and prediction task. If we're predicting whether or not an image contains a cat, we may be OK returning the label "cat" even when our model is only 60% confident. However, if we have a model that predicts whether or not a medical image contains a disease, we'd likely want our threshold to be much higher.

The What-If Tool helps us choose a threshold based on various optimizations. Optimizing for "Demographic parity," for example, would ensure that our model approves the same percentage of applications for both HUD and non-HUD loans.[11] Alternatively, using an equality of opportunity[12] fairness metric will ensure that datapoints from both the HUD and non-HUD slice with a ground truth value of "approved" in the test dataset are given an equal chance of being predicted "approved" by the model.

Note that changing a model's prediction threshold is only one way to act on fairness evaluation metrics. There are many other approaches, including rebalancing training data, retraining a model to optimize for a different metric, and more.

---

11 This article (*https://oreil.ly/wFx_W*) provides more detail on the What-If Tool's options for fairness optimization strategies.

12 More details on equality of opportunity as a fairness metric can be found here (*https://oreil.ly/larIS*).

The What-If Tool is model agnostic and can be used for any type of model regardless of architecture or framework. It works with models loaded within a notebook or in TensorBoard (*https://oreil.ly/xWV4_*), models served via TensorFlow Serving, and models deployed to Cloud AI Platform Prediction. The What-If Tool team also created a tool for text-based models called the Language Interpretability Tool (LIT) (*https://oreil.ly/CZ60B*).

Another important consideration for post-training evaluation is testing our model on a balanced set of examples. If there are particular slices of our data that we anticipate will be problematic for our model—like inputs that could be affected by data collection or representation bias—we should ensure our test set includes enough of these cases. After splitting our data, we'll use the same type of analysis we employed in the "Before training" part of this section on *each* split of our data: training, validation, and test.

As seen from this analysis, there is no one-size-fits-all solution or evaluation metric for model fairness. It is a continuous, iterative process that should be employed throughout an ML workflow—from data collection to deployed model.

Trade-Offs and Alternatives

There are many ways to approach model fairness in addition to the pre- and post-training techniques discussed in the Solution section. Here, we'll introduce a few alternative tools and processes for achieving fair models. ML fairness is a rapidly evolving area of research—the tools included in this section aren't meant to provide an exhaustive list, but rather a few techniques and tools currently available for improving model fairness. We'll also discuss the differences between the Fairness Lens and Explainable Predictions design patterns, as they are related and often used together.

Fairness Indicators

Fairness Indicators (*https://github.com/tensorflow/fairness-indicators*) (FI) are a suite of open source tools designed to help in understanding a dataset's distribution before training, and evaluating model performance using fairness metrics. The tools included in FI are TensorFlow Data Validation (TFDV) and TensorFlow Model Analysis (TFMA). Fairness Indicators are most often used as components in TFX pipelines (see "Design Pattern 25: Workflow Pipeline" on page 282 in Chapter 6 for more details) or via TensorBoard. With TFX, there are two pre-built components that utilize Fairness Indicator tools:

- ExampleValidator for data analysis, detecting drift, and training–serving skew with TFDV.

- Evaluator uses the TFMA library to evaluate a model across subsets of a dataset. An example of an interactive visualization generated from TFMA is shown in Figure 7-15. This looks at one feature in the data (height) and breaks down the model's false negative rate for each possible categorical value of that feature.

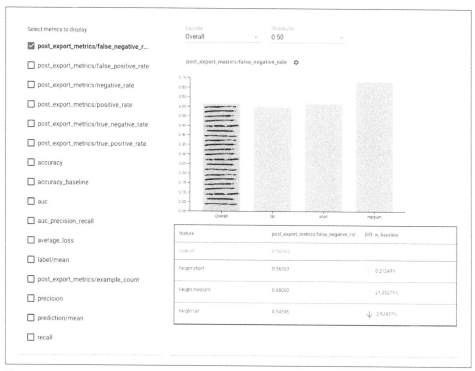

Figure 7-15. Comparing a model's false negative rate over different subsets of data.

From the Fairness Indicators Python package (*https://oreil.ly/pYM1j*), TFMA can also be used as a standalone tool that works with both TensorFlow and non-TensorFlow models.

Automating data evaluation

The fairness evaluation methods we discussed in the Solution section focused on manual, interactive data and model analysis. This type of analysis is important, especially in the initial phases of model development. As we operationalize our model and shift our focus to maintaining and improving it, finding ways to automate fairness evaluation will improve efficiency and ensure that fairness is integrated throughout our ML process. We can do this through "Design Pattern 18: Continued Model Evaluation" on page 220 discussed in Chapter 5, or with "Design Pattern 25: Workflow Pipeline" on page 282 in Chapter 6 using components like those provided by TFX for data analysis and model evaluation.

Allow and disallow lists

When we can't find a way to fix inherent bias in our data or model directly, it's possible to hardcode rules on top of our production model using allow and disallow lists. This applies mostly to classification or generative models, when there are labels or words we don't want our model to return. As an example, gendered words such as "man" and "woman" were removed (*https://oreil.ly/WY2vp*) from Google Cloud Vision API's label detection feature. Because gender cannot be determined by appearance alone, it would have reinforced unfair biases to return these labels when the model's prediction is based solely on visual features. Instead, the Vision API returns "person." Similarly, the Smart Compose feature in Gmail avoids the use of gendered pronouns (*https://oreil.ly/dtMhK*) when completing sentences such as "I am meeting an investor next week. Do you want to meet ___?"

These allow and disallow lists can be applied in one of two phases in an ML workflow:

Data collection
> When training a model from scratch or using the Transfer Learning design pattern to add our own classification layer, we can define our model's label set in the data collection phase, before a model has been trained.

After training
> If we're relying on a pre-trained model for predictions, and are using the same labels from that model, an allow and disallow list can be implemented in production—after the model returns a prediction but before those labels are surfaced to end users. This could also apply to text generation models, where we don't have complete control of all possible model outputs.

Data augmentation

In addition to the data distribution and representation solutions discussed earlier, another approach to minimizing model bias is to perform data *augmentation*. Using this approach, data is changed before training with the goal of removing potential sources of bias. One specific type of data augmentation is known as ablation, and is especially applicable in text models. In a text sentiment analysis model, for example, we could remove identity terms from text to ensure they don't influence our model's predictions. Building on the ice cream example we used earlier in this section, the sentence "Mint chip is their best ice cream flavor" would become "BLANK is their best ice cream flavor" after applying ablation. We'd then replace all other words throughout the dataset that we didn't want to influence the model's sentiment prediction with the same word (we used BLANK here, but anything not present in the rest of the text data will work). Note that while this ablation technique works well for many text models, it's important to be careful when removing areas of bias from tabular datasets, as mentioned in the Problem section.

Another data augmentation approach involves generating new data, and it was used by Google Translate to minimize gender bias (*https://oreil.ly/3Rkdr*) when translating text to and from gender-neutral and gender-specific languages. The solution involved rewriting translation data such that when applicable, a provided translation would be offered in both the feminine and masculine form. For example, the gender-neutral English sentence "We are doctors" would yield two results when being translated to Spanish, as seen in Figure 7-16. In Spanish, the word "we" can have both a feminine and masculine form.

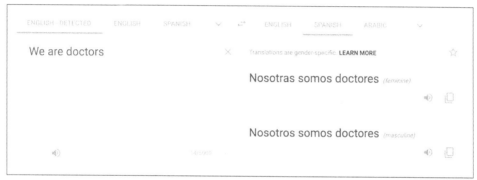

Figure 7-16. When translating a gender-neutral word in one language (here, the word "we" in English) to a language where that word is gender-specific, Google Translate now provides multiple translations to minimize gender bias.

Model Cards

Originally introduced in a research paper (*https://oreil.ly/OAIcs*), Model Cards provide a framework for reporting a model's capabilities and limitations. The goal of Model Cards is to improve model transparency by providing details on scenarios where a model should and should not be used, since mitigating problematic bias only works if a model is used in the way it was intended. In this way, Model Cards encourage accountability for using a model in the correct context.

The first Model Cards (*https://oreil.ly/OwiJY*) released provide summaries and fairness metrics for the Face Detection and Object Detection features in Google Cloud's Vision API. To generate Model Cards for our own ML models, TensorFlow provides a Model Card Toolkit (*https://github.com/tensorflow/model-card-toolkit*) (MCT) that can be run as a standalone Python library or as part of a TFX pipeline. The toolkit reads exported model assets and generates a series of charts with various performance and fairness metrics.

Fairness versus explainability

The concepts of fairness and explainability in ML are sometimes confused since they are often used together and are both part of the larger initiative of Responsible AI.

Fairness applies specifically to identifying and removing bias from models, and explainability is *one* approach for diagnosing the presence of bias. For example, applying explainability to a sentiment analysis model might reveal that the model is relying on identity terms to make its prediction when it should instead be using words like "worst," "amazing," or "not."

Explainability can also be used outside the context of fairness to reveal things like why a model is flagging particular fraudulent transactions, or the pixels that caused a model to predict "diseased" in a medical image. Explainability, therefore, is a method for improving model transparency. Sometimes transparency can reveal areas where a model is treating certain groups unfairly, but it can also provide higher-level insight into a model's decision-making process.

Summary

While Peter Parker may not have been referring to machine learning when he said, "With great power comes great responsibility," the quote certainly applies here. ML has the power to disrupt industries, improve productivity, and generate new insights from data. With this potential, it's especially important that we understand how our models will impact different groups of stakeholders. Model stakeholders could include varying demographic slices of model users, regulatory groups, a data science team, or business teams within an organization.

The Responsible AI patterns outlined in this chapter are an essential part of every ML workflow—they can help us better understand the predictions generated by our models and catch potential adverse behavior before models go to production. Starting with the *Heuristic Benchmark* pattern, we looked at how to identify an initial metric for model evaluation. This metric is useful as a comparison point for understanding subsequent model versions and summarizing model behavior for business decision makers. In the *Explainable Predictions* pattern, we demonstrated how to use feature attributions to see which features were most important in signaling a model's prediction. Feature attributions are one type of explainability method and can be used for both evaluating the prediction on a single example or over a group of test inputs. Finally, the *Fairness Lens* design pattern presented tools and metrics for ensuring a model's predictions treat all groups of users in a way that is fair, equitable, and unbiased.

## CHAPTER 8

Connected Patterns

We set out to create a catalog of machine learning design patterns, solutions to recurring problems when designing, training, and deploying machine learning models and pipelines. In this chapter, we provide a quick reference to this inventory of patterns.

We organized the patterns in the book in terms of where they would be used in a typical ML workflow. Thus, we had a chapter on input representation and another on model selection. We then discussed patterns that modify the typical training loop and make inference more resilient. We ended with patterns that promote a responsible use of ML systems. This is akin to organizing a recipe book with separate sections on appetizers, soups, entrees, and desserts. Such an organization, however, can make it hard to determine when to choose which soup and what desserts go well with some entree. Therefore, in this chapter, we also draw out how the patterns are related to one another. Finally, we also put together "meal plans" by discussing how the patterns interact for common categories of ML tasks.

Patterns Reference

We've discussed a lot of different design patterns and how they can be used to address common challenges that arise in machine learning. Here is a summary.

Chapter	Design pattern	Problem solved	Solution
Data Representation	Hashed Feature	Problems associated with categorical features such as incomplete vocabulary, model size due to cardinality, and cold start.	Bucket a deterministic and portable hash of string representation and accept the trade-off of collisions in the data representation.
	Embeddings	High-cardinality features where closeness relationships are important to preserve.	Learn a data representation that maps high-cardinality data into a lower-dimensional space in such a way that the information relevant to the learning problem is preserved.
	Feature Cross	Model complexity insufficient to learn feature relationships.	Help models learn relationships between inputs faster by explicitly making each combination of input values a separate feature.
	Multimodal Input	How to choose between several potential data representations.	Concatenate all the available data representations.
Problem Representation	Reframing	Several problems including confidence for numerical prediction, ordinal categories, restricting prediction range, and multitask learning.	Change the representation of the output of a machine learning problem; for example, representing a regression problem as a classification (and vice versa).
	Multilabel	More than one label applies to a given training example.	Encode the label using a multi-hot array, and use k sigmoids as the output layer.
	Ensembles	Bias–variance trade-off on small- and medium-scale problems.	Combine multiple machine learning models and aggregate their results to make predictions.
	Cascade	Maintainability or drift issues when a machine learning problem is broken into a series of ML problems.	Treat an ML system as a unified workflow for the purposes of training, evaluation, and prediction.
	Neutral Class	The class label for some subset of examples is essentially arbitrary.	Introduce an additional label for a classification model, disjoint from the current labels.
	Rebalancing	Heavily imbalanced data.	Downsample, upsample, or use a weighted loss function depending on different considerations.

Chapter	Design pattern	Problem solved	Solution
Patterns That Modify Model Training	Useful Overfitting	Using machine learning methods to learn a physics-based model or dynamical system.	Forgo the usual generalization techniques in order to intentionally overfit on the training dataset.
	Checkpoints	Lost progress during long-running training jobs due to machine failure.	Store the full state of the model periodically, so that partially trained models are available and can be used to resume training from an intermediate point, instead of starting from scratch.
	Transfer Learning	Lack of large datasets that are needed to train complex machine learning models.	Take part of a previously trained model, freeze the weights, and use these nontrainable layers in a new model that solves a similar problem.
	Distribution Strategy	Training large neural networks can take a very long time, which slows experimentation.	Carry the training loop out at scale over multiple workers, taking advantage of caching, hardware acceleration, and parallelization.
	Hyperparameter Tuning	How to determine the optimal hyperparameters of a machine learning model.	Insert the training loop into an optimization method to find the optimal set of model hyperparameters.
Resilience	Stateless Serving Function	Production ML system must be able to synchronously handle thousands to millions of prediction requests per second.	Export the machine learning model as a stateless function so that it can be shared by multiple clients in a scalable way.
	Batch Serving	Carrying out model predictions over large volumes of data using an endpoint that is designed to handle requests one at a time will overwhelm the model.	Use software infrastructure commonly used for distributed data processing to carry out inference asynchronously on a large number of instances at once.
	Continued Model Evaluation	Model performance of deployed models degrades over time either due to data drift, concept drift or other changes to the pipelines which feed data to the model.	Detect when a deployed model is no longer fit-for-purpose by continually monitoring model predictions and evaluating model performance.
	Two-Phase Predictions	Large, complex models must be kept performant when they are deployed at the edge or on distributed devices.	Split the use case into two phases with only the simpler phase being carried out on the edge.
	Keyed Predictions	How to map the model predictions that are returned to the corresponding model input when submitting large prediction jobs.	Allow the model to pass through a client-supported key during prediction that can be used to join model inputs to model predictions.

Chapter	Design pattern	Problem solved	Solution
Reproducibility	Transform	The inputs to a model must be transformed to create the features the model expects and that process must be consistent between training and serving.	Explicitly capture and store the transformations applied to convert the model inputs into features.
	Repeatable Splitting	When creating data splits, it's important to have a method that is lightweight and repeatable regardless of the programming language or random seeds.	Identify a column that captures the correlation relationship between rows and use the Farm Fingerprint hashing algorithm to split the available data into training, validation, and testing datasets.
	Bridged Schema	As new data becomes available, any changes to the data schema could prevent using both the new and old data for retraining.	Adapt the data from its older, original data schema to match the schema of the newer, better data.
	Windowed Inference	Some models require an ongoing sequence of instances to run inference, or features must be aggregated across a time window in such a way that avoids training–serving skew.	Externalize the model state and invoke the model from a stream analytics pipeline to ensure that features calculated in a dynamic, time-dependent way can be correctly repeated between training and serving.
	Workflow Pipeline	When scaling the ML workflow, run trials independently and track performance for each step of the pipeline.	Make each step of the ML workflow a separate, containerized service that can be chained together to make a pipeline that can be run with a single REST API call
	Feature Store	The ad hoc approach to feature engineering slows model development and leads to duplicated effort between teams as well as work stream inefficiency.	Create a feature store, a centralized location to store and document feature datasets that will be used in building machine learning models and can be shared across projects and teams.
	Model Versioning	It is difficult to carry out performance monitoring and split test model changes while having a single model in production or to update models without breaking existing users.	Deploy a changed model as a microservice with a different REST endpoint to achieve backward compatibility for deployed models.

Chapter	Design pattern	Problem solved	Solution
Responsible AI	Heuristic Benchmark	Explaining model performance using complicated evaluation metrics does not provide the intuition that business decision makers need.	Compare an ML model against a simple, easy-to-understand heuristic.
	Explainable Predictions	Sometimes it is necessary to know why a model makes certain predictions either for debugging or for regulatory and compliance standards.	Apply model explainability techniques to understand how and why models make predictions and improve user trust in ML systems.
	Fairness Lens	Bias can cause machine learning models to not treat all users equally and can have adverse effects on some populations.	Use tools to identify bias in datasets before training and evaluate trained models through a fairness lens to ensure model predictions are equitable across different groups of users and different scenarios.

Pattern Interactions

Design patterns don't exist in isolation. Many of them are closely related to one another either directly or indirectly and often complement one another. The interaction diagram in Figure 8-1 summarizes the interdependencies and some relationships between different design patterns. If you find yourself using a pattern, you might benefit from thinking how you could incorporate other patterns that are related to it.

Here, we'll highlight some of the ways in which these patterns are related and how they can be used together when developing a full solution. For example, when working with categorical features, the Hashed Feature design pattern may be combined with the Embeddings design pattern. These two patterns work together to address high-cardinality model inputs, such as working with text. In TensorFlow, this is demonstrated by wrapping a `categorical_column_with_hash_bucket` feature column with an `embedding` feature column to convert the sparse, categorical text input to a dense representation:

```
import tensorflow.feature_column as fc
keywords = fc.categorical_column_with_hash_bucket("keywords",
    hash_bucket_size=10K)
keywords_embedded = fc.embedding_column(keywords, num_buckets=16)
```

We saw when discussing Embeddings that this technique is recommended when using the Feature Cross design pattern. Hashed Features go hand in hand with the Repeatable Splitting design pattern since the Farm Fingerprint hashing algorithm can be used for data splitting. And, when using the Hashed Features or Embeddings design pattern, it's common to turn to concepts within Hyperparameter Tuning to determine the optimal number of hash buckets or the right embedding dimension to use.

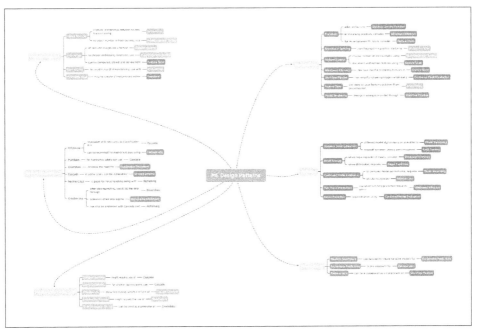

Figure 8-1. Many of the patterns discussed in this book are related or can be used together. This image is available in the GitHub repository (https://github.com/Google CloudPlatform/ml-design-patterns) for this book.

In fact, the Hyperparameter Tuning design is a common part of the machine learning workflow and is often used in conjunction with other patterns. For example, we might use hyperparameter tuning to determine the number of older examples to use if we're implementing the Bridged Schema pattern. And, when using hyperparameter tuning, it's important to keep in mind how we've set up model Checkpoints using virtual epochs and Distributed Training. Meanwhile, the Checkpoints design pattern naturally connects to Transfer Learning since earlier model checkpoints are often used during fine-tuning.

Embeddings show up throughout machine learning, so there are many ways in which the Embeddings design pattern interacts with other patterns. Perhaps the most notable is Transfer Learning since the output generated from the intermediate layers of a pre-trained model are essentially learned feature embeddings. We also saw how incorporating the Neutral Class design pattern in a classification model, either naturally or through the Reframing pattern, can improve those learned embeddings. Further downstream, if those embeddings are used as features for a model, it could be advantageous to save them using the Feature Store pattern so they can be easily accessed and versioned. Or, in the case of Transfer Learning, the pre-trained model output could be viewed as the initial output of a Cascade pattern.

We also saw how the Rebalancing pattern could be approached by combining two other design patterns: Reframing and Cascade. Reframing would allow us to represent the imbalanced dataset as a classification of either "normal" or "outlier." The output of that model would then be passed to a secondary regression model, which is optimized for prediction on either data distribution. These patterns will likely also lead to the Explainable Predictions pattern, since when dealing with imbalanced data, it is especially important to verify that the model is picking up on the right signals for prediction. In fact, it's encouraged to consider the Explainable Predictions pattern when building a solution involving a cascade of multiple models, since this can limit model explainability. This trade-off of model explainability shows up again with the Ensemble and Multimodel Input patterns since these techniques also don't lend themselves well to some explainability methods.

The Cascade design pattern might also be helpful when using the Bridged Schema pattern and could be used as an alternative pattern by having a preliminary model that imputes missing values of the secondary schema. These two patterns might then be combined to save the resulting feature set for later use as described in the Feature Store pattern. This is another example which highlights the versatility of the Feature Store pattern and how it is often combined with other design patterns. For example, a feature store provides a convenient way to maintain and utilize streaming model features that may arise through the Windowed Inference pattern. Feature stores also work hand in hand with managing different datasets that might arise in the Reframing pattern, and provide a reusable version of the techniques that arise when using the Transform pattern. The feature versioning capability as discussed in the Feature Store pattern also plays a role with the Model Versioning design pattern.

The Model Versioning pattern, on the other hand, is closely related to the Stateless Serving Function and Continued Model Evaluation patterns. In Continued Model Evaluation, different model versions may be used when assessing how a model's performance has degraded over time. Similarly, the different model signatures of the serving function provide an easy means of creating different model versions. This approach to model versioning via the Stateless Serving Function pattern can be connected back to the Reframing pattern where two different model versions could provide their own REST API endpoints for the two different model output representations.

We also discussed how, when using the Continued Model Evaluation pattern, it's often advantageous to explore solutions presented in the Workflow Pipeline pattern as well, both to set up triggers that will initiate the retraining pipeline as well as maintain lineage tracking for various model versions that are created. Continued Model Evaluation is also closely connected to the Keyed Predictions pattern since this can provide a mechanism for easily joining ground truth to the model prediction outputs. In the same way, the Keyed Predictions pattern is also intertwined with the Batch Serving pattern. By the same token, the Batch Serving pattern is often used in conjunction with the Stateless Serving Function pattern to carry out prediction jobs at scale which, in turn, relies on the Transform pattern under the hood to maintain consistency between training and serving.

Patterns Within ML Projects

Machine learning systems enable teams within an organization to build, deploy, and maintain machine learning solutions at scale. They provide a platform for automating and accelerating all stages of the ML life cycle, from managing data, to training models, evaluating performance, deploying models, serving predictions, and monitoring performance. The patterns we have discussed in this book show up throughout any machine learning project. In this section, we'll describe the stages of the ML life cycle and where many of these patterns are likely to arise.

ML Life Cycle

Building a machine learning solution is a cyclical process that begins with a clear understanding of the business goals and ultimately leads to having a machine learning model in production that benefits that goal. This high-level overview of the ML life cycle (see Figure 8-2) provides a useful roadmap designed to enable ML to bring value to businesses. Each of the stages is equally important, and failure to complete any one of these steps increases the risk in later stages of producing misleading insights or models of no value.

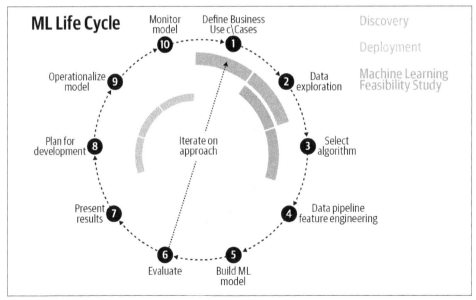

Figure 8-2. The ML life cycle begins with defining the business use case and ultimately leads to having a machine learning model in production that benefits that goal.

The ML life cycle consists of three stages, as shown in Figure 8-2: discovery, development, and deployment. There is a canonical order to the individual steps of each stage. However, these steps are completed in an iterative manner and earlier steps may be revisited depending on the outcomes and insights gathered from later stages.

Discovery

Machine learning exists as a tool to solve a problem. The discovery stage of an ML project begins with defining the business use case (Step 1 of Figure 8-2). This is a crucial time for business leaders and ML practitioners to align on the specifics of the problem and develop an understanding of what ML can and cannot do to achieve that goal.

It is important to keep sight of the business value through each stage of the life cycle. Many choices and design decisions must be made throughout the various stages, and often there is no single "right" answer. Rather, the best option is determined by how the model will be used in support of the business goal. While a feasible goal for a research project could be to eke out 0.1% more accuracy on a benchmark dataset, this is not acceptable in industry. For a production model built for a corporate organization, success is governed by factors more closely tied to the business, like improving customer retention, optimizing business processes, increasing customer engagement, or decreasing churn rates. There could also be indirect factors related to the business use case that influence development choices, like speed of inference, model size, or model interpretability. Any machine learning project should begin with a thorough understanding of the business opportunity and how a machine learning model can make a tangible improvement on current operations.

A successful discovery stage requires collaboration between the business domain experts as well as machine learning experts to assess the viability of an ML approach. It is crucial to have someone who understands the business and the data collaborating with teams that understand the technical challenges and the engineering effort that would be involved. If the overall investment of development resources outweighs the value to the organization, then it is not a worthwhile solution. It is possible that the technical overhead and cost of resources for productionization exceed the benefit provided by a model that ultimately improves churn prediction by only 0.1%. Or maybe not. If an organization's customer base has 1 billion people, then 0.1% is still 1 million happier customers.

During the discovery phase, it is important to outline the business objectives and scope for the task. This is also the time to determine which metrics will be used to measure or define success. Success can look different for different organizations, or even within different groups of the same organization. See, for example, the discussion on multiple objectives in "Common Challenges in Machine Learning" on page 11 in Chapter 1. Creating well-defined metrics and key performance indicators

(KPIs) at the onset of an ML project can help to ensure everyone is aligned on the common goal. Ideally there is already some procedure in place that provides a convenient baseline against which to measure future progress. This could be a model already in production, or even just a rules-based heuristic that is currently in use. Machine learning is not the answer to all problems, and sometimes a rule-based heuristic is hard to beat. Development shouldn't be done for development's sake. A baseline model, no matter how simple, is helpful to guide design decisions down the road and understand how each design choice moves the needle on that predetermined evaluation metric. In Chapter 7, we discussed the role of a Heuristic Benchmark as well as other topics related to Responsible AI that often come up when communicating the impact and influence of machine learning with business stakeholders.

Of course, these conversations should also take place in the context of the data. A business deep dive should go hand in hand with a deep dive of data exploration (Step 2 of Figure 8-2). As beneficial as a solution might be, if quality data is not available, then there is no project. Or perhaps the data exists, but because of data privacy reasons, it cannot be used or must be scrubbed of relevant information needed for the model. In any case, the viability of a project and the potential for success all rely on the data. Thus, it is essential to have data stewards within the organization involved in these conversations early.

The data guides the process and it's important to understand the quality of the data that is available. What are the distributions of the key features? How many missing values are there? How will missing values be handled? Are there outliers? Are any input values highly correlated? What features exist in the input data and which features should be engineered? Many machine learning models require a massive dataset for training. Is there enough data? How can we augment the dataset? Is there bias in the dataset? These are important questions, and they only touch the surface. One possible decision at this stage is that more data, or data of a specific scenario, needs to be collected before the project can proceed.

Data exploration is a key step in answering the question of whether data of sufficient quality exists. Conversation alone is rarely a substitute for getting your hands dirty and experimenting with the data. Visualization plays an important role during this step. Density plots and histograms are helpful to understand the spread of different input values. Box plots can help to identify outliers. Scatter plots are useful for discovering and describing bivariate relationships. Percentiles can help identify the range for numeric data. Averages, medians, and standard deviations can help to describe central tendency. These techniques and others can help determine which features are likely to benefit the model as well as further understanding of which data transformations will be needed to prepare the data for modeling.

Within the discovery stage, it can be helpful to do a few modeling experiments to see if there really is "signal in the noise." At this point, it could be beneficial to perform a machine learning feasibility study (Step 3). Just as it sounds, this is typically a short technical sprint spanning only a few weeks whose goal is to assess the viability of the data for solving the problem. This provides a chance to explore options for framing the machine learning problem, experiment with algorithm selection, and learn which feature engineering steps would be most beneficial. The feasibility study step in the discovery stage is also a good point at which to create a Heuristic Benchmark (see Chapter 7).

Development

After agreeing on key evaluation metrics and business KPIs, the development stage of the machine learning life cycle begins. The details of developing an ML model are covered in detail in many machine learning resources. Here, we highlight the key components.

During the development stage, we begin by building data pipelines and engineering features (Step 4 of Figure 8-2) to process the data inputs that will be fed to the model. The data collected in real-world applications can have many issues such as missing values, invalid examples, or duplicate data points. Data pipelines are needed to pre-process these data inputs so that they can be used by the model. Feature engineering is the process of transforming raw input data into features that are more closely aligned with the model's learning objective and expressed in a format that can be fed to the model for training. Feature engineering techniques can involve bucketizing inputs, converting between data formats, tokenizing and stemming text, creating categorical features or one-hot encoding, hashing inputs, creating feature crosses and feature embeddings, and many others. Chapter 2 of this book discusses Data Representation design patterns and covers many data aspects that arise during this stage of the ML life cycle. Chapter 5 and Chapter 6 describe patterns related to resilience and reproducibility in ML systems, which help in building data pipelines.

This step may also involve engineering the labels for the problem and design decisions related to how the problem is represented. For example, for time-series problems, this may involve creating feature windows and experimenting with lag times and the size of label intervals. Or perhaps it's helpful to reframe a regression problem as a classification and change the representation of the labels entirely. Or maybe it is necessary to employ rebalancing techniques, if the distribution of output classes is overrepresented by a single class. Chapter 3 of this book is focused on problem representation and addresses these and other important design patterns that are related to problem framing.

The next step (Step 5 in Figure 8-2) of the development stage is focused on building the ML model. During this development step, it is crucial to adhere to best practices of capturing ML workflows in a pipeline: see "Design Pattern 25: Workflow Pipeline" on page 282 in Chapter 6. This includes creating repeatable splits for training/validation/test sets before any model development has begun to ensure there is no data leakage. Different model algorithms or combinations of algorithms may be trained to assess their performance on the validation set and to examine the quality of their predictions. Parameter and hyperparameters are tuned, regularization techniques are employed, and edge cases are explored. The typical ML model training loop is described in detail at the beginning of Chapter 4 where we also address useful design patterns for changing the training loop to attain specific objectives.

Many steps of the ML life cycle are iterative, and this is particularly true during model development. Many times, after some experimentation, it may be necessary to revisit the data, business objectives, and KPIs. New data insights are gleaned during the model development stage and these insights can shed additional light on what is possible (and what is not possible). It is not uncommon to spend a long time in the model development phase, particularly when developing a custom model. Chapter 6 addresses many other reproducibility design patterns that address challenges that arise during this iterative phase of model development.

Throughout development of the model, each new adjustment or approach is measured against the evaluation metrics that were set in the discovery stage. Hence, successful execution of the discovery stage is crucial, and it is necessary to have alignment on the decisions made during that stage. Ultimately, model development culminates in a final evaluation step (Step 6 of Figure 8-2). At this point, model development ceases and the model performance is assessed against those predetermined evaluation metrics.

One of the key outcomes of the development stage is to interpret and present results (Step 7 of Figure 8-2) to the stakeholders and regulatory groups within the business. This high-level evaluation is crucial and necessary to communicate the value of the development stage to management. This step is focused on creating numbers and visuals for initial reports that will be brought to stakeholders within the organization. Chapter 7 discusses some of the common design patterns that ensure AI is being used responsibly and can help with stakeholder management. Typically, this is a key decision point in determining if further resources will be devoted to the final stage of the life cycle, machine learning productionization and deployment.

Deployment

Assuming successful completion of the model development and evidence of promising results, the next stage is focused on productionization of the model, with the first step (Step 8 in Figure 8-2) being to plan for deployment.

Training a machine learning model requires a substantial amount of work, but to fully realize the value of that effort, the model must run in production to support the business efforts it was designed to improve. There are several approaches that achieve this goal and deployment can look different among different organizations depending on the use case. For example, productionized ML assets could take the form of interactive dashboards, static notebooks, code that is wrapped in a reusable library, or web services endpoints.

There are many considerations and design decisions for productionizing models. As before, many of the decisions that are made during the discovery stage guide this step as well. How should model retraining be managed? Will input data need to stream in? Should training happen on new batches of data or in real time? What about model inference? Should we plan for one-off batch inference jobs each week or do we need to support real-time prediction? Are there special throughput or latency issues to consider? Is there a need to handle spiky workloads? Is low latency a priority? Is network connectivity an issue? The design patterns in Chapter 5 touch on some of the issues that arise when operationalizing an ML model.

These are important considerations, and this final stage tends to be the largest hurdle for many businesses, as it can require strong coordination among different parts of the organization and integration of a variety of technical components. This difficulty is also in part due to the fact that productionization requires integrating a new process, one that relies on the machine learning model, into an existing system. This can involve dealing with legacy systems that were developed to support a single approach, or there could be complex change control and production processes to navigate within the organization. Also, many times, existing systems do not have a mechanism for supporting predictions coming from a machine learning model, so new applications and workflows must be developed. It is important to anticipate these challenges, and developing a comprehensive solution requires significant investment from the business operations side to make the transition as easy as possible and increase the speed to market.

The next step of the deployment stage is to operationalize the model (Step 9 in Figure 8-2). This field of the practice is typically referred to as MLOps (ML Operations) and covers aspects related to automating, monitoring, testing, managing, and maintaining machine learning models in production. It is a necessary component for any company hoping to scale the number of machine learning–driven applications within their organization.

One of the key characteristics of operationalized models is automated workflow pipelines. The development stage of the ML life cycle is a multistep process. Building pipelines to automate these steps enables more efficient workflows and repeatable processes that improve future model development, and allows for increased agility in solving problems that arise. Today, open source tools like Kubeflow (*https://oreil.ly/*

I_cJf) provide this functionality and many large software companies have developed their own end-to-end ML platforms, like Uber's Michelangelo (*https://oreil.ly/se4G9*) or Google's TFX (*https://oreil.ly/OznI3*), which are also open source.

Successful operationalization incorporates components of continuous integration and continuous delivery (CI/CD) that are the familiar best practices of software development. These CI/CD practices are focused on reliability, reproducibility, speed, security, and version control within code development. ML/AI workflows benefit from the same considerations, though there are some notable differences. For example, in addition to the code that is used to develop the model, it is important to apply these CI/CD principles to the data, including data cleaning, versioning, and orchestration of data pipelines.

The final step to be considered in the deployment stage is to monitor and maintain the model. Once the model has been operationalized and is in production, it's necessary to monitor the model's performance. Over time, data distributions change, causing the model to become stale. This model staleness (see Figure 8-3) can occur for many reasons, from changes in customer behavior to shifts in the environment. For this reason, it is important to have in place mechanisms to efficiently monitor the machine learning model and all the various components that contribute to its performance, from data collection to the quality of the predictions during serving. The discussion of "Design Pattern 18: Continued Model Evaluation" on page 220 in Chapter 5 covers this common problem and its solution in detail.

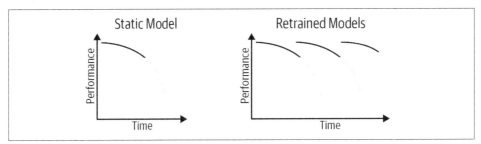

Figure 8-3. Model staleness can occur for many reasons. Retraining models periodically can help to improve their performance over time.

For example, it is important to monitor the distribution of feature values to compare against the distributions that were used during the development steps. It is also important to monitor the distribution of label values to ensure that some data drift hasn't caused an imbalance or shift in label distribution. Oftentimes, a machine learning model relies on data collected from an outside source. Perhaps our model relies on a third-party traffic API to predict wait times for car pickups or uses data from a weather API as input to a model that predicts flight delays. These APIs are not managed by our team. If that API fails or its output format changes in a significant way, it will have consequences for our production model. In this case, it is important to set up monitoring to check for changes in these upstream data sources. Lastly, it is important to set up systems to monitor prediction distributions and, when possible, measure the quality of those predictions in the production environment.

Upon completion of the monitoring step, it can be beneficial to revisit the business use case and objectively, accurately assess how the machine learning model has influenced business performance. Likely, this will lead to new insights and the start of new ML projects, and the life cycle begins again.

AI Readiness

We find that different organizations working on building machine learning solutions are at different stages of AI Readiness. According to a white paper published by Google Cloud (*https://oreil.ly/5GljC*), a company's maturity in incorporating AI into the business can typically be characterized into three phases: tactical, strategic, and transformational. Machine learning tools in these three phases go from involving primarily manual development in the tactical phase, to using pipelines in the strategic phase, to being fully automated in the transformational phase.

Tactical phase: Manual development

The tactical phase of AI Readiness is often seen in organizations just beginning to explore the potential for AI to deliver, with focus on short-term projects. Here, the AI/ML use cases tend to be more narrow, focusing more on proofs of concept or prototypes; a direct link to the business goals may not always be clear. In this stage, organizations recognize the promise of advanced analytics work, but the execution is driven primarily by individual contributors or outsourced entirely to partners; access to large-scale, quality datasets within the organization can be difficult.

Typically, in this phase, there is no process to scale solutions consistently, and the ML tools used (see Figure 8-4) are developed on an ad hoc basis. Data is warehoused offline or in isolated data islands and accessed manually for data exploration and analysis. There are no tools in place to automate the various phases of the ML development cycle and there is little attention paid to developing repeatable processes of the workflow. This makes it difficult to share assets within members of the organization, and there is no dedicated hardware for development.

The extent of MLOps is limited to a repository of trained models, and there is little distinction between testing and production environments where the final model may be deployed as an API-based solution.

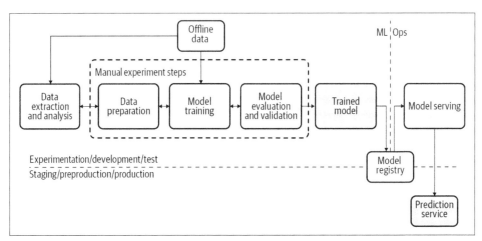

Figure 8-4. Manual development of AI models. Figure adapted from Google Cloud documentation (https://oreil.ly/aC1HP).

Strategic phase: Utilizing pipelines

Organizations in the strategic phase have aligned AI efforts with business objectives and priorities, and ML is seen as a pivotal accelerator for the business. As such, there is often senior executive sponsorship and dedicated budget for ML projects that are executed by skilled teams and strategic partners. There is infrastructure in place for these teams to easily share assets and develop ML systems that leverage both ready-to-use and custom models. There is a clear distinction between development and production environments.

Teams typically already have skills in data wrangling with expertise in descriptive and predictive analytics. Data is stored in an enterprise data warehouse, and there is a unified model for centralized data and ML asset management. The development of ML models occurs as an orchestrated experiment. The ML assets and source code for these pipelines is stored in a centralized source repository and easily shared among members of the organization.

The data pipelines for developing ML models are automated utilizing a fully managed, serverless data service for ingestion and processing and are either scheduled or event driven. Additionally, the ML workflow for training, evaluation, and batch prediction is managed by an automated pipeline so that the stages of the ML life cycle, from data validation and preparation to model training and validation (see Figure 8-5), are executed by a performance monitoring trigger. These models are stored in a centralized trained models registry and able to be deployed automatically based on predetermined model validation metrics.

There may be several ML systems deployed and maintained in production with logging, performance monitoring, and notifications in place. The ML systems leverage a model API that is capable of handling real-time data streams both for inference and to collect data that is fed into the automated ML pipeline to refresh the model for later training.

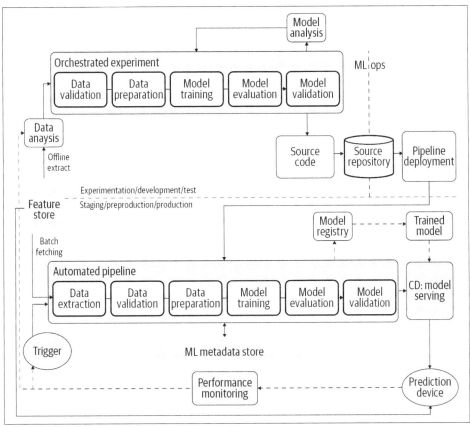

Figure 8-5. Pipelines phase of AI development. Figure adapted from Google Cloud documentation (https://oreil.ly/sMNo7).

Transformational phase: Fully automated processes

Organizations in the transformational phase of AI Readiness are actively using AI to stimulate innovation, support agility, and cultivate a culture where experimentation and learning is ongoing. Strategic partnerships are used to innovate, co-create, and augment technical resources within the company. Many of the design patterns related to reproducibility and resilience in Chapters 5 and 6 arise in this phase of AI Readiness.

In this phase, it is common to have product-specific AI teams embedded into the broader product teams and supported by the advanced analytics team. In this way, ML expertise is able to diffuse across various lines of business within the organization. The established common patterns and best practices as well as standard tools and libraries for accelerating ML projects are shared easily among different groups within the organization.

Datasets are stored in a platform that is accessible to all teams, making it easy to discover, share, and reuse datasets and ML assets. There are standardized ML feature stores, and collaborations across the entire organization are encouraged. Fully automated organizations operate an integrated ML experimentation and production platform where models are built and deployed and ML practices are accessible to everyone in the organization. That platform is supported by scalable and serverless computation for batch and online data ingestion and processing. Specialized ML accelerators such as GPUs and TPUs are available on demand and there are orchestrated experiments for end-to-end data and ML pipelines.

The development and production environments are similar to the pipeline stage (see Figure 8-6) but have incorporated CI/CD practices into each of the various stages of their ML workflow as well. These CI/CD best practices focus on reliability, reproducibility, and version control for the code to produce the ML models as well as the data and the data pipelines and their orchestration. This allows for building, testing, and packaging of various pipeline components. Model versioning is maintained by an ML Model Registry that also stores necessary ML metadata and artifacts.

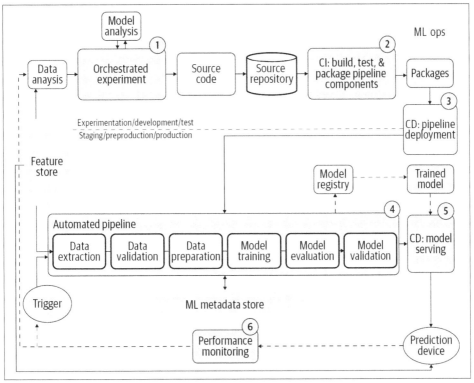

Figure 8-6. Fully automated processes support AI development. Figure adapted from Google Cloud documentation. (https://oreil.ly/VX31C)

Common Patterns by Use Case and Data Type

Many of the design patterns discussed in this book are utilized throughout the course of any machine learning development cycle and will likely be used regardless of the production use case—for example, Hyperparameter Tuning, Heuristic Benchmark, Repeatable Splitting, Model Versioning, Distributed Training, Workflow Pipelines, or Checkpoints. Other design patterns, you might find, are particularly useful for certain scenarios. Here, we'll group together commonly used design patterns according to popular machine learning use cases.

Natural Language Understanding

Natural language understanding (NLU) is a branch of AI that focuses on training a machine to understand the meaning behind text and language. NLU is used by speech agents like Amazon's Alexa, Apple's Siri, and Google's Assistant to understand sentences like, "What is the weather forecast this weekend?" There are many use cases that fall under the umbrella of NLU and it can be applied to a lot of processes, such as text classification (email filtering), entity extraction, question answering, speech recognition, text summarization, and sentiment analysis.

- Embeddings
- Hashed Feature
- Neutral Class
- Multimodal Input
- Transfer Learning
- Two-Phase Predictions
- Cascade
- Windowed Inference

Computer Vision

Computer vision is the broad parent name for AI that trains machines to understand visual input, such as images, videos, icons, and anything where pixels might be involved. Computer vision models aim to automate any task that might rely on human vision, from using an MRI to detect lung cancer to self-driving cars. Some classical applications of computer vision are image classification, video motion analysis, image segmentation, and image denoising.

- Reframing
- Neutral Class
- Multimodal Input
- Transfer Learning
- Embeddings
- Multilabel
- Cascade
- Two-Phase Predictions

Predictive Analytics

Predictive modeling uses historical data to find patterns and determine the likelihood of a certain event occurring in the future. Predictive models can be found across many different industry domains. For example, businesses might use predictive models to forecast revenue more accurately or anticipate future demand for products. In medicine, predictive models might be used to assess the risk of a patient developing a chronic disease or predict when a patient might not show up for a scheduled appointment. Other examples include energy forecasting, customer churn prediction, financial modeling, weather prediction, and predictive maintenance.

- Feature Store
- Feature Cross
- Embeddings
- Ensemble
- Transform
- Reframing
- Cascade
- Multilabel
- Neutral Class
- Windowed Inference
- Batch Serving

IoT analytics is also a broad category that sits within predictive analytics. IoT models rely on data collected by internet-connected sensors called IoT devices. Consider a commercial aircraft that has thousands of sensors on it collecting more than 2 TB of data per day. Machine learning of IoT sensor device data can provide predictive models to warn against equipment failure before it happens.

- Feature Store
- Transform
- Reframing
- Hashed Feature
- Cascade
- Neutral Class
- Two-Phase Predictions
- Stateless Serving Function
- Windowed Inference

Recommendation Systems

Recommender systems are one of the most widespread applications of machine learning in business and they often arise whenever users interact with items. Recommender systems capture features of past behavior and similar users and recommend items most relevant for a given user. Think of how YouTube will recommend a series of videos for you to watch based on your watch history, or Amazon may recommend purchases based on items in your shopping cart. Recommendation systems are popular throughout many businesses, particularly for product recommendation, personalized and dynamic marketing, and streaming video or music platforms.

- Embeddings
- Ensemble
- Multilabel
- Transfer Learning
- Feature Store
- Hashed Feature
- Reframing
- Transform
- Windowed Inference
- Two-Phase Predictions
- Neutral Class
- Multimodal Input
- Batch Serving

Fraud and Anomaly Detection

Many financial institutions use machine learning for fraud detection to keep their consumers' accounts safe. These machine learning models are trained to flag transactions that appear fraudulent based on certain characteristics or patterns that have been learned in the data.

More broadly, anomaly detection is a technique used to find abnormal behavior or outlier elements in a dataset. Anomalies can arise as spikes or dips that deviate from the normal patterns, or they can be longer-term abnormal trends. Anomaly detection shows up through many different use cases in machine learning and might even be used in conjunction with a separate use case. For example, consider a machine learning model that identifies anomalous train tracks based on images.

- Rebalancing
- Feature Cross
- Embeddings
- Ensemble
- Two-Phase Predictions
- Transform

- Feature Store
- Cascade
- Neutral Class
- Reframing

Index

A

ablation, 356
AdaBoost, 102, 107
AdaNet, 108
AI Platform Notebooks, 336
AI Platform Pipelines, 230, 288
AI Platform Prediction, 9, 288, 315, 354
AI Platform pusher component, 288
AI Platform Training, 9, 197
AI readiness, 373-376
Alexander, Christopher, 1-2
all-reduce algorithm, 176
anomaly detection, 132-136, 243, 380
Apache Airflow, 284, 292
Apache Beam, 217, 256, 276-278, 297, 309
Apache Flink, 297, 309
Apache Spark, 217, 297, 309
Apigee, 314
application-specific integrated circuit (see
    ASIC)
ARIMA, 79, 275
arrays, 27-28
ASIC, 184, 239
asynchronous serving, 247
asynchronous training, 179-181
attribution values, 136
autoencoders, 49-51
AutoML Tables, 339
AutoML technique, 108
autoregressive integrated moving average (see
    ARIMA)
AWS Lambda, 206, 220, 228, 314
Azure, 184, 314
Azure Functions, 206, 228
Azure Machine Learning, 314
Azure ML Pipelines, 288

B

bag of words approach (see BOW encoding)
bagging, 100-101, 104-105, 108
baseline, 330-336
    (see also informative baseline, uninforma-
        tive baseline)
batch prediction, 7, 241, 247
batch serving, 218-220
Batch Serving design pattern, 201, 213-220, 365
batch size, 185-186
batching, 208
Bayesian optimization, 193-194
beam search algorithm, 79
BERT, 50-51, 174, 186
bias
    data, 339, 344, 350-356
        (see also data collection bias, data distri-
            bution bias, data representation bias,
            experimenter bias, implicit bias,
            problematic bias, proxy bias, report-
            ing bias)
    human, 12-13, 343
        (see also implicit bias, problematic bias,
            proxy bias,)
    model, 72, 100, 106, 128, 326, 343, 345-346,
        356-358
        (see also label bias)
    unfair, 356
bias-variance tradeoff, 100, 108
Bidirectional Encoding Representations from
    Transformers (see BERT)

BigQuery
    about, 2, 9
    features of, 217, 220, 297, 309
    performance of, 280
    uses of, 33, 52, 54-55, 91, 95, 214, 285
BigQuery Machine Learning (see BigQuery
    ML)
BigQuery ML
    about, 9
    features of, 127, 251-252
    performance of, 57
    uses of, 55-56, 132, 250
BigQueryExampleGen component, 284, 286
BigTable, 309
binary classification, 93-99, 322
binary classifier, 98, 117, 121, 153
binary encoding, 37
boolean variables, 19
boosting, 102-102, 104, 105, 108
bootstrap aggregating (see bagging)
bottleneck layer, 163-168
BOW encoding, 67-70
Box-Cox transform, 27
Bridged Schema design pattern, 250, 266-273,
    364, 365
bucketing, 30

C

CAIP (see Cloud AI Platform)
capacity, 100, 157
Cartesian product, 62
cascade, 110, 129-132, 271
Cascade design pattern, 79, 97, 108-117, 272,
    364-365
Cassandra, 297, 309
categorical data, 6
categorical inputs, 28-31
CBOW, 50
CentralStorageStrategy, 177
centroid, 133-134
chaos theory, 145
checkpoint selection, 155-157
checkpointing, 151
checkpoints, 151-155
Checkpoints design pattern, 149-161, 364
CI/CD, 290, 291-292, 308, 372, 376
Civil Comments dataset, 350
classification models, 6
classification threshold, 353

clipping, 23-24
closeness relationships, 40-41, 60
Cloud AI Platform, 2, 9, 222, 223
Cloud AI Platform Pipelines, 284, 287
Cloud AI Platform Predictions, 220
Cloud AI Platform Training, 285, 287
Cloud Build, 292
Cloud Composer/Apache Airflow, 230
Cloud Dataflow, 220
Cloud Functions, 228, 291
Cloud Run, 206, 315
Cloud Spanner, 219
clustering, 5
clustering models, 5
CNN, 72, 169-171
cold start, 32, 35
combinatorial explosion, 189
completeness, 12
components, definition of, 284
computer vision, 378
concept drift, 220, 231
confidence, 98, 120, 223
confusion matrix, 123, 225
consistency, 12-13
containers, 282, 284, 289
context language models, 50-51
    (see also BERT, Word2Vec)
Continued Model Evaluation design pattern,
    201, 220-231, 314, 320, 355, 365
Continuous Bag of Words (see CBOW)
continuous evaluation, 247-248
continuous integration and continuous delivery
    (see CI/CD)
convolutional neural network (see CNN)
Coral Edge TPU, 239
counterfactual analysis, 339-342
counterfactual reasoning, 224
cryptographic algorithms, 38
custom serving function, 209

D

DAG, 289, 292
Darwin, Charles, 197
data accuracy, 11
data analysts, 10
data augmentation, 356
data collection bias, 348, 350
data distribution bias, 344
data drift, 14-15, 220, 231, 243, 310

data engineers, 9, 16, 297
data parallelism, 175-176, 178, 181, 184
data preprocessing, 6
    (see also data transformation, feature engi-
      neering)
data representation, 20-21
data representation bias, 348
data scientists
    role of, 9, 16-17, 207, 327
    tasks of, 283, 295, 311
data transformation, 7
data validation, 7, 231
data warehouses, 51-52
dataset-level transformations, 255
datasets, definition of, 6
Datastore, 219
decision trees, 5, 19-21, 108, 135, 139, 327
Deep Galerkin Method, 147-148
deep learning, 4-5, 77
deep neural network (see DNN model)
default, definition of, 312
Dense layers, 64, 76
design patterns, definition of, 1-2
Design Patterns: Elements of Reusable Object-
    Oriented Software, 2
developers, 10, 16
dimensionality reduction, 5
directed acyclic graph (see DAG)
discrete probability distribution, 81, 82, 84
distributed data processing infrastructure, 214
DistributedDataParallel, 178
Distribution Strategy design pattern, 175-187
DNN model, 44, 57, 107, 139
Docker container, 287, 315
downsampling, 123, 125-127, 134-135
dropout technique, 107
dummy coding, 29

E

early stopping, 155
edge, 232-233, 239
Embedding design pattern, 20-21, 39-52, 62-65,
    66, 363-364
embeddings, 167
    (see also bottleneck layer)
embeddings, as similarity, 47
Ensemble design pattern, 79-80, 99-108, 110,
    134, 365
ensemble methods, 100

    (see also bagging, boosting, stacking)
epochs
    training, 6, 155
    using, 140, 150, 159-160
    virtual, 160, 180, 364
evaluation, definition of, 7
example-based explanation, 339-342
ExampleGen components, 284-284
ExampleValidator, 284
experimenter bias, 345
explainability, 310, 327, 329, 335, 357-358, 365
    (see also deep learning, post hoc explaina-
      bility method)
Explainable AI, 9, 136, 335, 339
Explainable Predictions design pattern, 320,
    326-343, 365
exported model, 150

F

Facets, 231
Fairness Indicators, 354
Fairness Lens design pattern, 320, 343-358
Farm Fingerprint hashing algorithm, 259, 263,
    265
FarmHash, 33
Feast, 298-309
feature attributions, 329-339
feature columns, 33, 39, 42, 252-255
Feature Cross design pattern, 21, 52-62, 363
feature cross, cardinality, 61
feature engineering, 6, 20, 257, 295, 368
    (see also data preprocessing)
feature extraction, 21, 172-173, 260
Feature Store design pattern, 250, 257, 295-310,
    364-365
feature, definition of, 7, 20
FeatureSet, 299-302
feed-forward neural networks (see neural net-
    works)
field-programmable gate array (FPGA), 184
fine-tuning, 157, 172-173, 229
    (see also progressive fine-tuning)
fingerprint hashing algorithm, 38
fitness function, 198
flat approach, 97
Flatten layer, 71
FPGA (field-programmable gate array), 184
fraud detection, 122-126, 134, 220, 224, 263,
    265

G

Gamma, Erich, 2
Gaussian process, 194
genetic algorithms, 194, 197-198
GitHub Actions, 292
GitLab Triggers, 292
GKE, 284, 287
GLoVE, 51
Google App Engine, 206
Google Bolo, 242
Google Cloud Functions, 206
Google Cloud Public Datasets, 9
Google Container Registry, 287
Google Kubernetes Engine (see GKE)
Google Translate, 241
GPU, 162, 175-178, 184, 186, 214, 287, 376
Gradient Boosting Machines, 102
gradient descent (see SGD)
graphics processing unit (see GPU)
grid search, 188-190, 192
Grid-SearchCV, 189
ground truth label, 7, 12, 224-227

H

hash buckets
    collisions, 35
    empty, 39
    heuristic to choose numbers, 34
Hashed Feature design pattern, 21, 32-39, 363
Helm, Richard, 2
Heroku, 206
heuristic benchmark , 321-324, 333, 369
Heuristic Benchmark design pattern, 320-325
hidden layers, 4
high cardinality, 32, 35, 40
histogram equalization, 27
Hive, 297, 309
Hopsworks, 309
hyperparameter tuning, 37, 160
Hyperparameter Tuning design pattern,
    187-198, 363
hyperparameters, 6, 187

I

idioms, 22, 28, 31
IG, 336, 337
image embeddings, 45
ImageDataGenerator, 237

ImageNet

ImageNet, 45, 49, 97, 162, 163, 165
implicit bias, 345
imputation, 270-273
Inception, 45
inference, 8, 143
    (see also ML approximation)
informative baseline, 330-333
input, definition of, 7, 20
instance, definition of, 7
instance-level transformations, 255
integrated gradients (see IG)
interpretability (see explainability)
interpretable by design, 327
IoT analytics, 379
irreducible error, 100

J

Jetson Nano, 239
Johnson, Ralph, 2
JSON, 208

K

k-nearest neighbors (kNN), 105
Kaggle, 235
Kale, 293
Keras
    about, 2, 4, 128
    features of, 127, 152, 167, 177, 190, 236
    uses of, 42-45, 64-65, 71, 73, 94, 252-255
Keras ImageDataGenerator, 129
Keras Sequential API, 92
Keras Training Loop, 140
kernel size, 72
key performance indicator (see KPI)
Keyed Predictions design pattern, 2, 201,
    244-248, 365
keys, 244-248
KFP (see Kubeflow Pipelines)
kNN (k-nearest neighbors), 105
KPI , 367-370
Kubeflow Pipelines, 113, 284, 288, 292

L

label bias, 89
label, definition of, 7
    (see also ground truth label, prediction)
labeling, 13, 118, 224, 324, 345
labels, overlapping, 97-99

LAMB, 186
Lambda architecture, 219
Language Interpretability Tool, 354
library function, 213
Light on Two Sides of Every Room pattern, 1-2
lineage tracking, 294
linear models, 5, 322
long short-term memory model (see LSTM)
low latency, 8, 209-209, 215, 237, 296-298, 307
LSTM, 135, 275, 281

M

machine learning engineers (see ML engineers)
machine learning feasibility study, 369
machine learning framework, 14
machine learning life cycle (see ML life cycle)
machine learning models, 4
machine learning problems (see supervised
    learning; unsupervised learning)
machine learning, definition of, 4
MAE (mean absolute error), 320
MAP (mean average precision), 321
MapReduce, 216
matrix factorization, 79
MD5 hash, 38
mean absolute error (MAE), 320
mean average precision (MAP), 321
Mesh TensorFlow, 184
mesh-free approximation, 147
microservices architecture, 283
min-max scaling, 23-25
Mirrored Variable, 176
MirroredStrategy, 177, 179
Mixed Input Representation, 115
ML approximation, 143, 145
ML engineers
    role of, 10, 16, 207, 319, 327
    tasks of, 295, 297, 311, 314
ML life cycle, 366-367, 369-373
ML Operations (see MLOps)
ML pipelines, 8
ML researchers, 319
MLflow, 284
MLOps, 371, 373
MNIST dataset, 71, 74
MobileNetV2, 236, 238
Mockus, Jonas, 193
model builders, 319
    (see also data scientists, ML researchers)

Model Card Toolkit, 357
Model Cards, 357
model evaluation, 8, 123, 294, 343, 345
    (see also Continued Model Evaluation
        design pattern)
model parallelism, 175, 183-184
model parameters, 187-188
model understanding (see explainability)
Model Versioning design pattern, 250, 310-317,
    365
model, pre-trained, 167-169, 173, 319, 364
model, text classification, 203, 209, 250, 316
monolithic applications, 283
Monte Carlo approach, 146-147
multi-hot encoding, 31
multiclass classification problems, 90
multilabel classification, 93-95, 322
Multilabel design pattern, 79, 90-99
multilabel, multiclass classification (see Multi-
    label design pattern)
Multimodal Input design pattern, 62-77, 365
multimodal inputs, definition of, 65
MultiWorkerMirroredStrategy, 177, 179
MySQL, 219
MySQL Cluster, 309

N

naive Bayes, 105
natural language understanding (NLU), 377
Netflix Prize, 107
Neural Machine Translation, 183
neural networks, 4, 147
Neutral Class design pattern, 80, 117-122, 320,
    364
NLU (natural language understanding), 377
NNLM, 51
nonlinear transformations, 26-27
numerical data, 6-7

O

objective function, 193
OCR (optical character recognition), 116
one versus rest approach, 98
one-hot encoding, 29-30, 39-40, 48, 267
OneDeviceStrategy, 179, 180
online machine learning, 230
online prediction, 7, 247
online update, 279
ONNX, 205

optical character recognition (OCR), 116
orchestration, definition of, 293
outliers, 24
output layer bias, 128
overfit model, 100, 142
  (see also physics-based model)
overfitting, 148-149

P

parameter server architecture, 179
parameter sharing, 89-90
ParameterServerStrategy, 180
partial differential equation (see PDE)
Parzen estimator, 194
Pattern Language, A, 1
PCA, 41, 49
PDE, 141-143, 146, 147
PDF, 82, 84, 87
physics-based model, 142
pipeline, 284
pixel values, 71
post hoc explainability method, 329
posterior probability distribution, 82
precision, 124
prediction, 7, 110, 213
  (see also batch prediction, inference, online
    prediction)
predictive modeling, 378
principal components analysis (see PCA)
probability density function (see PDF)
problematic bias, 343-346
productionizing models, 371
progressive fine-tuning, 172
proxy bias, 345
Pusher component, 285
PyTorch, 152, 175, 178, 196

Q

quantile regression, 83, 86
quantization, 233, 237, 239
quantization aware training, 241

R

random forest, 108, 189
random search, 190, 192
random seed, 258-259
RandomForestRegressor, 188
RandomizedSearchCV, 190

ratings, representation of, 66
ray-tracing model, 143
Rebalancing design pattern, 109, 115, 122-136,
  350, 365
recall, 124
recommendation systems
  reframing as regression, 84
  uses for, 82, 89, 379
Redis, 297, 309
reducible error, 100
reframing , 123, 129-132
Reframing design pattern, 79-90, 331, 364-365
regression models, 6, 322
regularization, 107, 141, 146, 149, 155-157, 271
relative frequency, 31
repeatability, 13
Repeatable Splitting design pattern, 250,
  258-265, 320, 363, 370
reporting bias, 344
reproducibility, 13-14
research scientists, 10
ResNet, 45
responsible AI, 320, 357, 370
REST API, for model serving, 213
retraining trigger, 228
roles, 9-10
roles, impact of team size, 10
Runge-Kutta methods, 146
runs, definition of, 286

S

SageMaker, 207, 288, 314
salt, 38
Sampled Shapley, 336
SavedModel, 205, 212, 220
  (see also saved_model_cli)
saved_model_cli, 205
scaling, 16, 22-23
SchemaGen, 284
scikit-learn, 14, 22, 69, 107, 135, 189, 190
sentence embeddings, 174
Sequential API, 71, 73
serverless, 8, 315
serverless triggers, 228
serving, definition of, 7
SGD, 139-140, 176
SHAP, 136, 333-336
Shapley Value, 330, 333
sigmoid, 94, 99-99

(see also sigmoid activation)
sigmoid activation, 91-95
Six-Foot Balcony pattern, 1-2
skip-gram model, 50
Smart Compose, 356
SMOTE, 128-129, 134
softmax, 45, 90, 94
  (see also softmax activation)
softmax activation, 92
software reliability engineer (SRE), 209
spurious correlation, 37
SRE (software reliability engineer), 209
Stack Overflow, 67-69, 70, 91, 95
stacking, 103-104, 106-106
stakeholders, 319, 327
stateful stream processing, 275
stateful vs. stateless components, 202-203
stateless functions, 202-203
Stateless Serving Function design pattern,
  201-213, 316, 365
StatisticsGen component, 284
stochastic gradient descent (see SGD)
stratified split, 264
streaming, definition of, 8
structured data, 6
  (see also categorical data, numerical data,
    tabular data)
supervised learning, 5
support vector machine (see SVM)
surrogate function, 194
survival of the fittest theory, 197
SVM, 105, 139
Swivel, 52
synchronous training, 176-181
Synthetic Minority Over-sampling Technique
  (see SMOTE)

T
TabNet, 50, 174
tabular data
  about, 6
    (see also structured data)
  applications for, 91, 173-174
  representation of, 65-66
tensor processing unit (see TPU)
TensorBoard, 354
TensorFlow
  about, 2, 4, 107
  features of, 135, 152, 155, 175, 176, 252, 335

uses of, 13, 33, 42, 56
TensorFlow Data Validation, 231, 354
TensorFlow dataset, 158
TensorFlow Extended, 256, 284
TensorFlow hub, 52, 167, 169, 335
Tensorflow Lite, 233, 237
TensorFlow Lite, 233
TensorFlow Model Analysis (see TFMA)
TensorFlow Probability, 86
TensorFlow Serving, 207, 256, 315, 354
TensorFlow Transform method, 114
  (see also Transform design pattern)
test data, 6, 140, 258, 264
testing dataset (see test data)
text embeddings, 42-45
TF Hub (see TensorFlow hub)
TF Lite Interpreter, 237-238
TFMA, 354-355
TFX, 231, 284-292, 354
threshold selection, 96
threshold, definition of, 95
time-windowed average, 218
timeliness, 13
tokenization, 42-45
TorchServe, 207
TPAClusterResolver, 185
TPU
  about, 2, 186, 376
  features of, 150, 287
  uses of, 184, 214
TPUStrategy, 185
Trainer component, 285, 288
training data, 6
training examples, 7
training loop, 139-141, 150, 155
  (see also well-behaved training loop)
training, definition of, 7
training, synchronous vs. asynchronous,
  179-181
training-serving skew, 251, 257, 309
Transfer Learning design pattern, 161-174, 356,
  364-364
Transform component, 284
Transform design pattern, 2, 56, 250-258, 309,
  365
trials, definition of, 188
Tweedie distribution, 81
Two-Phase Predictions design pattern, 201,
  232-243, 282

U

underfit model, 100
Uniform Approximation Theorem, 144-145
uninformative baseline, 330-332
Universal Sentence Encoder, 174
unsampled data, 125
unstructured data, 6, 265
unsupervised learning, 5, 132
upsampling, 123, 128-129
Useful Overfitting design pattern, 141-149

V

validation data, 6, 140
validation dataset (see validation data)
VGG, 164-168, 239
Vision API, 356, 357
Vlissides, John, 2
vocabulary, 29-33, 34-35, 42, 67-70

W

well-behaved training loop, 155, 157

What-If Tool, 136, 339, 347-354
Wheeler, David, 209
Windowed Inference design pattern, 250, 273-282
winsorizing, 24-25
word index, 68
Word2Vec, 50-51
Workflow Pipeline design pattern, 112-113, 228, 250, 282-294, 355, 365

X

XGBoost, 69, 102, 107, 135, 315
XRAI, 337, 337

Z

z-score normalization, 23-25

About the Authors

Valliappa (Lak) Lakshmanan is Global Head for Data Analytics and AI Solutions on Google Cloud. His team builds software solutions for business problems using Google Cloud's data analytics and machine learning products. He founded Google's Advanced Solutions Lab ML Immersion program. Before Google, Lak was a Director of Data Science at Climate Corporation and a Research Scientist at NOAA.

Sara Robinson is a Developer Advocate on Google's Cloud Platform team, focusing on machine learning. She inspires developers and data scientists to integrate ML into their applications through demos, online content, and events. Sara has a bachelor's degree from Brandeis University. Before Google, she was a Developer Advocate on the Firebase team.

Michael Munn is an ML Solutions Engineer at Google where he works with customers of Google Cloud on helping them design, implement, and deploy machine learning models. He also teaches an ML Immersion Program at the Advanced Solutions Lab. Michael has a PhD in mathematics from the City University of New York. Before joining Google, he worked as a research professor.

Colophon

The animal on the cover of *Machine Learning Design Patterns* is a sunbittern *(Eurypyga helias)*, a bird found in tropical regions of the Americas, from Guatemala to Brazil. The sunbittern's closest living relative is the kagu, a bird found only in New Caledonia, an archipelago in the southwest Pacific Ocean.

Sunbitterns are cryptic, meaning their coloration of subtle black, gray, and brown patterns acts as camouflage in their environment. Their flight feathers are red, yellow, and black, and with their wings fully spread, these feathers look like eyespots. These spots are shown in courtship and threat displays and used to startle predators. The birds have powder down, a specialized down found in only a few types of birds.

Male and female sunbitterns take turns incubating eggs and feeding their chicks. The diet consists of a wide range of animals, including insects, crustaceans, fish, and amphibians. Although only observed in captive sunbitterns, the birds have been seen fishing with lures to attract prey within striking distance.

The sunbittern's conservation status is least concern. Many of the animals on O'Reilly covers are endangered; all of them are important to the world.

The cover illustration is by Karen Montgomery, based on a black and white engraving from *Elements of Ornithology*. The cover fonts are Gilroy Semibold and Guardian Sans. The text font is Adobe Minion Pro; the heading font is Adobe Myriad Condensed; and the code font is Dalton Maag's Ubuntu Mono.

O'REILLY®

There's much more where this came from.

Experience books, videos, live online training courses, and more from O'Reilly and our 200+ partners—all in one place.

Learn more at oreilly.com/online-learning

Milton Keynes UK
Ingram Content Group UK Ltd.
UKHW012305210924
448614UK00001B/5